ALABAMA WILDLIFE

~VOLUME 2~

Imperiled Aquatic Mollusks and Fishes

RALPH E. MIRARCHI
JEFFREY T. GARNER
MAURICE F. (SCOTT) METTEE
PATRICK E. O'NEIL

Published for, and in cooperation with,

The Division of Wildlife and Freshwater Fisheries
Department of Conservation and Natural Resources

and

The School of Forestry and Wildlife Sciences and
The Alabama Agricultural Experiment Station, Auburn University

by

The University of Alabama Press
Tuscaloosa and London

Manufactured in Korea by Pacifica Communications

∞
The paper on which this book is printed meets the minimum requirements of American National Standard for Information Science–Permanence of Paper for Printed Library Materials, ANSI Z39.48-1984.

Library of Congress Cataloging-in-Publication Data

Alabama wildlife / [edited by] Ralph E. Mirarchi ... [et al.].
p. cm.
Updates and expands: Vertebrate wildlife of Alabama. 1984 andVertebrate animals of Alabama in need of special attention. 1986.
"Published for, and in cooperation with, the Division of Wildlife and Freshwater Fisheries, Department of Conservation and Natural Resources and the School of Forestry and Wildlife Sciences and the Alabama Experiment Station, Auburn University."
Includes bibliographical references.
Contents:
— v.1. A checklist of vertebrates and selected invertebrates: Aquatic mollusks, fishes, amphibians, reptiles, birds, and mammals. ISBN 0-8173-5130-2 (pbk. : alk paper)
— v.2. Imperiled aquatic mollusks and fishes ISBN 0-8173-5131-0
— v.3. Imperiled amphibians, reptiles, birds, and mammals. ISBN 0-8173-5132-9 (pbk. : alk paper)
— v.4.Conservation and management recommendations for imperiled wildlife. ISBN 0-8173-5133-7 (pbk. : alk paper)
1. Vertebrates—Alabama. 2. Endangered species—Alabama. I. Mirarchi, R. E. (Ralph Edward), 1950– II. Alabama. Division of Wildlife and Freshwater Fisheries. III. Auburn University. School of Forestry and Wildlife Sciences. IV. Alabama Agricultural Experiment Station. V. Vertebrate wildlife of Alabama. VI. Vertebrate animals of Alabama in need of special attention.

AL606.52.U6A58 2004
596'.09761—dc22

2003064557

The University of Alabama Press
Tuscaloosa, Alabama 35487-0380

TABLE OF CONTENTS

FISHES

LIST OF FIGURES

Counties of Alabama

Major Rivers, Basins, and Dams of Alabama

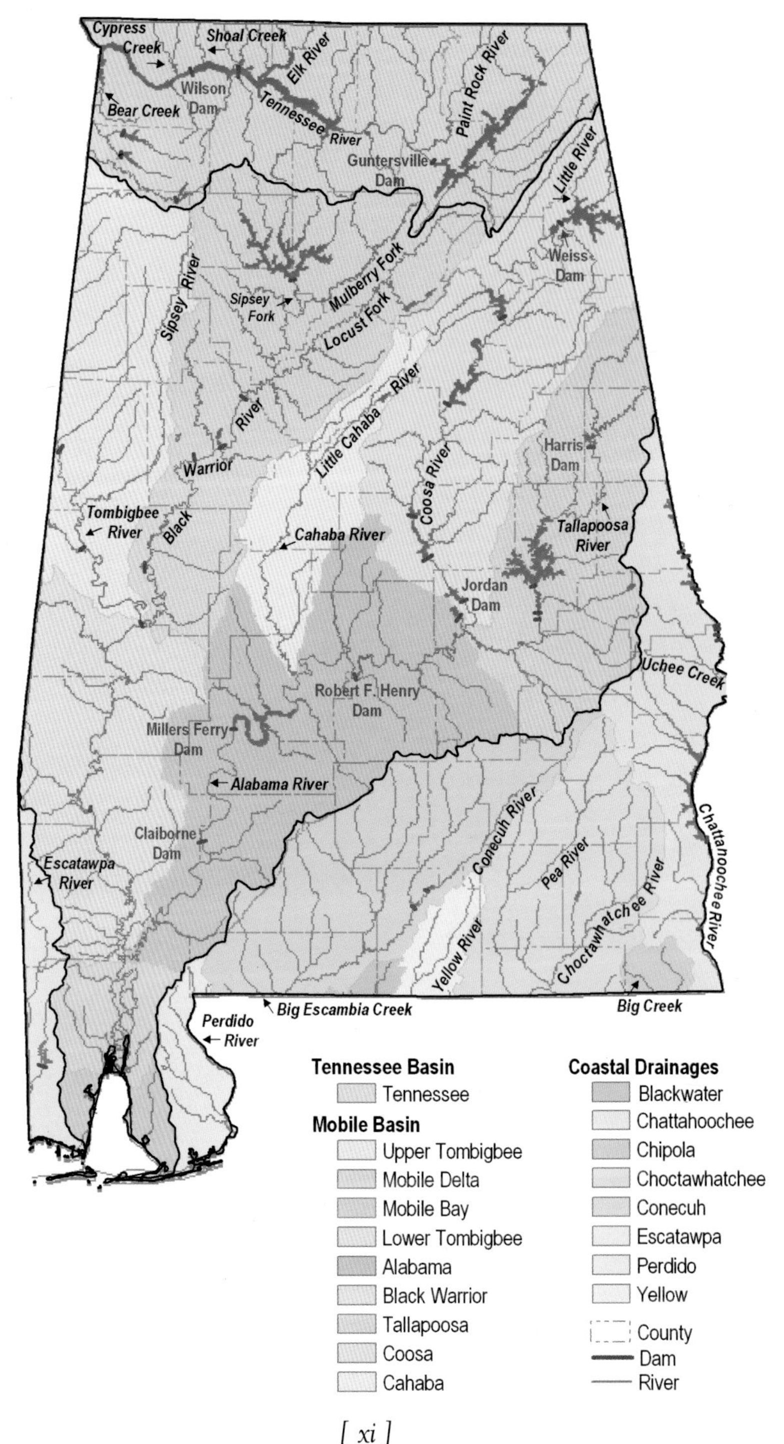

Map Symbols

Approximates
Fall Line
State
Rivers/streams
Historic fish distribution
Current fish distribution
Mussel/snail historic distribution
Snail spring distribution
Mussel/snail current distribution
National
National distribution

Foreword

Alabama is blessed with tremendous natural diversity that spans terrestrial habitats from the Gulf beaches to the lower Appalachian Mountains. The state also contains a wealth of water and wetland resources. Indeed, the Mobile-Tensaw Delta is recognized as one of the most significant and important delta complexes in the nation. This great physical diversity produces numerous habitat types and the abundance of wildlife species they support.

The volumes in this collection present detailed information about the known vertebrate and freshwater mussel and snail taxa in Alabama. Many are common on the landscape and well recognized by the public. Others are endemic, exist in very restrictive habitats, and are unknown to many. Regardless of their abundance, each species is part of the gift that has been bestowed upon us. Each has its place, role, and function. As the most intelligent beings on this planet, and those capable of impacting it the most, we have been given the responsibility to allow other organisms their space and to understand their roles and functions as best we can.

As we begin this century, these volumes are a good indicator of what is known about these species and what is yet to be learned. Some past actions were taken without an understanding of the consequences, and some species have suffered. We are better today at predicting consequences, both beneficial and negative. As our knowledge increases, so will our understanding of the outcome of our actions.

Human populations will continue to grow, effectively shrinking the natural world. Economic needs and demands will increase. The more we understand the natural world, the better we will be prepared to meet the needs of the various species. Economic development and growth can occur without creating drastic negative consequences for wildlife. It will be important to continually create opportunities for the public to enjoy the wildlife in our state and to become more aware of each species' habitat requirements.

We must recognize that the majority of wildlife habitat in Alabama is, and always will be, in the hands of private landowners. It will be their management decisions that determine the future for much of the wildlife in our state. Through developing a climate of cooperation and trust between private landowners and government agencies, wildlife will benefit.

The efforts of all the scientists and interested individuals who participated in the nongame wildlife conference and who contributed information for this publication are appreciated. The chief editor, Dr. Ralph Mirarchi, has provided great leadership and resolve to create these volumes. A special thank you for his efforts is deserved. In addition, all of the scientists who toiled in quest of gaining an answer to an often self-induced question about our natural world deserve recognition. It is the body of knowledge they created that is studied and built upon. I hope that these volumes provide understanding and direction to scientists as they seek answers to the many questions that still exist.

Gary H. Moody, Chief
Wildlife Section
Division of Wildlife and Freshwater Fisheries

Dedication

These volumes are dedicated to the memory of

Robert "Bob" C. McCollum, III
September 13, 1963 – January 22, 2002

Robert Clyde McCollum, III, son of Dr. Tom P. and Patsy L. Watson and the late Robert C. McCollum, Jr., was born in Albany, Georgia. Bob graduated from Hawkinsville High School, Hawkinsville, Georgia, in 1981. He attended both Middle Georgia College and the Georgia Institute of Technology prior to deciding on a career in wildlife management and enrolling at Auburn University in January 1987. He received a B.S. in Wildlife Science from the then-Department of Zoology and Wildlife Sciences, Auburn University, in June 1989 and an M.S. from the same program, under the auspices of the Alabama Cooperative Fish and Wildlife Research Unit, in August 1992. Bob worked as the Nongame Wildlife Coordinator for the Alabama Department of Conservation and Natural Resources, Division of Wildlife and Freshwater Fisheries from July 22, 1996, until his death.

Bob believed deeply in the conservation of nongame species and their associated habitats. He always endeavored to demonstrate the benefits of nonconsumptive activities associated with wildlife conservation to the public, whether through his daily routine, in developing birding trails, or by writing articles on wildlife. Bob loved all aspects of the outdoors. He enjoyed fly-fishing for trout and squirrel hunting as much as providing habitat in his yard for a variety of wildlife species. He was the moving force behind, and actively involved in, the initial planning for the Second Nongame Wildlife Conference at the time of his passing. He will be sorely missed, both as a wildlife professional and as a friend.

EDITOR'S NOTES

Authors making reference to this publication in its entirety should cite it as follows:

Mirarchi, R. E., J. T. Garner, M. F. Mettee, P. E. O'Neil, eds. 2004. Alabama wildlife. Volume 2. Imperiled aquatic mollusks and fishes. The University of Alabama Press, Tuscaloosa, AL. 255 pp.

Authors making reference to individual species accounts should cite them as per the following example:

McGregor, S. W. 2004. Elktoe *Alasmidonta marginata*. Pp. 16 *in* R. E. Mirarchi, J. T. Garner, M. F. Mettee, P. E. O'Neil, eds. Alabama wildlife. Volume 2. Imperiled aquatic mollusks and fishes. The University of Alabama Press, Tuscaloosa, AL.

ACKNOWLEDGMENTS

Those individuals who provided photographs for this volume are identified with each accompanying photograph. Special thanks to the personnel of the Geological Survey of Alabama for providing the base map of the river systems of Alabama used throughout. Numerous individuals, agencies, and organizations whose names do not appear elsewhere, or whose contributions exceed normal acknowledgment in this publication, deserve recognition for efforts that aided in the completion of this particular volume. They include: Gary H. Moody, Keith D. Guyse, Mark S. Sasser, Ericha S. Shelton, Alabama Department of Conservation and Natural Resources (ADCNR), Division of Wildlife and Freshwater Fisheries; James B. Grand, Elise R. Irwin, Michael S. Mitchell, Amy L. Silvano, Cari-Ann Hayer, and Benton D. Taylor, Alabama Cooperative Fish and Wildlife Research Unit, U.S. Geological Survey (USGS), Alabama Gap Analysis Team; Catherine L. Jackson, Teresa E. Rodriguez, Leigh Ann Stribling, Jamie Creamer, Editorial Staff, Alabama Agricultural Experiment Station; Alabama Power Company (APC); Alabama Wildlife Federation; Auburn University (AU) Student Chapter of The Wildlife Society; AU Student Chapter of the American Fisheries Society; Black Warrior-Cahaba Rivers Land Trust; International Paper Company; MeadWestvaco Corporation; William R. and Fay Ireland Distinguished Professorship Endowment Fund; The Nature Conservancy of Alabama's Natural Heritage Program; ADCNR, State Lands' Natural Heritage Section; Sherry Bostick, USGS; David Campbel and Jeff Sides, Univ. of Alabama-Tuscaloosa; Laura Mirarchi and Mike Gangloff, AU; Randy Haddock, Cahaba River Society; Bob Jones, Miss. Museum of Natural Science; Paul Parmalee, McClung Museum, Univ. of Tennessee; Malcolm Pierson, APC; Amy Sides, Alabama Rivers Alliance; and Richard Brinker, Dean, School of Forestry and Wildlife Sciences, AU, for supporting the chief editor's request for professional improvement leave to complete this project.

PREFACE

This volume is the second of four that updates and expands the coverage provided by *Vertebrate Wildlife of Alabama* (Mount 1984) and *Vertebrate Animals of Alabama in Need of Special Attention* (Mount 1986). Volume 1 (*see* Mirarchi 2004) provides a brief review of the historical process and key people associated with the initiation of protection of the most sensitive taxa within Alabama, as well as a detailed annotated list of the aquatic mollusks and all the vertebrate wildlife of the state. In Volume 2, we specifically begin to focus on the most critically imperiled forms. Herein we update the fish portion of *Vertebrate Animals in Need of Special Attention* (Mount 1986) and add an entirely new section on the imperiled freshwater mussels and snails of the state. This is the first treatment of that particular group of organisms in almost 30 years (*see* Stansbery 1976 and Stein 1976), and is long overdue.

The tremendous abundance and diversity of the aquatic mollusks of the pre-European human-dominated Alabama landscape surely would have awed malacologists (those who study mollusks) of today. Unfortunately, much of this bounty was lost forever before it could be studied in depth due to alteration of the natural waterways of the state via impoundments, channelization, canalization, and sedimentation. Up until a few years ago, progress in acquiring and consolidating the knowledge we need to conserve these resources had been slow. Even though early descriptions of the mussel and snail fauna of the state date back to Conrad and Lea during the mid-1800s, some 150 years later Stansbery (1976:39) lamented, ".....When one examines the information available concerning the present status of the freshwater bivalve mollusks of Alabama one is impressed by how little we know concerning the current distribution and abundance of most of the species recorded from this state...." Similarly, Stein (1976:19), commenting on the snail fauna at that time, wrote, "The absence of adequate data on the present abundance and distribution of most of these species precludes any definitive evaluation of their current status." As such, she assigned no categories of endangerment in her listing because she "believed that all of the forms included are in danger of becoming extinct if present trends continue." Interestingly, some 30 years later, there still is no comprehensive text (although one is being prepared) on either the mussels or snails of a state reputed to have the richest molluscan fauna in the United States. Fortunately, systematic coverage of our remaining mollusks has improved in recent years, and Jeff Garner herein provides us with the necessary details on those that are imperiled in the state. Sadly, as he indicates in his introductory comments, only about 73 percent of the snail and mussel species historically occurring in the state are still alive, and only about 19 percent of those are known to be secure.

The convergence of abundant surface water, diverse physiographic regions, and moderate climate of Alabama also has allowed an extremely abundant and diverse fish fauna to evolve. Ichthyologists (those who study fish) rank Alabama among the richest states in terms of fish diversity, with new forms still being described. Fortunately, despite being challenged by the same stressors faced by the aquatic mollusks, the fishes of Alabama have fared somewhat better. Although the early research on this group started approximately during the same period as the mollusks some 150 years ago with published papers by Storer and Agassiz, progress in amassing and consolidating information has come about more quickly. This resulted in the first comprehensive treatment of the state's fishes, entitled *Freshwater Fishes of Alabama* (Vaniz-Smith 1968), more than 30 years ago. More recently, that treatise has been updated and expanded beautifully and cogently into the *Fishes of Alabama and the Mobile Basin* (Mettee *et al.* 1996). Another in-depth treatment of life history and ecology of these fishes apparently also is in progress (Boschung and Mayden, in press). Whether this wealth of information on fishes, when compared with the mollusk group, is reflective of more research interest or better research funding is open to conjecture. Nonetheless, while not suffering the cataclysmic declines in diversity of species suffered by the mollusks, this diverse and beautiful group of organisms still faces stiff challenges in the future. As Maurice Mettee forthrightly points out in his introductory comments, there is both good and bad news regarding the fishes of Alabama. The good news is that a comparison of the current status of the fishes with those listed in the 1986 publication indicates no new taxa have been added to the state's list of extinct or extirpated species; the bad news is that the status of 39 of the state's most imperiled taxa remains unchanged and that the current list has grown to the largest in the state's history (47). Hence, we would do well to heed his call for cooperation of all stakeholders to promote full recovery of these imperiled species.

Much has happened in the wildlife conservation arena in the 18 years since the publication of the last texts on Alabama's imperiled wildlife. The human population of the state has continued to expand, and the resulting development pressures have wrought numerous changes to the native landscape. Demands for food, fiber, housing, jobs, water, energy, recreation, and the infrastructure (highways, bridges, dams, power plants, fiber mills, shopping malls, sewage treatment facilities, and gas, power, and sewage line right-of-ways) to support these demands all continue to impact the land and the river systems that drain it. Of all the organisms that have been negatively impacted by these human activities, those associated with our state's waterways have fared the worst. Part of this is due to the fact that Alabama was so richly blessed with surface water and had such a tremendous diversity of aquatic organisms that inhabited them. However, we must also acknowledge that we have squandered much of this wealth of water and the wildlife resources that depend upon it. All too frequently, in the name of "progress," we laid waste to communities of organisms living wild and free because they appeared either inexhaustible or were deemed economically and aesthetically less important than inanimate objects that brought us more immediate gratification. A quick tally of the excruciatingly long list of Alabama's freshwater mollusks and fish that have become extinct (71) or have been extirpated (31) in the last 75 years, along with those that are otherwise currently imperiled (96) or on the Alabama Watch List teetering on the abyss of imperilment (128), should force even the most ecologically ignorant among us to pause and think. For those of us provided with an ecological education, it is almost too much to bear. Perhaps this text will make more of the uninformed among us ask if one more place to shop, one more new restaurant in which to eat, or one more lake upon which to jet-ski is worth the cost of further increasing the length of these lists. Perhaps it will make us ask if the uncontrolled demands of an ever-increasing human population are worth the cost in terms of unique and beautiful organisms never to be enjoyed again by future generations of Alabamians. I certainly hope so.

This volume is much more detailed in its coverage than Volume 1 of the series, and includes only those imperiled members among both the aquatic mussels and snails and the fishes of Alabama. A list of extinct species from each group precedes the appropriate section. A detailed account of each imperiled taxon assigned a priority designation (Extirpated, Extirpated/Conservation Action Underway, Priority 1, or Priority 2) immediately follows that list and contains information on other common names for the taxon, a physical description (including photographs), the distribution (including maps) of the taxon nationally and in Alabama, the habitat where it is found, any key life history and ecological information, and the basis for its status classification. Definitions of the various Priority Designations are contained in Volume 1 and also are repeated at the beginning of each group's Priority Designation accounts. The detailed accounts are followed by tables that consolidate critical information relevant to the text, and lists of scientific names of plants and animals associated with the imperiled taxa that are keyed to the text via common names. Finally, an Alabama Watch List consisting of Priority 3 Conservation Designees—those taxa that could easily slip into one of the imperiled categories—and references to support information in the text completes the coverage of each group. An extensive glossary of technical terms, designed to assist the average lay reader, is used to minimize verbiage throughout the text and is found at the end of the volume. Detailed coverage of the amphibians, reptiles, birds, and mammals is contained in Volume 3. Conservation and management recommendations for all taxa, along with general conservation recommendations for the physical communities in which they reside, are included in Volume 4.

The maps on pages x through xii are provided to assist the reader in locating many of the geographic, ecological, and specific site locations mentioned in the text. Only major rivers, and those creeks and dams frequently mentioned in the text, were labeled. All dams on major rivers were included (Mettee *et al.* 1993), although not all were labeled. These decisions were made by the senior editor in an attempt to avoid unnecessary cluttering of the map.

It is our hope that a wide cross section of Alabamians will take the time to read these volumes and learn more about these unique parts of our heritage, and that others from across the country will introduce themselves to an important part of what makes Alabama a special place to live.

Ralph E. Mirarchi

References

Boschung, H. T., and R. L. Mayden. In press. *The Fishes of Alabama*. Smithsonian Institution Press, Washington, D.C.

Mettee, M. F., P. E. O'Neil, and J. M. Pierson. 1996. Fishes of Alabama and the Mobile Basin. Geological Surv. Alabama Monograph 15. Oxmoor House, Birmingham, AL. 820 pp.

Mirarchi, R. E., ed. 2004. Alabama wildlife. Volume 1. A checklist of vertebrates and selected invertebrates: aquatic mollusks, fishes, amphibians, reptiles, birds, and mammals. The Univ. of Alabama Press, Tuscaloosa, AL. 209 pp.

Mount, R. H., ed. 1984. Vertebrate wildlife of Alabama. Ala. Agric. Expt. Sta., Auburn Univ., Auburn Univ., Auburn, AL. 44 pp.

_____. 1986. Vertebrate animals of Alabama in need of special attention. Ala. Agric. Expt. Sta., Auburn Univ., Auburn, AL. 124 pp.

Smith-Vaniz, W. F. 1968. Freshwater fishes of Alabama. Ala. Agric. Expt. Sta., Auburn Univ., Auburn, AL. 211 pp.

Stansbery, D. H. 1976. Naiad mollusks. Pp. 38-48 *in* H. T. Boschung, ed. Endangered and threatened plants and animals of Alabama. Bull. Ala. Mus. Nat. 2. 92 pp.

Stein, C. B. 1976. Gastropods. Pp. 17-37 *in* H. T. Boschung, ed. Endangered and threatened plants and animals of Alabama. Bull. Ala. Mus. Nat. 2. 92 pp.

FRESHWATER MUSSELS AND SNAILS

Photo: Malcolm Pierson

Finerayed Pocketbook
Lampsilis altilis

INTRODUCTION

Freshwater mollusks (i.e., mussels and snails) live sedentary, cryptic lives in our streams, rivers, and lakes. Whether a mussel lying buried in the bottom of a turbid coastal plain river or a snail crawling across a slab of bedrock in a clear mountain stream, mollusks are seldom noticed at all and almost never recognized as the unique, and often beautiful, creatures that they are. The diversity in form and function displayed by this group of animals is astonishing. Alabama is known for its multiplicity of aquatic habitats, and mollusks have adapted to, and reside in, practically every one, resulting in more species of freshwater mussels and snails than any other state and many entire countries. One hundred seventy-five freshwater mussels and 174 freshwater snails have been reported from Alabama.

Unfortunately, freshwater mollusks have suffered greatly with modern alterations to rivers and streams. The greatest diversity of species, both mussels and snails, occurred in riffles and shoals of medium to large streams and rivers. Impoundment of long stretches of river by construction of dams eliminated many species from much of their distributions and isolated remaining population fragments. These smaller population fragments were, in turn, more vulnerable to further habitat degradation from stream channelization, sedimentation, streambed instability, and water pollution. These factors have resulted in freshwater mollusks being one of the most imperiled groups of animals in the world. This document reflects their dire circumstances. Only 75 percent of the snail species and 71 percent of the mussel species historically occurring in Alabama are still alive there at the time of this writing. This percentage actually may be somewhat lower for mussels, and considerably lower for snails, since distributions of some species are poorly known. Of all species known to occur in the state, only 17 percent of snails and 21 percent of mussels are believed to be secure. The remainder are imperiled to various degrees, ranging from relict populations that are no longer reproducing to species that are still widespread but appear to be suffering from declining population trends.

Although neglect of our molluscan fauna has resulted in their dismal predicament, they are finally getting some long overdue interest from government agencies and conservation groups. Nationwide, 32 snails and 70 mussels have been listed as threatened or endangered by the U.S. Fish and Wildlife Service under the Endangered Species Act since 1976. With such protection come detailed recovery plans and funding for continued monitoring and recovery work. As a result, mussel and snail propagation techniques are being developed and offer promise for some species. However, suitable habitat is essential for recovery efforts to proceed and be successful. The federal Clean Water Act and other momentous legislative measures have resulted in greatly improved habitat since the dark days of unregulated pollution during the middle of the twentieth century. Inventive new tools for recovery, such as nonessential experimental populations of federally endangered species, also will play a major role in future recovery efforts.

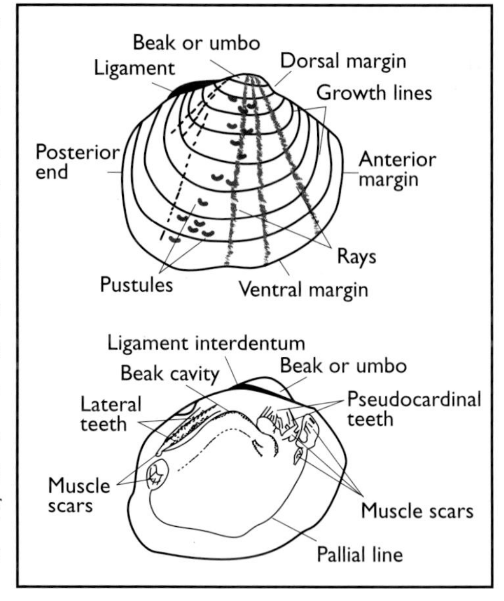

Figure 1, above. Morphology of the exterior and interior of a typical freshwater mussel. Figure 2, below. Life cycle of a typical mussel.

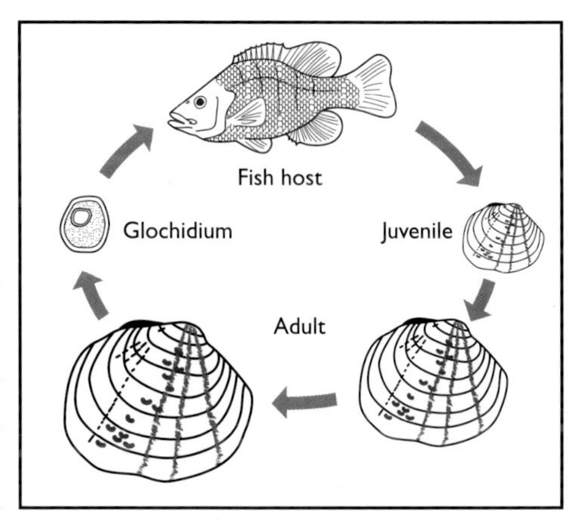

The morphology of a typical freshwater mussel, and the terminology used to describe it, are unique; they have two valves (hence the name bivalves) that can open and close for a variety of functions, from locomotion to reproduction. (Fig. 1). Their life cycle begins when fertilized eggs develop into larva called glochidia within the gills of the female. When released, they must come in contact with a fish and parasitize its gills, fins, or body (Fig. 2). A few days or weeks later, the glochida drop off and settle to the bottom of the waterway in which they reside to begin life as juvenile mussels. They then may take several years to reproduce and may live more than 70 years. Snails, on the other hand, have shells that are all one piece, albeit with a variety of different shaped whorls and color patterns. Their shells have only one opening at the base of the shell through which the various bodily functions take place. They also reproduce differently than mussels in that there is no parasitic phase; they can either lay eggs that eventually hatch into the immature forms or bear their young alive.

Work toward a better understanding of taxonomic relationships among mollusk taxa continues using modern genetic comparisons in conjunction with shell morphology and soft tissue anatomy. Thus, the names of species are in a state of flux. Those used here are current, but some changes are likely in the near future.

This publication serves not only as a status report for Alabama's molluscan species, but as a "report card" for the citizens of Alabama regarding the protection of our wildlife resources. Although mussel and snail recovery efforts are well underway, the task at hand is daunting. Increased funding and effort, as well as additional propagation facilities, will be necessary to progress beyond our current state of affairs. Also, continued legislative vigilance will be necessary, lest important legislation that protects our waterways is weakened in the name of economic "progress."

Jeffrey T. Garner

AUTHORS

Mr. Jeffrey T. Garner, Division of Wildlife and Freshwater Fisheries, Alabama Department of Conservation and Natural Resources, 350 County Road 275, Florence, AL 35633, Co-editor/Co-compiler

Mrs. Holly Blalock-Herod, U.S. Fish and Wildlife Service, 1601 Balboa Avenue, Panama City, FL 32405

Dr. Arthur E. Bogan, North Carolina State Museum of Natural Sciences, Research Laboratory, 4301 Reedy Creek Road, Raleigh, NC 27607

Mr. Robert S. Butler, U.S. Fish and Wildlife Service, 160 Zillicoa Street, Asheville, NC 28801

Dr. Wendell R. Haag, U.S.D.A. Forest Service, Center for Bottomland Hardwoods Research, 1000 Front Street, Oxford, MS 38655

Mr. Paul D. Hartfield, U.S. Fish and Wildlife Service, 6578 Dogwood View Parkway, Jackson, MS 39213

Mr. Jeffrey J. Herod, U.S. Fish and Wildlife Service, Jackson Guard at Eglin Air Force Base, 107 Highway 85 North, Niceville, FL 32578

Dr. Paul D. Johnson, Tennessee Aquarium Research Institute, 5385 Red Clay Road, Cohutta Springs, GA 30710

Mr. Stuart W. McGregor, Geological Survey of Alabama, P.O. Box 869999, Tuscaloosa, AL 35486

Freshwater Mussels

EXTINCT

Taxa that historically occurred in Alabama, but are no longer alive anywhere within their former distribution.

Order Unionoida
Family Unionidae

COOSA ELKTOE *Alasmidonta mccordi*
WINGED SPIKE *Elliptio nigella*
SUGARSPOON *Epioblasma arcaeformis*
ANGLED RIFFLESHELL *Epioblasma biemarginata*
LEAFSHELL *Epioblasma flexuosa*
YELLOW BLOSSOM *Epioblasma florentina florentina*
ACORNSHELL *Epioblasma haysiana*
NARROW CATSPAW *Epioblasma lenior*
FORKSHELL *Epioblasma lewisii*
ROUNDED COMBSHELL *Epioblasma personata*
TENNESSEE RIFFLESHELL *Epioblasma propinqua*
CUMBERLAND LEAFSHELL *Epioblasma stewardsonii*
TUBERCLED BLOSSOM *Epioblasma torulosa torulosa*
TURGID BLOSSOM *Epioblasma turgidula*
LINED POCKETBOOK *Lampsilis binominata*
HADDLETON LAMPMUSSEL *Lampsilis haddletoni*
HIGHNUT *Pleurobema altum*
HAZEL PIGTOE *Pleurobema avellanum*
YELLOW PIGTOE *Pleurobema flavidulum*
BROWN PIGTOE *Pleurobema hagleri*
GEORGIA PIGTOE *Pleurobema hanleyianum*
ALABAMA PIGTOE *Pleurobema johannis*
COOSA PIGOE *Pleurobema murrayense*
LONGNUT *Pleurobema nucleopsis*
WARRIOR PIGTOE *Pleurobema rubellum*
TRUE PIGTOE *Pleurobema verum*
STIRRUPSHELL *Quadrula stapes*

Freshwater Mussels

EXTIRPATED

Taxa that historically occurred in Alabama, but are now absent; may be rediscovered in the state, or be reintroduced from populations existing outside the state.

Order Unionoida
Family Unionidae

PHEASANTSHELL *Actinonaias pectorosa*
ELKTOE *Alasmidonta marginata*
SOUTHERN ELKTOE *Alasmidonta triangulata*
CHIPOLA SLABSHELL *Elliptio chipolaensis*
BROTHER SPIKE *Elliptio fraterna*
UPLAND COMBSHELL *Epioblasma metastriata*
CATSPAW *Epioblasma obliquata obliquata*
SOUTHERN ACORNSHELL *Epioblasma othcaloogensis*
GREEN FLOATER *Lasmigona subviridis*
SCALESHELL MUSSEL *Leptodea leptodon*
COOSA MOCCASINSHELL *Medionidus parvulus*
HICKORYNUT *Obovaria olivaria*
LITTLEWING PEARLYMUSSEL *Pegias fabula*
CLUBSHELL *Pleurobema clava*
BLACK CLUBSHELL *Pleurobema curtum*
FLAT PIGTOE *Pleurobema marshalli*
FLUTED KIDNEYSHELL *Ptychobranchus subtentum*
CUMBERLAND BEAN *Villosa trabalis*

PHEASANTSHELL

Actinonaias pectorosa (Conrad)

OTHER NAMES. None.

Photo—From Parmalee and Bogan 1998

DESCRIPTION. Has moderately thick shell (max. length = 150 mm [5 7/8 in.]), elongate oval to elliptical in outline, and inflated. Dorsal and ventral margins slightly convex to almost straight; anterior margin broadly rounded and posterior margin narrowly rounded to bluntly pointed. Females may be somewhat more rounded posteriorly than males, and have slight marsupial swelling. Posterior ridge rounded and posterior slope steep dorsally, becoming more flat ventrally. Shell disk and posterior slope without sculpture. Umbos moderately inflated, but elevated only slightly above hinge line. Periostracum tawny to light greenish brown and usually marked with faint, wide, broken, green rays. Pseudocardinal teeth erect and triangular; lateral teeth short and straight. A long, moderately wide interdentum separates them, and the umbo cavity is open and moderately deep. Shell nacre white, but may have a light salmon wash, particularly in the umbo cavity. (Modified from Simpson 1914, Parmalee and Bogan 1998)

DISTRIBUTION. Endemic to Tennessee and Cumberland River systems (Burch 1975, Parmalee and Bogan 1998). Historically occurred in Tennessee River downstream to Muscle Shoals, but not reported there since the river was impounded (Garner and McGregor 2001). Also occurred in Paint Rock River (Isom and Yokley 1973), and possibly other tributaries in Alabama, but no recent reports of continued existence there.

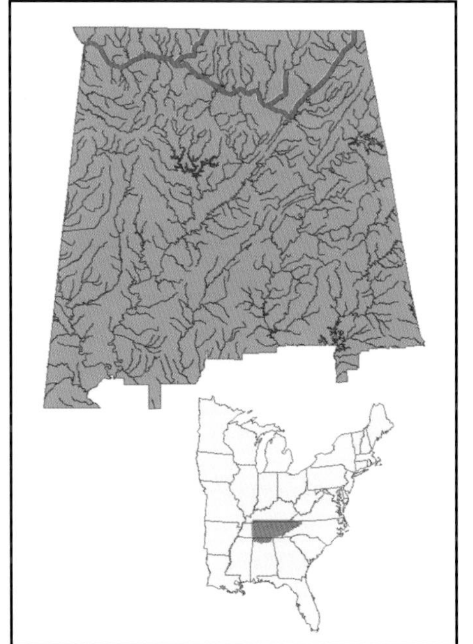

HABITAT. Shoals, usually at shallow depths (Parmalee and Bogan 1998).

LIFE HISTORY AND ECOLOGY. A long-term brooder, with gravid specimens reported in September and May (Ortmann 1921). Hosts of glochidia unknown.

BASIS FOR STATUS CLASSIFICATION. Apparently unable to adapt to impounded habitats. Its disappearance from Paint Rock River and possibly other tributaries likely caused by degradation of habitat from channelization, sedimentation, and pollution from poor agricultural practices. Listed as a species of special concern throughout its distribution (Williams *et al.* 1993), and considered endangered (Stansbery 1976*a*) and possibly extirpated (Lydeard *et al.* 1999) in Alabama.

Prepared by: **Jeffrey T. Garner**

ELKTOE

Alasmidonta marginata Say

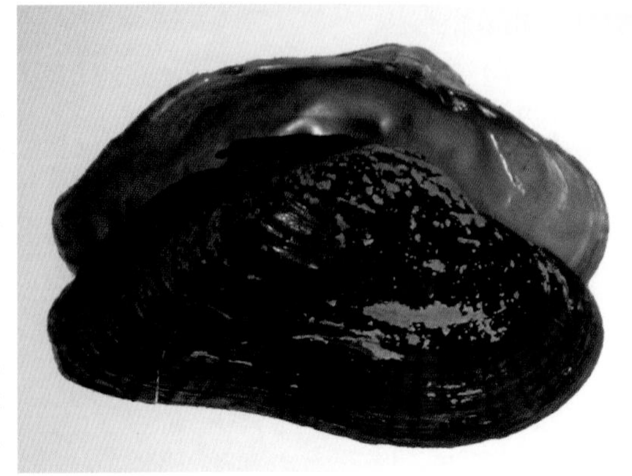

Photo—From Parmalee and Bogan 1998

OTHER NAMES. None.

DESCRIPTION. Shell elongate, somewhat rhomboid and inflated, thin when young, but thickening and becoming solid with age (max. length = 75 mm [2 15/16 in.]). Anterior margin sharply rounded and posterior margin straight, but obliquely oriented. Ventral margin may be straight, slightly convex, or slightly concave. Posterior ridge high and sharp, with numerous fine, radial ridges on posterior slope that extend upward toward the margin. Shell disk smooth. Umbos large, inflated, elevated above the hinge line, and almost centrally positioned. Umbo sculpture consists of a few strong, usually double-looped, corrugations. Periostracum tawny or greenish, usually with numerous green or black rays, plus additional dark spots that appear in connection with the rays. Pseudocardinal teeth thin, low, and elongate; lateral teeth manifested as a thickened hinge line. No interdentum present, and umbo cavity moderately deep. Shell nacre bluish white, occasionally with shades of pink. (Modified from Simpson 1914, Parmalee and Bogan 1998)

DISTRIBUTION. Encompasses much of the Interior Basin, including the Ohio, Cumberland, and Tennessee River systems. Also in St. Lawrence River drainage, and Susquehanna River drainage. Historically known from the Tennessee River at Muscle Shoals and large Tennessee River tributaries (Clarke 1981, Parmalee and Bogan 1998).

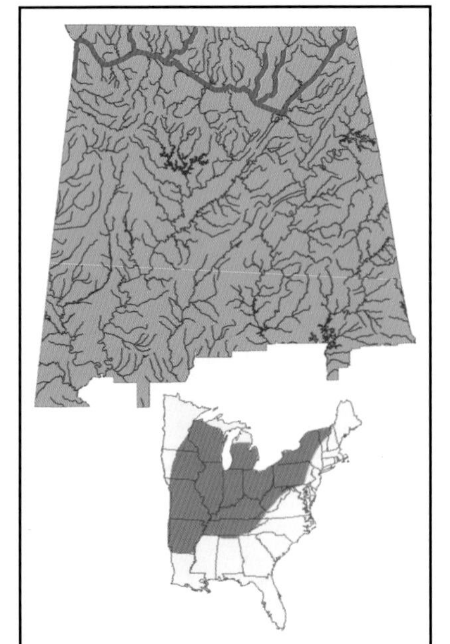

HABITAT. Seems to prefer small, shallow rivers with moderately swift current and a mixture of fine gravel and sand (Parmalee and Bogan 1998). However, its presence at Muscle Shoals prior to impoundment of the Tennessee River attests to its ability to inhabit large rivers under some conditions.

LIFE HISTORY AND ECOLOGY. A long-term brooder, with a breeding season extending from mid-July to mid-June. Reported hosts of glochidia include white sucker, northern hog sucker, shorthead redhorse, rock bass, and warmouth (Parmalee and Bogan 1998).

BASIS FOR STATUS CLASSIFICATION. Has disappeared from much of former distribution and extant populations isolated, resulting in classification as a species of special concern (Williams *et al.* 1993). Has not been reported from Alabama for several decades, although listed as a species of special concern there (Stansbery 1976*a*, Lydeard *et al.* 1999).

Prepared by: **Stuart W. McGregor**

SOUTHERN ELKTOE

Alasmidonta triangulata (I. Lea)

OTHER NAMES. None.

Photo—Mike Gangloff

DESCRIPTION. Shell moderately thin, inflated, subtriangular in outline, with a rounded anterior margin and bluntly pointed posterior margin (max. length = 70 mm [2 3/4 in.]). Posterior ridge pronounced, steep, and finely rugose. Shell disk without sculpture, but lower posterior slope scaly. Umbos high and full, positioned anterior of the center. Umbo sculpture consists of strong, concentric ridges, with radial lirae in front and behind. Periostracum of young specimens dark yellowish green to bluish green, and strongly rayed when viewed with transmitted light. Adults glossy brownish black and appear finely rayed with transmitted light. Right valve has single pseudocardinal tooth and left valve has two short, pointed pseudocardinals; lateral teeth rudimentary. Shell nacre bluish white to salmon. (Modified from Simpson 1914, Clench and Turner 1956, Brim Box and Williams 2000)

DISTRIBUTION. Endemic to Apalachicola Basin, including Chattahoochee River system in Alabama. Last reported from Uchee and Little Uchee Creeks in Russell and Lee Counties (Brim Box and Williams 2000).

HABITAT. Seems to prefer sandy mud substrata, particularly in vicinity of rocks in larger creeks and rivers. Also occurs in stable sand and gravel bars (Clench and Turner 1956, Brim Box and Williams 2000).

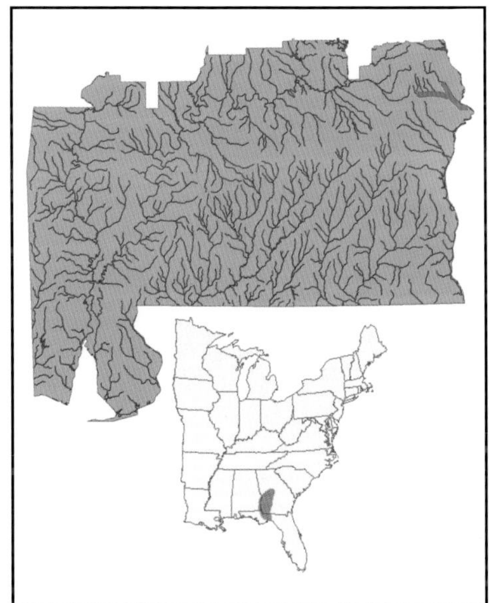

LIFE HISTORY AND ECOLOGY. A long-term brooder with glochidia reported in the marsupia year round. Hosts of glochidia unknown (Brim Box and Williams 2000).

BASIS FOR STATUS CLASSIFICATION. Regarded as rare (Clench and Turner 1956), and classified as endangered throughout its distribution (Williams *et al.* 1993) based on significant reductions in distribution and habitat degradation and fragmentation (Brim Box and Williams 2000). Listed as a species of special concern (Stansbery 1976*a*) and as imperiled in Alabama (Lydeard *et al.* 1999).

Prepared by: **Stuart W. McGregor**

CHIPOLA SLABSHELL

Elliptio chipolaensis (Walker)

OTHER NAMES. None.

Photo—Paul Johnson

DESCRIPTION. Has moderately thin, inflated shell (max. length = 85 mm [3 3/8 in.]), ovate to subelliptical in outline, with slightly convex dorsal and ventral margins, rounded anterior margin, and a posterior margin that is slightly biangulate. Posterior ridge rounded dorsally, but flattens with a ventral progression, forming the posterior biangulation at distal end. Posterior slope wide, shallow, and slightly concave. Shell disk and posterior slope unsculptured. Umbos prominent and elevated above the hinge line. Periostracum chestnut brown, often darker in the umbo area and with alternating light and dark bands in a ventral progression. Pseudocardinal teeth compressed and crenulate; lateral teeth long, slender, and slightly curved. Umbo cavity deep. Shell nacre salmon. (Modified from Simpson 1914, Williams and Butler 1994)

DISTRIBUTION. Thought to be endemic to Chipola River drainage, generally in the main stem and lower ends of larger tributaries (van der Schalie 1940, Clench and Turner 1956, Williams and Butler 1994), until a museum record from a Chattahoochee River tributary was reported (Brim Box and Williams 2000). Species was not reported from Alabama (Brim Box and Williams 2000) and appears extirpated from the state.

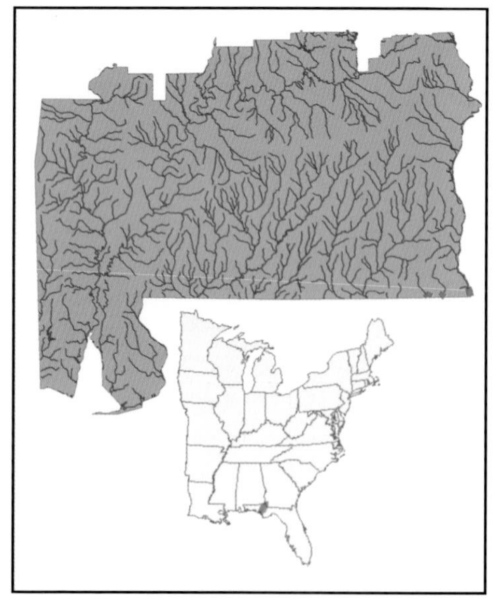

HABITAT. Silty sand substrata in large creeks and rivers with slow to moderate flow (van der Schalie 1940, Williams and Butler 1994, Brim Box and Williams 2000).

LIFE HISTORY AND ECOLOGY. Little known, although presumably a short-term brooder. Gravid females collected in late June (Brim Box and Williams 2000).

BASIS FOR STATUS CLASSIFICATION. Has a narrowly restricted distribution and could suffer severely from habitat degradation. Classified as threatened throughout distribution (Williams *et al.* 1993), and listed as extirpated from Alabama (Lydeard *et al.* 1999). **Listed as *threatened* by the U.S. Fish and Wildlife Service in 1998.**

Prepared by: **Stuart W. McGregor**

BROTHER SPIKE

Elliptio fraterna (I. Lea)

Photo—Paul Johnson

OTHER NAMES. None.

DESCRIPTION. Has fairly thin, but solid, shell (max. length = 100 mm [3 15/16 in.]), elongate and somewhat compressed, subrhomboidal in outline. Anterior margin rounded and posterior margin slightly biangulate. Dorsal and ventral margins straight to slightly convex. Posterior ridge well defined near umbo, but becomes less distinct ventrally. A faint secondary ridge usually present dorsally, resulting in the biangulation of the posterior margin. Posterior slope adorned with radially oriented wrinkles that extend from posterior ridge to dorsal margin. Umbos moderately full, with little elevation above the hinge line. Periostracum smooth and shiny, reddish to yellowish brown in young specimens, often with fine green rays, but darkening with age to become dark reddish brown. Pseudocardinal teeth moderately thick, low, and stumpy; lateral teeth moderately long and slightly curved, and with no interdentum, remote from the pseudocardinals. Umbo cavity very shallow. Shell nacre varies from white to purple. (Modified from Simpson 1914, Johnson 1970, Brim Box and Williams 2000)

DISTRIBUTION. Known only from Chattahoochee and Flint Rivers of Alabama and Georgia and Savannah River drainage of South Carolina. Alabama records limited to the main stem Chattahoochee River (Brim Box and Williams 2000).

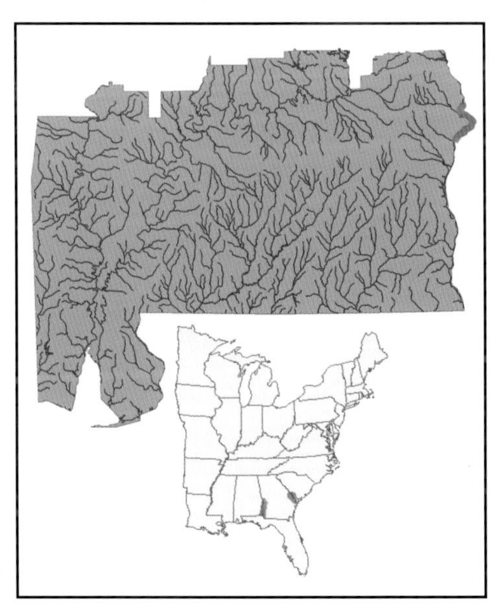

HABITAT. Rivers and large tributaries in shallow lotic areas with sandy substrata (Britton and Fuller 1979, Brim Box and Williams 2000).

LIFE HISTORY AND ECOLOGY. No information available, but presumably a short-term brooder like its congeners.

BASIS FOR STATUS CLASSIFICATION. Appears to have been rare historically, based on a paucity of specimens in the known few collections. Has not been reported from Apalachicola Basin since 1929. Based on museum records and failure to collect this species during recent surveys, Brim Box and Williams (2000) classified it as extirpated from the basin. Also classified as endangered throughout its distribution (Williams *et al.* 1993), and listed as imperiled and possibly extinct in Alabama (Lydeard *et al.* 1999).

***Prepared by:* Stuart W. McGregor**

UPLAND COMBSHELL

Epioblasma metastriata (Conrad)

OTHER NAMES. None.

Female upland combshell. Photo—Mike Gangloff

DESCRIPTION. Has moderately solid and inflated shell (max. length = 40 mm [1 5/8 in.]), elongate and subtriangular to subquadrate in outline. Anterior margin rounded. Posterior margin of females rounded, but males are truncate posteriorly. Ventral margin broadly rounded, but in adult females interrupted by a marsupial swelling toward posterior end. A slight sulcus is located anterior to the marsupial swelling. Posterior slope low and evenly rounded in males, but more distinct in females, being incorporated into the marsupial swelling. Fine, radial striations adorn the shell posteriorly, generally stronger on the marsupial swelling of females. Umbos moderately full and elevated above the hinge line. Periostracum smooth and shiny, yellowish to greenish yellow, with numerous narrow, light green rays. Pseudocardinal teeth somewhat elongate and doubled in both valves; lateral teeth short and almost straight. A short, narrow interdentum separates them. Umbo cavity very shallow. Shell nacre white, but may have a bluish tinge. (Modified from Simpson 1914, Parmalee and Bogan 1998)

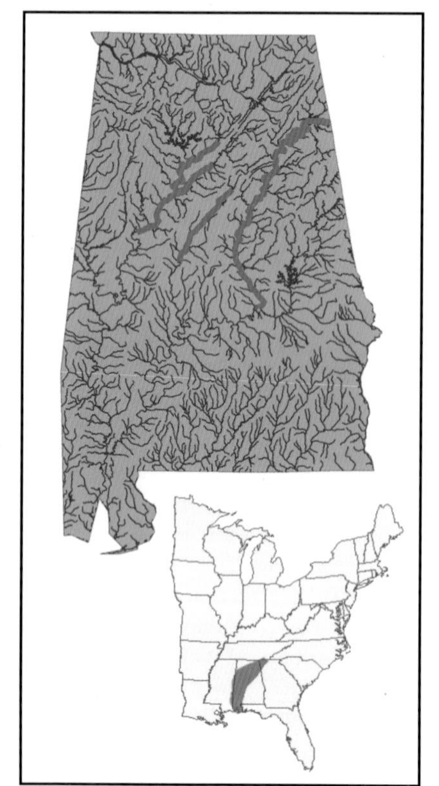

DISTRIBUTION. Endemic to Mobile Basin in Alabama, Georgia, and Tennessee. Known from Coosa, Cahaba, and Black Warrior River systems (USFWS 2000).

HABITAT. Lotic areas in medium to large rivers.

LIFE HISTORY AND ECOLOGY. Unknown. Other species of *Epioblasma* are long-term brooders and use darters (Percidae) as glochidial hosts (Yeager and Saylor 1995, Rogers *et al.* 2001).

BASIS FOR STATUS CLASSIFICATION. Suffered drastic declines with modern perturbations to habitat. Extirpated from most of its former distribution and may be extinct. Was classified as endangered throughout its distribution (Williams *et al.* 1993) and threatened in (Stansbery 1976*a*) and possibly extirpated from Alabama (Lydeard *et al.* 1999). **Listed as *endangered* by the U.S. Fish and Wildlife Service in 1993.**

***Prepared by:* Wendell R. Haag**

CATSPAW

Epioblasma obliquata obliquata (Rafinesque)

OTHER NAMES. Peewee, Cat's Claw.

Above– female, below– male.
Photo—From Parmalee and Bogan 1998

DESCRIPTION. Shells solid and trapezoidal to subovate in outline, with males somewhat truncate anteriorly (max. length = 70 mm [2 3/4 in.]) and females rounded anteriorly (max. length = 50 mm [2 in.]). Males bluntly pointed posteriorly and females of reproductive age irregularly ovate, with a marsupial swelling producing an emargination. Marsupial swelling inflated, rounded, radially striate, and separated from posterior ridge by a strong sulcus. Posterior ridge of males low and not well developed, but it is doubled. In females, posterior ridge somewhat obscured by the sulcus. Shell disk without sculpture. Umbos prominent and elevated above the hinge line. Umbo sculpture consists of a few faint corrugations, which may be broken or double looped. Periostracum smooth and somewhat shiny, yellowish green to brown, usually with numerous fine, wavy green rays. Pseudocardinal teeth subtriangular and ragged, separated from the short, straight lateral teeth by a short, wide interdentum. Umbo cavity shallow. Shell nacre usually purple, but may be white. (Modified from Simpson 1914, Johnson 1978, and Parmalee and Bogan 1998)

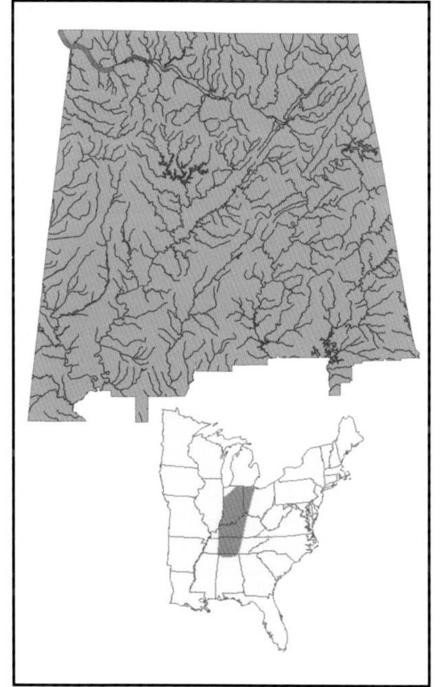

DISTRIBUTION. Historically, throughout much of the eastern Interior Basin, including parts of the Ohio and Cumberland Rivers and up the Tennessee River as far as Muscle Shoals (Parmalee and Bogan, 1998). Although listed as extinct by Turgeon *et al.* (1998), a viable population is extant in Killbuck Creek of the Muskingum River system in Ohio (R.S. Butler, USFWS, pers. comm.).

HABITAT. Lotic areas in medium to large rivers (Parmalee and Bogan 1998).

LIFE HISTORY AND ECOLOGY. Nothing known. Probably a long-term brooder, like its congeners.

BASIS FOR STATUS CLASSIFICATION. Was classified as endangered throughout its distribution (Williams *et al.* 1993), and endangered but extirpated from Alabama (Lydeard *et al.* 1999). **Listed as *endangered* by the U.S. Fish and Wildlife Service in 1990.**

***Prepared by:* Jeffrey T. Garner**

SOUTHERN ACORNSHELL

Epioblasma othcaloogensis (Lea)

OTHER NAMES. None.

Photo—Mike Gangloff

DESCRIPTION. Has moderately solid and inflated shell (max. length = 25 mm [1 in.]), quadrate in outline, with anterior margin rounded, posterior margin truncate above and rounded below, dorsal margin slightly convex, and ventral margin straight. Females may have a slight marsupial swelling and may be somewhat more truncate posteriorly than males. Posterior ridge rounded. Weak, radial sculpture may be visible on posterior half of shell. Umbos moderately inflated and elevated well above hinge line. Periostracum smooth and shiny, yellowish, and without rays. Pseudocardinal teeth elevated, ragged, and slightly compressed; lateral teeth short and straight to slightly curved. No interdentum present, and umbo cavity shallow. Shell nacre white. (Modified from Simpson 1914, Parmalee and Bogan 1998)

DISTRIBUTION. Endemic to upper Coosa River system in Alabama, Georgia, and Tennessee (Parmalee and Bogan 1998), but has not been reported in more than 20 years.

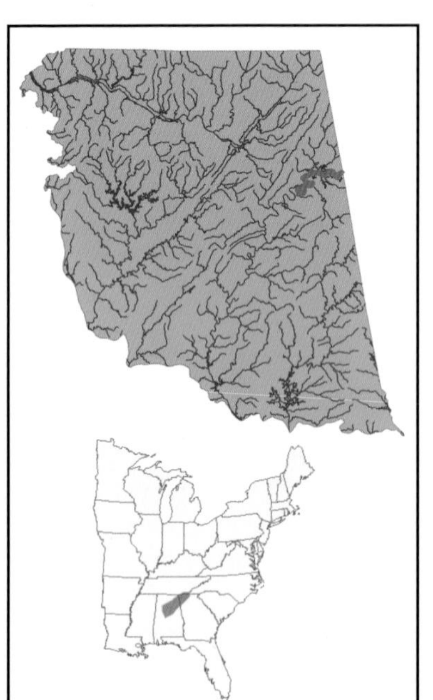

HABITAT. Small to medium rivers in fine gravel substrata (Parmalee and Bogan 1998).

LIFE HISTORY AND ECOLOGY. Unknown. Other species of *Epioblasma* are long-term brooders and use darters (Percidae) as glochidial hosts (Yeager and Saylor 1995, Rogers *et al.* 2001).

BASIS FOR STATUS CLASSIFICATION. Has not been seen alive in more than 20 years. Was classified as endangered throughout its distribution (Williams *et al.* 1993), in Alabama (Stansbery 1976*a*), and Lydeard *et al.* (1999) suggested that it was extirpated from state. **Listed as *endangered* by the U.S. Fish and Wildlife Service in 1993.**

***Prepared by:* Wendell R. Haag**

GREEN FLOATER

Lasmigona subviridis (Conrad)

OTHER NAMES. None.

DESCRIPTION. Has thin, slightly inflated, subovate shell (max. length approx. 60 mm [2 3/8 in.]), narrower anteriorly, higher posteriorly, with a dorsal margin that forms a blunt angle with posterior margin. Posterior ridge low and rounded, appearing more as a slight swelling than a ridge. Umbos low and not elevated above hinge line. Periostracum color varies from dull yellow to tan or brownish green, with variable dark green rays. Pseudocardinal teeth lamellate and directed forward of beak, almost parallel with hinge line; lateral teeth long, straight, and thin. Shell nacre dull, bluish white, often with mottled shades or tints of salmon in umbo cavity. (Modified from Parmalee and Bogan 1998)

Photo—From Parmalee and Bogan 1998

DISTRIBUTION. Atlantic Coast drainages, but also occurs in Apalachicola Basin of Alabama and Georgia (Brim Box and Williams 2000). Found as far north as St. Lawrence-Hudson River system. Also reported from Watauga River of upper Tennessee River system (Parmalee and Bogan 1998). In Alabama, reported only from Chattahoochee River near Columbus, Georgia (Brim Box and Williams 2000).

HABITAT. Primarily quiet reaches of streams out of main current, in pools or eddies with gravel and sand bottoms, but also may occur in canals (Ortmann 1919). Reported from pockets of sand and gravel among boulders in East Tennessee (Parmalee and Bogan 1998).

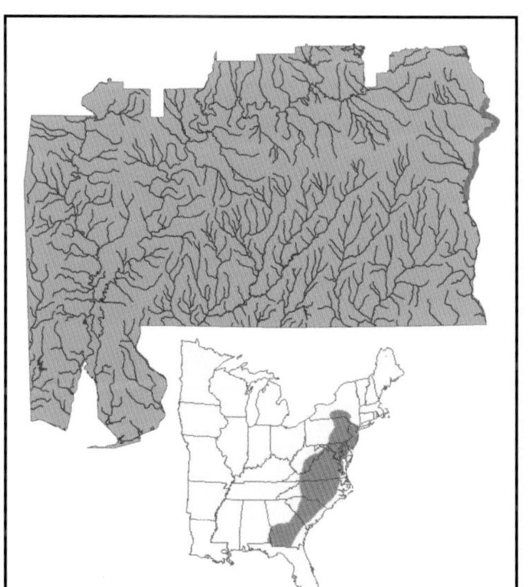

LIFE HISTORY AND ECOLOGY. A long-term brooder, gravid from September to June in Pennsylvania, and typically hermaphroditic (Ortmann 1919). Hosts of glochidia unknown.

BASIS FOR STATUS CLASSIFICATION. Has not been reported from Chattahoochee River system since the nineteenth century (Brim Box and Williams 2000). Classified as threatened throughout distribution (Williams *et al.* 1993) and deemed extirpated from Alabama (Lydeard *et al.* 1999).

Prepared by: **Stuart W. McGregor**

SCALESHELL MUSSEL

Leptodea leptodon (Rafinesque)

Photo—From Parmalee and Bogan 1998

OTHER NAMES. None.

DESCRIPTION. Has thin, fragile, and compressed shell (max. length approx. 120 mm [4 3/4 in.]), elongate and ovate to rhomboidal in shape, with a rounded anterior margin, bluntly pointed posterior margin and nearly straight ventral margin. Posterior ridge distinct, but low and rounded. Shell disk and posterior ridge unsculptured. Umbos small, compressed, placed anteriorly, and not elevated above hinge line. Periostracum yellowish or olive green, often with numerous wide, faint green rays. Pseudocardinal teeth reduced to very small, tubercular swellings; lateral teeth long and appear as swellings of hinge line anteriorly. Interdentum absent and umbo cavity very shallow or absent. Shell nacre purplish or salmon on upper half, with remainder bluish and iridescent. (Modified from Parmalee and Bogan 1998)

DISTRIBUTION. Historically, from upper Mississippi drainage south to Tennessee River system, and from western New York to southern Michigan and southern Manitoba (Parmalee and Bogan 1998). In Alabama, was confined to Tennessee River at Muscle Shoals (Ortmann 1925). Not reported from Alabama since impoundment of the Tennessee River (Garner and McGregor 2001).

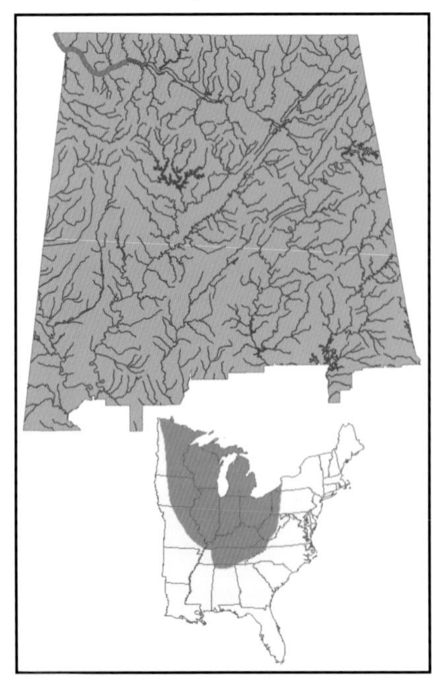

HABITAT. A shoal species, found in clear, unpolluted water with good current, and known to burrow down into substratum several centimeters (inches) (Oesch 1995).

LIFE HISTORY AND ECOLOGY. A long-term brooder (Baker 1928) with glochidia reported in marsupia in September to November and March (Gordon 1991). Only reported host for glochidia is the freshwater drum (USFWS 2001).

BASIS FOR STATUS CLASSIFICATION. Was classified as endangered throughout its distribution (Williams *et al.* 1993) and deemed extirpated from Alabama (Stansbery 1976*a*, Lydeard *et al.* 1999). **Listed as *endangered* by the U.S. Fish and Wildlife Service in 2001.**

***Prepared by:* Stuart W. McGregor**

COOSA MOCCASINSHELL

Medionidus parvulus (Lea)

OTHER NAMES. None.

Photo—From Parmalee and Bogan 1998

DESCRIPTION. Has thin and slightly compressed shell (max. length = 40 mm [1 5/8 in.]), subrhomboid in outline, with a rounded anterior margin, bluntly pointed posterior margin, slightly convex dorsal margin, and straight to somewhat concave ventral margin. Males often slightly arcuate and females slightly more inflated than males posteriorly. Posterior ridge rounded. Low, curved folds or plications adorn posterior slope and may extend onto posterior ridge. Remainder of shell disk unsculptured. Umbos moderately inflated, but hardly elevated above hinge line. Periostracum greenish to yellowish, with numerous, faint green wavy lines that sometimes form wider rays. Pseudocardinal teeth small, compressed, and erect; lateral teeth long and thin, straight to slightly curved. No interdentum present and umbo cavity very shallow. Shell nacre bluish white. (Modified from Simpson 1914, Parmalee and Bogan 1998)

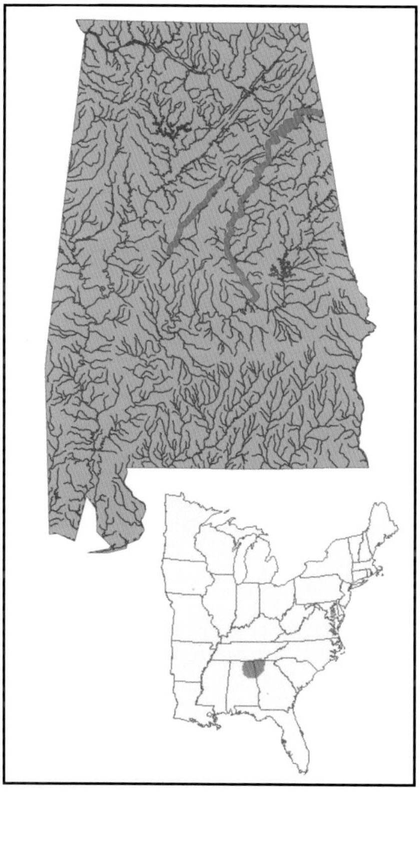

DISTRIBUTION. Endemic to Coosa River system in Alabama, Georgia, and Tennessee (Parmalee and Bogan 1998). Apparently extant only in some headwater rivers in Georgia.

HABITAT. Historically, in sand, gravel, and cobble substrata (Parmalee and Bogan 1998) of small streams to large rivers, mostly above the Fall Line.

LIFE HISTORY AND ECOLOGY. Unknown. Other species of genus are long-term brooders and use darters (Percidae) as glochidial hosts (Zale and Neves 1982*a*, Haag and Warren 1997; O'Brien and Williams 2002; Haag and Warren 2003).

BASIS FOR STATUS CLASSIFICATION. Has experienced a dramatic reduction in distribution and today occurs only in a few widely scattered localities. Classified as endangered throughout its distribution (Williams *et al.* 1993) and in Alabama (Lydeard *et al.* 1999). **Listed as *endangered* by the U.S. Fish and Wildlife Service in 1993.**

***Prepared by:* Wendell R. Haag**

HICKORYNUT

Obovaria olivaria (Rafinesque)

OTHER NAMES. None.

Photo—From Parmalee and Bogan 1998

DESCRIPTION. Has thick, solid, inflated shell (max. length = approx. 100 mm [3 3/4 in.]). Males generally larger than females. Ovate or elliptical in outline, with anterior and ventral margins broadly rounded. Posterior margin broadly pointed in males and more rounded in females. Posterior ridge rounded and barely perceptible. Shell disk and posterior slope smooth. Umbos inflated, directed forward, and elevated above hinge line. Umbo sculpture consists of a few somewhat double-looped bars. Periostracum olive green to yellowish brown, becoming very dark brown with age. Young shells have distinct, fine green rays, but often obscured in old specimens. Pseudocardinal teeth heavy, roughened, and triangular, with small, thin accessory teeth; lateral teeth long, raised, and striate. Interdentum short and wide and umbo cavity shallow. Shell nacre white, but often has a pink or cream wash in center of valve. (Modified from Simpson 1914, Parmalee and Bogan 1998)

DISTRIBUTION. Throughout most of the Mississippi River drainage from Pennsylvania and New York to Minnesota and Kansas, south to Louisiana. Also occurs in St. Lawrence River Basin (Parmalee and Bogan 1998). In the Tennessee River, it apparently only occurred as far upstream as Muscle Shoals (Ortmann 1925), but has not been collected from there in recent years (Garner and McGregor 2001).

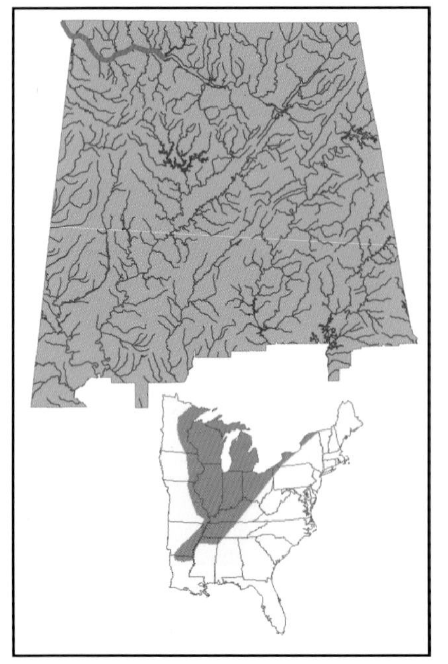

HABITAT. Seems to prefer sand or gravel substrata in water generally more than two meters (6 1/2 feet) deep, with good current. Often found in mussel beds in midstream of larger rivers (Parmalee and Bogan 1998).

LIFE HISTORY AND ECOLOGY. A long-term brooder with eggs and developing glochidia reported in marsupia from August to June. Only reported host of glochidia is shovelnose sturgeon (Coker *et al.* 1921).

BASIS FOR STATUS CLASSIFICATION. Has not been reported from Alabama since 1970s (Gooch *et al.* 1979, Garner and McGregor 2001). While Stansbery (1976*a*) considered it endangered in Alabama, more recently listed as currently stable throughout its distribution (Williams *et al.* 1993) and in Alabama (Lydeard *et al.* 1999). Thought extremely rare in, or extirpated from, entire Tennessee River (Parmalee and Bogan 1998).

***Prepared by:* Stuart W. McGregor**

LITTLEWING PEARLYMUSSEL

Pegias fabula (Lea)

OTHER NAMES. None.

DESCRIPTION. Has oblong shell (rarely exceeding 35 mm [1 3/8 in.]), somewhat triangular in outline, with a rounded anterior margin and posterior margin biangulate in males and truncate in females. Posterior ridge sharp and well defined with a medial ridge anterior to posterior ridge, and a wide, radial depression separating them. Depression ends at marginal biangulation on posterior ventral margin. Biangulation weakened in females, which are more swollen posteriorly and have a more truncate posterior margin. Shell disk and posterior slope without sculpture. Umbos elevated above hinge line. Umbo sculpture consists of heavy, subconcentric rings. Shells almost always heavily eroded, often with no vestige of periostracum. When present, periostracum tawny to brown, with brown or olive green rays. Pseudocardinal teeth irregular and triangular; lateral teeth appear as short, faint ridges. A rudimentary interdentum may or may not be present. Umbo cavity deep and impressed. Shell nacre thickened anteriorly and tan or salmon in umbo cavity, but white elsewhere. (Modified from Simpson 1914, Parmalee and Bogan 1998)

Male littlewing pearlymussel.
Photo—From Parmalee and Bogan 1998

DISTRIBUTION. Endemic to Cumberland and Tennessee River systems (Stansbery 1976*b*, Ahlstedt and Saylor 1995). In Alabama, historically found in Bluewater Creek, Lauderdale County (Ortmann 1925), and possibly other Tennessee River tributaries.

HABITAT. Riffles, often in high gradient tributary streams (Ahlstedt and Saylor 1995). Apparently also occurred in streams of moderate to low gradient prior to modern perturbation of these systems (*e.g.*, Bluewater Creek, Lauderdale County; Ortmann 1925).

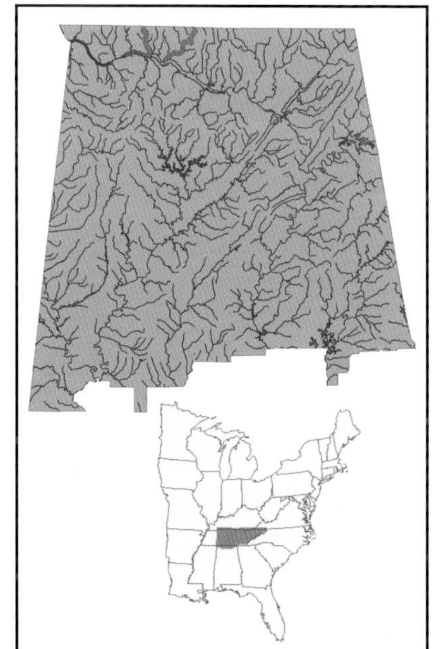

LIFE HISTORY AND ECOLOGY. A long-term brooder. Gravid females reported as early as September (Ortmann 1914, 1921; Ahlstedt and Saylor 1995). Greenside and emerald darters identified as glochidial hosts (Layzer and Madison 1995). Although remaining buried in substrata or under flat rocks for much of their lives, gravid females may emerge and lie on surface during autumn, presumably as part of reproductive activities such as glochidia discharge (Ahlstedt and Saylor 1995).

BASIS FOR STATUS CLASSIFICATION. Suffered severe distribution reductions during twentieth century. Limited distribution, rarity, and declining population trend make it vulnerable to extinction. Classified as endangered throughout its distribution (Stansbery 1970, Williams *et al.* 1993) and extirpated from Alabama (Stansbery 1976*a*, Lydeard *et al.* 1999). **Listed as *endangered* by the U.S. Fish and Wildlife Service in 1976.**

***Prepared by:* Jeffrey T. Garner**

CLUBSHELL

Pleurobema clava (Lamarck)

Photo—From Parmalee and Bogan 1998

OTHER NAMES. Northern Clubshell.

DESCRIPTION. Has solid and moderately inflated shell (max. length = 65 mm [2 5/8 in.]), elongate and triangular in outline, with a broadly rounded anterior margin, obliquely truncate or broadly pointed posterior margin, straight dorsal margin, and slightly convex to straight ventral margin. Posterior ridge rounded, but usually prominent, with a wide, shallow sulcus located anteriorly. Shell disk and posterior slope without sculpture. Umbos full and elevated above hinge line, located and oriented anteriorly, and often project past anterior margin of shell. Umbo sculpture consists of a few irregular, often broken, ridges. Periostracum yellowish or yellowish brown, usually with dark green rays that may be broken into irregular blotches and most prominent on umbo. Pseudocardinal teeth triangular, serrate, and erect, usually parallel to hinge line; lateral teeth long, thin, elevated, and may be straight or slightly curved. Interdentum narrow and umbo cavity shallow. Shell nacre white. (Modified from Simpson 1914, Parmalee and Bogan 1998)

DISTRIBUTION. Historically, from the Tennessee and Cumberland Rivers, north to New York and Minnesota and west to Nebraska (Parmalee and Bogan 1998). However, has disappeared from much of its distribution and appears to be extirpated from Alabama (Garner and McGregor 2001).

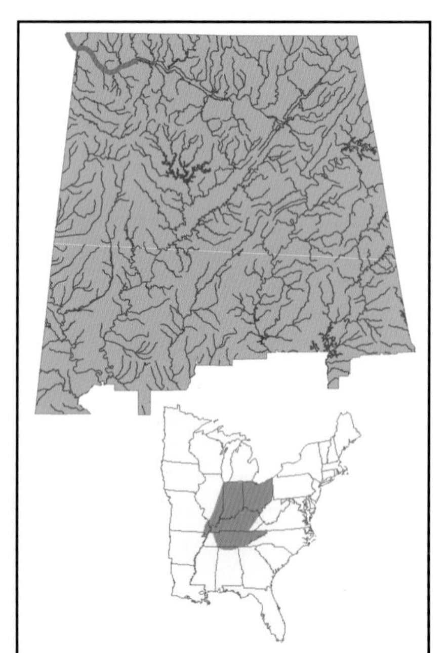

HABITAT. Medium to large rivers, generally occurring in shoal habitats (Parmalee and Bogan 1998).

LIFE HISTORY AND ECOLOGY. A short-term brooder, with gravid females from May through July (Ortmann 1919). Hosts of glochidia unknown.

BASIS FOR STATUS CLASSIFICATION. Suffered drastic habitat loss and population fragmentation during twentieth century. Has been classified as endangered in Alabama (Stansbery 1976*a*) and throughout its distribution (Williams *et al.* 1993), and more recently listed as extirpated from the state (Lydeard *et al.* 1999). **Listed as *endangered* by the U.S. Fish and Wildlife Service in 1993.**

Prepared by: **Jeffrey T. Garner**

BLACK CLUBSHELL

Pleurobema curtum (Lea)

Photo—Bob Jones

OTHER NAMES. None.

DESCRIPTION. Has moderately thick and inflated shell (max. length = 50 mm [2 in.]), elongate triangular in outline, with a very broadly rounded to somewhat truncate anterior margin, bluntly pointed posterior margin, almost straight dorsal margin, and broadly convex ventral margin. Posterior ridge low, positioned near dorsal margin. A low, wide, radial swelling positioned just anterior of mid line. Shell disk and posterior slope without sculpture. Umbos full and elevated well above hinge line, positioned almost at anterior margin. Periostracum greenish in young specimens, darkening to greenish brown with age. Pseudocardinal teeth triangular and radially arranged; lateral teeth moderately long and almost straight. Interdentum short and narrow and umbo cavity shallow. Shell nacre bluish white. (Modified from Simpson 1914, USFWS 2000)

DISTRIBUTION. Endemic to Tombigbee River in Alabama and Mississippi (Stansbery 1976*a*, Williams 1982). Appears to have been extirpated from Alabama with construction of Tennessee-Tombigbee Waterway.

HABITAT. Was restricted to lotic habitat of main stem of Tombigbee River, where it occurred in gravel and sand substrata in moderate to swift current (Williams 1982).

LIFE HISTORY AND ECOLOGY. Unknown. Other species of *Pleurobema* short-term brooders and use cyprinids as host fishes (Haag and Warren 2003).

BASIS FOR STATUS CLASSIFICATION. Most of habitat destroyed by construction of Tennessee-Tombigbee Waterway. A small population persisted until mid-1980s in a short section of upper Tombigbee River in Mississippi not directly modified by waterway (Hartfield and Jones 1989). Extremely rare, if extant, anywhere within its limited historic distribution. Was listed as endangered (Stansbery 1976*a*) in Alabama and throughout its distribution (Williams *et al.* 1993), and most recently considered extirpated from Alabama (Lydeard *et al.* 1999). **Listed as *endangered* by the U.S. Fish and Wildlife Service in 1987.**

***Prepared by:* Wendell R. Haag**

FLAT PIGTOE

Pleurobema marshalli Frierson

Photo—Jim Williams

OTHER NAMES. None.

DESCRIPTION. Has solid, inflated shell (max. length = 60 mm [2 3/8 in.]), ovate to obliquely elliptical in outline, with broadly rounded anterior and posterior margins and gently convex dorsal and ventral margins. Posterior slope well-developed and may be slightly doubled. Although shell generally without a sulcus, usually with a flattened area just anterior to posterior ridge. Shell disk and posterior slope generally without sculpture, but there may be a few weak pustules ventrally. Umbos inflated and elevated well above hinge line, positioned near anterior end and turned slightly forward. Periostracum brown to reddish brown. Pseudocar-dinal teeth triangular and divergent; lateral teeth moderately long and slightly curved. Interdentum well developed, but umbo cavity shallow. Shell nacre white. (Modified from USFWS 2000)

DISTRIBUTION. Endemic to Tombigbee River in Alabama and Mississippi (Stansbery 1976*a*, Williams 1982). Extirpated from Tombigbee River proper, and may be extinct altogether, but could possibly exist in a large tributary of river.

HABITAT. Was restricted to lotic habitat of main channel of Tombigbee River. Occurred in gravel and sand substrata in moderate to swift current (Williams 1982).

LIFE HISTORY AND ECOLOGY. Unknown. Other species of *Pleurobema* short-term brooders and use cyprinids as hosts for their glochidia (Haag and Warren 2003).

BASIS FOR STATUS CLASSIFICATION. All known big river habitat was destroyed by construction of Tennessee-Tombigbee Waterway. Was classified as endangered in Alabama (Stansbery 1976*a*) and throughout its distribution (Williams *et al.* 1993), but more recently considered extirpated from state (Lydeard *et al.* 1999). **Listed as *endangered* by the U.S. Fish and Wildlife Service in 1987**.

***Prepared by:* Wendell R. Haag**

FLUTED KIDNEYSHELL

Ptychobranchus subtentum (Say)

OTHER NAMES. None.

Photo—From Parmalee and Bogan 1998

DESCRIPTION. Has solid, moderately inflated shell (max. length = 120 mm [4 3/4 in.], but usually 80-90 mm [3 1/8 - 3 1/2 in.] long), slightly obovate and subrhomboid in outline, with a broadly rounded anterior margin, straight or slightly concave ventral margin, and slightly convex dorsal margin. Posterior margin generally rounded in males, but females have a slightly pronounced basal region with posterior point raised a little above base line. Posterior ridge pronounced but rounded, sloping to a rounded margin. Posterior slope radially plicate or corrugated in most individuals. Shell disk unsculptured. Umbos fairly compressed, barely rising above hinge line. Periostracum smooth and slightly shiny in young individuals, but becomes duller and satin-like along ventral margins with age. Periostracum greenish yellow, becoming dull brown with age, generally with several broken, wide green rays that may be in form of square spots or zigzag markings. Pseudocardinal teeth stumpy and usually triangular; lateral teeth heavy and slightly curved. Interdentum narrow and curved. Umbo cavity shallow. Shell nacre varies from bluish white to tan or yellowish, often with a wash of salmon. (Modified from Parmalee and Bogan 1998)

DISTRIBUTION. From upper Cumberland and Tennessee Rivers, ranging downstream to Muscle Shoals in the Tennessee River (Ortmann 1925, Parmalee and Bogan 1998). Has not been reported from Alabama since impoundment of the Tennessee River (Garner and McGregor 2001).

HABITAT. Primarily streams and small rivers with fast current and a depth of less than or equal to one meter (3 1/4 feet), and sand or a mixture of sand and gravel substrata (Parmalee and Bogan 1998).

LIFE HISTORY AND ECOLOGY. A long-term brooder (Parmalee and Bogan 1998). Reported fish hosts include rainbow, redline, fantail, and barcheek darters, and banded sculpin (Luo 1993).

BASIS FOR STATUS CLASSIFICATION. Classified as a species of special concern throughout its distribution (Williams *et al.* 1993) and was listed as endangered in Alabama (Stansbery 1976*a*). More recently listed as extirpated from the state (Lydeard *et al.* 1999). **Recently elevated to a *candidate for listing as endangered* by the U.S. Fish and Wildlife Service.**

***Prepared by:* Stuart W. McGregor**

CUMBERLAND BEAN

Villosa trabalis (Conrad)

Photo—From Parmalee and Bogan 1998

OTHER NAMES. None.

DESCRIPTION. Has solid, elongate shell (max. length = 55 mm [2 1/8 in.]). Females grow slightly larger than males. Shell inflated, inequilateral, and irregularly oval, with a rounded anterior margin and ventral margin slightly rounded to straight, converging with posterior-dorsal surface in a rounded point. Males slightly narrowed at center and drawn out posteriorly, with posterior margin obliquely truncate above and ending in a rounded point below. Females higher and more evenly ovate and only slightly truncate behind and above posterior ridge. Posterior ridge full and rounded. Shell disk and posterior slope smooth. Umbos high, elevated slightly above hinge line, and positioned anterior of center and turned anteriorly. Periostracum dingy olive with numerous faint, wavy green rays. Pseudocardinal teeth heavy and triangular, right valve with accessory teeth; lateral teeth long, straight, and heavy. Interdentum narrow and umbo cavity shallow. Shell nacre bluish white, or white with a bluish posterior iridescence. (Modified from Simpson 1914, Parmalee and Bogan 1998)

DISTRIBUTION. Endemic to upper Cumberland River system in Kentucky and Tennessee River system from headwaters downstream to Muscle Shoals (Parmalee and Bogan 1998). Not reported from Alabama since impoundment of Tennessee River.

HABITAT. Most often in swift riffles of small rivers and streams, with gravel or a mixture of sand and gravel substrata (Parmalee and Bogan 1998). Although it was rare, its presence in main stem Tennessee River prior to its impoundment indicates its ability to live in large rivers under some conditions (Ortmann 1925).

LIFE HISTORY AND ECOLOGY. A long-term brooder. Fishes reportedly serving as hosts of glochidia include arrow, barcheek, fantail, Johnny, rainbow, snubnose, dirty, striped, and stripetail darters (Parmalee and Bogan 1998).

BASIS FOR STATUS CLASSIFICATION. Has disappeared from much of its former distribution and populations have become fragmented. Limited distribution and rarity make it vulnerable to extinction. Classified as endangered throughout its distribution (Williams *et al.* 1993) and in Alabama (Lydeard *et al.* 1999). **Listed as *endangered* by the U.S. Fish and Wildlife Service in 1976.**

***Prepared by:* Stuart W. McGregor**

Freshwater Mussels

EXTIRPATED
CONSERVATION ACTION UNDERWAY

Taxa that historically occurred in Alabama, were absent for a period of time, and currently are being reintroduced, or have a plan for being reintroduced, into the state from populations outside the state.

Order Unionoida
Family Unionidae

DROMEDARY PEARLYMUSSEL *Dromus dromas*
OYSTER MUSSEL *Epioblasma capsaeformis*
BIRDWING PEARLYMUSSEL *Lemiox rimosus*
CUMBERLAND MONKEYFACE *Quadrula intermedia*

DROMEDARY PEARLYMUSSEL

Dromus dromas (Lea)

Photo—From Parmalee and Bogan 1998

OTHER NAMES. Camel Shell.

DESCRIPTION. Has solid shell (max. length = 100 mm [3 3/4 in.]), rounded to subtriangular in outline, with all margins convex. Shell compressed, without a distinct posterior ridge. A hump or irregular knob usually present on the median line, near the umbo, and may be an indistinct row of small, low knobs down the midline of shell. Umbo extends a little above the hinge line, and umbo sculpture consists of fine ridges parallel to growth annuli. Periostracum golden brown to reddish brown, with numerous thin rays consisting of fine dark green flecks, and occasional wider rays consisting of irregular blotches. Pseudocardinal teeth heavy, low, and divergent; lateral teeth short and straight, with a wide, flat interdentum between. Umbo cavity deep and compressed. Shell nacre varies from white to pink. (Modified from Simpson 1914, Parmalee and Bogan 1998)

DISTRIBUTION. Confined to upper and middle reaches of Cumberland and Tennessee Rivers (Parmalee and Bogan 1998). In Alabama, historically occurred in Tennessee River downstream to Muscle Shoals. Had not been reported from Alabama since 1930s (van der Schalie 1939) until recent reintroduction.

HABITAT. Lotic areas of medium to large rivers, generally in shoals with clean, mixed substrata ranging in size from sand to cobble (USFWS 1983, Neves 1991). However, may also occur in water deeper than three meters in riverine habitats of Cumberland River (USFWS 1983).

LIFE HISTORY AND ECOLOGY. A long-term brooder, with gravid specimens reported from September through April (Ortmann 1921, Bogan and Parmalee 1998, Jones and Neves 2002*a*). Hosts of glochidia include greenside, fantail, snubnose, tangerine, channel, gilt, and Roanoke darters; blotchside logperch; and logperch (Jones and Neves 2002*a*). However, the Roanoke darter does not occur sympatrically with this species, so not a natural host. Generally remains well buried in the substratum (Neves 1991). Based on archaeological evidence, appears to have been one of the most abundant species in middle reaches of Tennessee River prehistorically (Morrison 1942). Unclear whether *D. dromas* was already in decline when the river was impounded. It has not been reported from the Tennessee River in Alabama since the river was impounded (Garner and McGregor 2001).

BASIS FOR STATUS CLASSIFICATION. Imperiled because of its restricted distribution, rarity, and specialized habitat requirements. Has been classified as endangered throughout its distribution (Williams *et al.* 1993) and in Alabama (Stansbery 1976*a*), and is now listed as probably extirpated from Alabama (Lydeard *et al.* 1999). **Listed as *endangered* by the U.S. Fish and Wildlife Service in 1976.**

***Prepared by*: Jeffrey T. Garner**

OYSTER MUSSEL

Epioblasma capsaeformis (Lea)

OTHER NAMES. None.

DESCRIPTION. Has a moderately thin shell (max. length = 70 mm [2 3/4 in.]). Males elliptical in outline whereas females of reproductive age are ovate, with a large, rounded marsupial swelling posteriorly. Posterior ridge low and may be slightly double in males, giving the shell a slightly biangulate posterior margin. Marsupial swelling of females thin, often adorned with small marginal serrations, and usually separated from remainder of shell by slight sulci anteriorly and posteriorly. Shell disk and posterior slope without sculpture. Umbos moderately inflated and elevated slightly above the hinge line. Umbo sculpture consists of very weak parallel loops. Periostracum yellowish green, with thin green rays, although marsupial swelling of females is usually dark green, sometimes black. Pseudocardinal teeth small and triangular, separated from the short, slightly curved lateral teeth by a very narrow interdentum. Umbo cavity shallow. Shell nacre bluish white. (Modified from Simpson 1914, Johnson 1978, Parmalee and Bogan 1998)

Female oyster mussel.
Photo—From Parmalee and Bogan 1998

DISTRIBUTION. Endemic to Tennessee and Cumberland River systems. Historically occurred in Tennessee River downstream to Muscle Shoals, but naturally occurring populations not reported there since impoundment (Garner and McGregor 2001).

HABITAT. Lotic areas of small to large rivers, and large creeks. Although generally found in shoals with clean gravel substrata, also may inhabit quieter areas of shoals in substrata consisting of gravel and some mud (Neves 1991).

LIFE HISTORY AND ECOLOGY. A long-term brooder, holding glochidia from autumn through mid-May of the following year. Females display mantle tissue within their marsupial swelling during certain periods of the year. Display is bright blue in the Clinch River population and slate gray in the Duck River population. During display the shell is gaped to present the tissue, and small fish, including identified hosts, have been observed striking it. Glochidia discharged individually, as opposed to being discharged as part of host-attracting conglutinates. Fishes reported to serve as hosts for glochidia include spotted, redline, wounded, and dusky darters, as well as banded sculpin (Summarized from TVA 1986)

BASIS FOR STATUS CLASSIFICATION. Unable to cope with impoundment, and has disappeared from most of its historic distribution. Was classified as endangered throughout its distribution (Williams *et al.* 1993) and in Alabama (Lydeard *et al.* 1999). **Listed as *endangered* by the U.S. Fish and Wildlife Service in 1997**.

***Prepared by:* Jeffrey T. Garner**

BIRDWING PEARLYMUSSEL

Lemiox rimosus (Rafinesque)

OTHER NAMES. None.

DESCRIPTION. Has solid, slightly inflated shell (rarely > 50 mm [2 in.] long), subtriangular to subovate in outline, with a broadly rounded anterior margin and obliquely truncate posterior margin, that comes to a broad point ventrally; slightly convex dorsal margin and straight ventral margin. Posterior ridge somewhat rounded, but distinct. Males grow larger than females and often have a shallow, radial depression just anterior of posterior ridge. Females lack depression and often have a weak marsupial swelling posteriorly. Posterior half of shell marked by strong corrugated, subradial sculpture. Umbos high and turned anteriorly, barely elevated above hinge line. Umbo sculpture consists of distinct, double-looped bars. Periostracum dull green or yellowish green, and marked with weak rays in young specimens; darkens with age to almost black. Pseudocardinal teeth low and rugged; lateral teeth short and slightly curved and interdentum broad. Umbo cavity very shallow. Shell nacre white. (Modified from Simpson 1914, Parmalee and Bogan 1998)

Photo—From Parmalee and Bogan 1998

DISTRIBUTION. Historically occurred in Tennessee River system, from its headwaters downstream to Muscle Shoals. A disjunct population in the Duck River (USFWS 1983) may be the only viable population (Parmalee and Bogan 1998). Was extant, at least in small numbers, in the Clinch, Powell, and Elk Rivers as late as early 1980s (USFWS 1983).

HABITAT. Flowing water in small to large rivers (USFWS 1983, Parmalee and Bogan 1998). However, was extirpated from large rivers with their impoundment.

LIFE HISTORY AND ECOLOGY. A long-term brooder, gravid from mid-September through early July (Ortmann 1916, TVA 1986). Releases glochidia individually, with those aborted as conglutinates apparently not viable (TVA 1986). Fishes reportedly serving as hosts of glochidia are greenside and banded darters (TVA 1986).

BASIS FOR STATUS CLASSIFICATION. Classified as endangered throughout its distribution (Williams *et al.* 1993) and as extirpated from Alabama (Lydeard *et al.* 1999). **Listed as *endangered* by the U.S. Fish and Wildlife Service in 1976**.

***Prepared by:* Jeffrey T. Garner**

CUMBERLAND MONKEYFACE

Quadrula intermedia (Conrad)

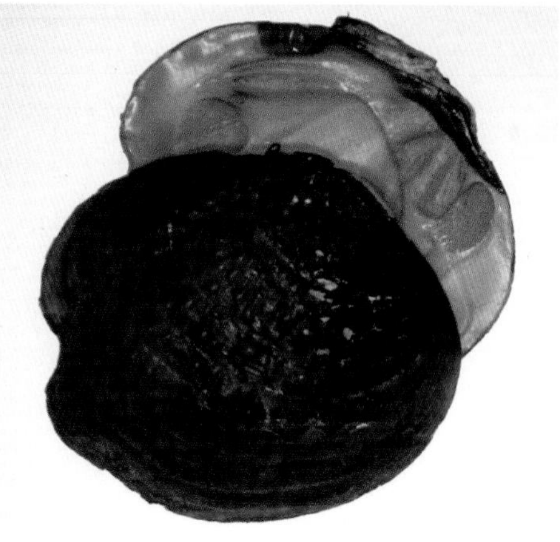

Photo—From Parmalee and Bogan 1998

OTHER NAMES. None.

DESCRIPTION. Has compressed and solid shell (max. length = 80 mm [3 1/8 in.]), suborbicular to subquadrate in outline, rounded anteriorly and posteriorly, but with posterior margin biangulate. Dorsal and ventral margins usually slightly rounded. Posterior ridge weak and posterior slope marked with a deep, wide, radial depression that gives biangulation to posterior margin. All but anterior one-third of shell surface covered with large tubercles. Umbos somewhat compressed and elevated slightly above hinge line. Periostracum greenish yellow, with variable, fine, angular green spots, chevrons, or zigzags, and may also have broken green rays. Pseudocardinal teeth radially striate and triangular; lateral teeth short, broad, and slightly curved. Interdentum wide and umbo cavity deep and compressed. Shell nacre white, often tinted with salmon, especially posteriorly. (Modified from Simpson 1914, Parmalee and Bogan 1998)

DISTRIBUTION. Endemic to Tennessee River system, historically found from its headwaters downstream to Muscle Shoals (Parmalee and Bogan 1998). Also occurred in some large tributaries such as the Elk River. Appears to be extirpated from Alabama.

HABITAT. Occurs in lotic habitat of small to medium rivers, typically in gravel substrata. Usually found well buried in substratum (Neves 1991). However, its presence at Muscle Shoals prior to impoundment of Tennessee River indicates that it can live in large rivers under some conditions (Ortmann 1925).

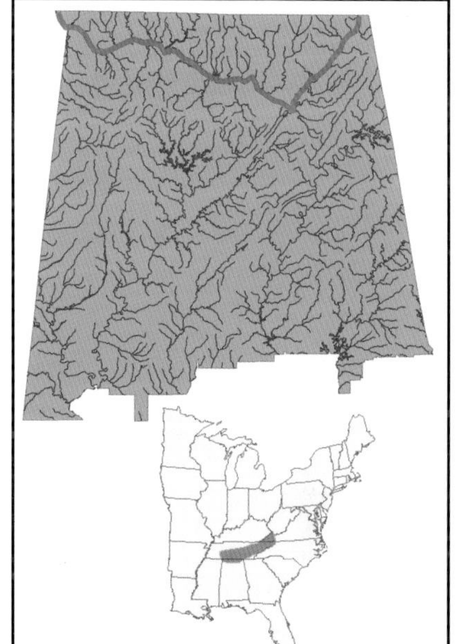

LIFE HISTORY AND ECOLOGY. A short-term brooder, with females gravid in May and June (TVA 1986). Fishes reported to serve as hosts for glochidia include streamline and blotched chubs (TVA 1986, Yeager and Saylor 1995).

BASIS FOR STATUS CLASSIFICATION. Suffered drastic reductions in distribution with impoundment of Tennessee River system. Limited distribution, rarity, and susceptibility to habitat degradation make it vulnerable to extinction. Classified as endangered throughout its distribution (Williams *et al.* 1993). Was listed as endangered in Alabama (Stansbery 1976*a*), but more recently listed as extirpated there (Lydeard *et al.* 1999). **Listed as *endangered* by the U.S. Fish and Wildlife Service in 1976.**

***Prepared by:* Jeffrey T. Garner**

Freshwater Mussels

PRIORITY I
HIGHEST CONSERVATION CONCERN

Taxa critically imperiled and at risk of extinction/extirpation because of extreme rarity, restricted distribution, decreasing population trend/population viability problems, and specialized habitat needs/habitat vulnerability due to natural/human-caused factors. Immediate research and/or conservation action required.

Order Unionoida
Family Margaritiferidae

SPECTACLECASE *Cumberlandia monodonta*
ALABAMA PEARLSHELL *Margaritifera marrianae*

Family Unionidae

MUCKET *Actinonaias ligamentina*
SLIPPERSHELL MUSSEL *Alasmidonta viridis*
FANSHELL *Cyprogenia stegaria*
ALABAMA SPIKE *Elliptio arca*
DELICATE SPIKE *Elliptio arctata*
SPIKE *Elliptio dilatata*
FLUTED ELEPHANTEAR *Elliptio mcmichaeli*
INFLATED SPIKE *Elliptio purpurella*
PURPLE BANKCLIMBER *Elliptoideus sloatianus*
CUMBERLAND COMBSHELL *Epioblasma brevidens*
SOUTHERN COMBSHELL *Epioblasma penita*
SNUFFBOX *Epioblasma triquetra*
SHINY PIGTOE *Fusconaia cor*
FINERAYED PIGTOE *Fusconaia cuneolus*
NARROW PIGTOE *Fusconaia escambia*
ROUND EBONYSHELL *Fusconaia rotulata*
LONGSOLID *Fusconaia subrotunda*
CRACKING PEARLYMUSSEL *Hemistena lata*
PINK MUCKET *Lampsilis abrupta*
SOUTHERN SANDSHELL *Lampsilis australis*
SHINYRAYED POCKETBOOK *Lampsilis subangulata*
ALABAMA LAMPMUSSEL *Lampsilis virescens*
SLABSIDE PEARLYMUSSEL *Lexingtonia dolabelloides*
CUMBERLAND MOCCASINSHELL *Medionidus*
GULF MOCCASINSHELL *Medionidus penicllatus*
RING PINK *Obovaria retusa*
ROUND HICKORYNUT *Obovaria subrotunda*
WHITE WARTYBACK *Plethobasus cicatricosus*
ORANGEFOOT PIMPLEBACK *Plethobasus cooperianus*
SHEEPNOSE *Plethobasus cyphyus*
PAINTED CLUBSHELL *Pleurobema chattanoogaense*
DARK PIGTOE *Pleurobema furvum*
SOUTHERN PIGTOE *Pleurobema georgianum*
TENNESSEE CLUBSHELL *Pleurobema oviforme*
OVATE CLUBSHELL *Pleurobema perovatum*
ROUGH PIGTOE *Pleurobema plenum*
OVAL PIGTOE *Pleurobema pyriforme*
PYRAMID PIGTOE *Pleurobema rubrum*
ROUND PIGTOE *Pleurobema sintoxia*
HEAVY PIGTOE *Pleurobema taitianum*
KIDNEYSHELL *Ptychobranchus fasciolaris*
TRIANGULAR KIDNEYSHELL *Ptychobranchus greenii*
SOUTHERN KIDNEYSHELL *Ptychobranchus jonesi*
RABBITSFOOT *Quadrula cylindrica cylindrica*
SCULPTURED PIGTOE *Quincuncina infucata*
CREEPER *Strophitus undulatus*
SOUTHERN PURPLE LILLIPUT *Toxolasma corvunculus*
PALE LILLIPUT *Toxolasma cylindrellus*
DEERTOE *Truncilla truncata*

SPECTACLECASE

Cumberlandia monodonta (Say)

Photo—From Parmalee and Bogan 1998

OTHER NAMES. None.

DESCRIPTION. Moderately solid shell (max. length = 180 mm [7 1/16 in.]) elongate and arcuate with rounded anterior and posterior margins; taller anteriorly than posteriorly and somewhat inflated from the umbo posteriorly, forming a broadly rounded posterior ridge. Posterior slope and disk of shell unsculptured. Umbo sculpture in the form of ridges parallel to growth annuli. Periostracum of young shells tan to light brown and may have a greenish tinge, darkening to black or dark brown with age. Pseudocardinal teeth rudimentary, consisting of a single peg-like tooth in right valve that fits into a depression in left valve. Pseudocardinals somewhat better developed in young shells; lateral teeth become indistinct with age. No interdentum, and umbo cavity shallow. Shell nacre white. (Modified from Simpson 1914, Parmalee and Bogan 1998, Baird 2000)

DISTRIBUTION. Throughout Mississippi River system, including Tennessee River of Alabama (Parmalee and Bogan 1998). Extant only in riverine reaches downstream of Wilson and Guntersville Dams (Garner and McGregor 2001).

HABITAT. A variety of substrata, including gravel, sand, and mud, in free-flowing, medium to large rivers, often in water more than four meters (13 feet) deep. Lives mostly among, and under, large rocks (Parmalee and Bogan 1998).

LIFE HISTORY AND ECOLOGY. Reproductive biology in Gasconade and Meramec Rivers described by Baird (2000). A short-term brooder, with gravid period in April and May. White, feather-shaped conglutinates observed, but not determined if they functioned as host attractants, or were simply aborted glochidial contents. Experimental determination of glochidial hosts inconclusive, but glochidia from natural infestations observed on bigeye chub and shorthead redhorse. Species often occurs in dense beds, with local densities reported as high as 120 per square meter in Gasconade and Meramec Rivers.

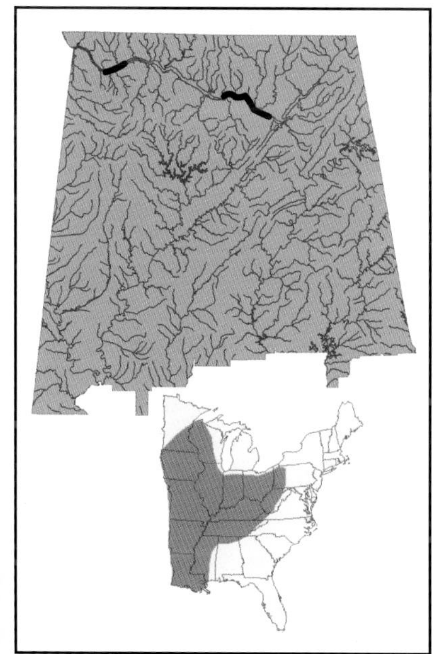

BASIS FOR STATUS CLASSIFICATION. Although widespread, has suffered extensive habitat loss and population fragmentation caused by impoundment of rivers and other human perturbations throughout Mississippi River drainage. Recently categorized as rare in the two areas where extant (Garner and McGregor 2001). Restricted distribution, specialized habitat requirements, and declining population trend make it vulnerable to extirpation from the state. Listed as a species of special concern (Stansbery 1976*a*) and considered imperiled in Alabama (Lydeard *et al.* 1999), and classified as threatened throughout its distribution (Williams *et al.* 1993). **Recently elevated to *candidate status* for protection by the U.S. Fish and Wildlife Service.**

Prepared by: **Jeffrey T. Garner**

ALABAMA PEARLSHELL

Margaritifera marrianae R. I. Johnson

OTHER NAMES. None.

Photo—Arthur Bogan

DESCRIPTION. Has moderately thick and compressed shell (max. length = 95 mm [3 3/4 in.]), oblong and subrhomboidal in outline, being rounded to somewhat truncate anteriorly and bluntly pointed posteriorly, with the point located toward the posterior-ventral margin. Ventral margin straight to slightly convex and dorsal margin convex and meeting the upper extent of posterior margin obliquely. Posterior ridge low and rounded, somewhat doubled, and adorned with strong corrugations that extend onto the posterior slope and for a short distance onto disk of shell. Periostracum smooth on disk, but roughened on corrugated areas; varies from olivaceous to very dark brown or black. Pseudocardinal teeth triangular, low, and stumpy; lateral teeth straight with no interdentum between. Umbo cavity wide and shallow. Shell nacre white, sometimes with pale purple cast. (Modified from Johnson 1983, Mott and Hartfield 1994)

DISTRIBUTION. Endemic to four-county area in south-central Alabama. Most of the area lies in the headwaters of the Escambia River drainage in Butler, Conecuh, and Crenshaw Counties, but a disjunct population is in Limestone Creek, a nearby tributary of the lower Alabama River in Monroe County (Mott and Hartfield 1994, Shelton 1997).

HABITAT. Shallow riffles and pool margins in substrata consisting of silty sand, sand, gravel, or a mixture of sand and gravel in headwater creeks (Mott and Hartfield 1994, Shelton 1997).

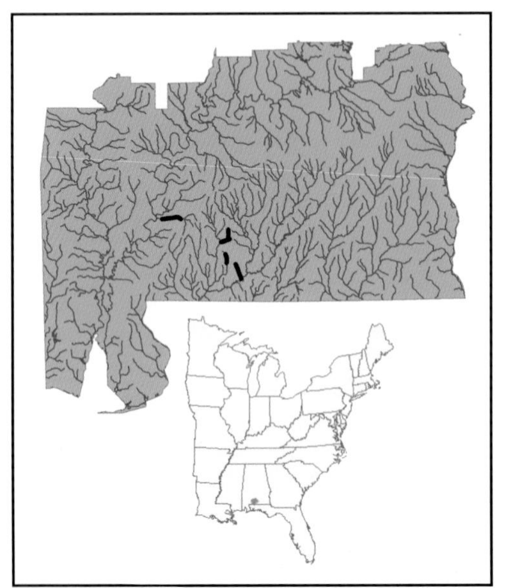

LIFE HISTORY AND ECOLOGY. Little known, although observed occupying streams in pairs, with male upstream of female (Shelton 1997). Females gravid in December, but not February, March, or July (Williams *et al.*, in review). Hosts of glochidia unknown.

BASIS FOR STATUS CLASSIFICATION. Restricted distribution, rarity, and declining population trend make it highly susceptible to extinction. Classified as endangered throughout its distribution (Williams *et al.* 1993), imperiled in Alabama (Lydeard *et al.* 1999), and **currently considered a *candidate for protection* by the U.S. Fish and Wildlife Service.**

Prepared by: **Stuart W. McGregor**

MUCKET

Actinonaias ligamentina (Lamarck)

Photo—From Parmalee and Bogan 1998

OTHER NAMES. Grass Mucket, Brass Mucket, Steamboat Mucket.

DESCRIPTION. Has solid shell (max. length = 140 mm [5 1/2 in.]), variable in outline, with adults usually being elongate oval to elliptical. Juveniles may be somewhat trapezoidal in outline. Dorsal and ventral margins generally slightly convex, but may be straight in subadults. Anterior margin broadly rounded and posterior margin narrowly to broadly rounded. Adults usually moderately inflated, but juveniles compressed. Females may be slightly swollen posteriorly. Posterior ridge low and rounded, but posterior slope may be moderately steep near the umbo. Shell disk and posterior slope without sculpture. Umbos broad and slightly inflated, barely elevated above the hinge line. Umbo sculpture consists of faint, irregular, double-looped ridges. Periostracum smooth and varies from tawny to greenish, sometimes with variable green rays, usually darkening to dark brown with age. Pseudocardinal teeth triangular and striate, and separated from slightly curved lateral teeth by a long, narrow interdentum. Umbo cavity open and moderately shallow. Shell nacre generally white, rarely light pink. (Modified from Simpson 1914, Parmalee and Bogan 1998)

DISTRIBUTION. Throughout the Mississippi River system, with exception of extreme southern and western reaches. Also occurs in St. Lawrence River Basin and tributaries of Lakes Erie, Michigan, and Ontario (Burch 1975, Parmalee and Bogan 1998). Apparently never common in lower bend of the Tennessee River, it has been reduced to no more than two tributary populations in Alabama, in Shoal and Second Creeks, Lauderdale County. Viability of both populations is questionable.

HABITAT. A variety of habitats, ranging from shoals with gravel and cobble bottoms to mud-bottomed pools (Parmalee and Bogan 1998).

LIFE HISTORY AND ECOLOGY. A long-term brooder, being gravid from August to May (Surber 1912). Glochidia reportedly use a variety of fish hosts, including rock bass, green sunfish, orangespotted sunfish, bluegill, smallmouth bass, largemouth bass, white crappie, black crappie, yellow perch, sauger, white bass, and banded killifish (Young 1911, Lefevre and Curtis 1910, Coker *et al.* 1921). American eels and tadpole madtoms natural hosts, but deemed of little significance (Coker *et al.* 1921).

BASIS FOR STATUS CLASSIFICATION. Restricted distribution, rarity, and declining population trend make species vulnerable to extirpation from state, where it has been listed as endangered (Stansbery 1976*a*) and imperiled (Lydeard *et al.* 1999). Williams *et al.* (1993) listed it as currently stable throughout most of its distribution.

***Prepared by:* Jeffrey T. Garner**

SLIPPERSHELL MUSSEL

Alasmidonta viridis (Rafinesque)

Photo—From Parmalee and Bogan 1998

OTHER NAMES. None.

DESCRIPTION. Has moderately thick shell (max. length = 55 mm [2 1/8 in.]), subrhomboidal in outline, with slightly convex dorsal and ventral margins, rounded anterior margin, and a posterior margin that is truncate or bluntly pointed ventrally. A well-developed, rounded posterior ridge extends to the point of the posterior margin. Shell disk and posterior slope unsculptured. A moderately inflated umbo is raised slightly above the hinge line and the umbo sculpture consists of irregular, heavy loops. Periostracum of young specimens greenish or yellowish, with numerous wavy green rays, darkening with age to a dark olive brown with rays often becoming indistinct. Pseudocardinal teeth triangular and somewhat rudimentary; lateral teeth indistinct. Interdentum narrow or absent and umbo cavity shallow. Shell nacre white. (Modified from Parmalee and Bogan 1998)

DISTRIBUTION. Widespread in eastern United States, and distributed from Lakes Huron, St. Clair, and Erie, and upper Mississippi River system, south to Ohio, Cumberland, and Tennessee River systems (Parmalee and Bogan 1998). In Alabama, confined to tributary streams of the Tennessee River system. Streams where recent collections have been made include Hurricane Creek in Paint Rock River system, Jackson County (Ahlstedt 1995, McGregor and Shelton 1995), and Fowler Creek in Flint River system, Madison County (McGregor and Shelton 1995).

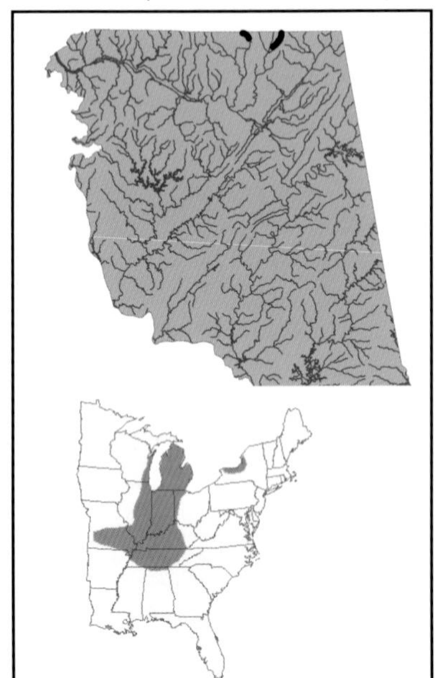

HABITAT. Often found in slow to moderate current, usually at depths of less than one meter (Neves 1991). May reside in substrata ranging from sand to gravel and appears somewhat tolerant of silt (Mathiak 1979). Often found associated with aquatic vegetation, such as water willow (*Justicia americana*) (Parmalee and Bogan 1998).

LIFE HISTORY AND ECOLOGY. A long-term brooder, becoming gravid in September (Ortmann 1921). Hosts of glochidia unknown.

BASIS FOR STATUS CLASSIFICATION. Restricted distribution, rarity, and declining population trend make species vulnerable to extirpation. Classified as a species of special concern in Alabama (Lydeard *et al.* 1999) and throughout its distribution (Williams *et al.* 1993).

Prepared by: **Stuart W. McGregor**

FANSHELL

Cyprogenia stegaria (Rafinesque)

Photo—From Parmalee and Bogan 1998

OTHER NAMES. Eastern Fanshell, Ohio Fanshell, Pimpleback, Ringed Wartyback.

DESCRIPTION. Has solid shell (max. length = 70 mm [2 3/4 in.]) rounded in outline, with all margins at least slightly convex, although the posterior margin may be somewhat truncate. Posterior ridge well developed and angular proximal to the umbo, but becomes narrowly rounded ventrally. Shell sculpture consists of numerous rounded and irregular pustules on the posterior two thirds, with pustules near the middle of the shell often forming regular rows. Umbos moderately inflated and raised above the hinge line. Strong, low, concentric ridges, indicative of growth rests, often present. Umbo sculpture rudimentary, consisting of weak ridges. Periostracum usually pale greenish yellow, but may darken to yellowish brown in older individuals. Pattern on periostracum consists of small, green flecks or dots, usually arranged as wide rays. Pseudocardinal teeth thick, rough, and divergent; lateral teeth short, heavy, and slightly curved. Interdentum wide and flat, and the umbo cavity shallow and compressed. Shell nacre white. (Modified from Simpson 1914, Parmalee and Bogan 1998)

DISTRIBUTION. Historically, the Ohio, Tennessee, and Cumberland River systems (Simpson 1914). In Alabama, historically found in the Tennessee River across northern Alabama, and probably the Elk River. Although extant, it is very rare in the tailwaters of Wilson Dam, but has not been collected in Guntersville Dam tailwaters since reported by Isom (1969). Viability of Wilson tailwater population questionable.

HABITAT. Lotic areas of medium to large rivers, generally in shoals with clean, coarse, sand and gravel substrata (Neves 1991). However, also occurs in water over three meters deep in riverine habitats downstream of Tennessee River Dams (*e.g.*, Wilson Dam).

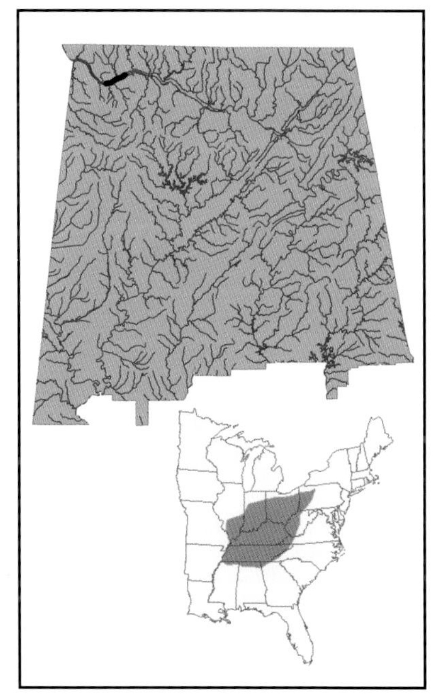

LIFE HISTORY AND ECOLOGY. Reproductive biology from Tennessee River headwaters (Jones and Neves 2002*b*) indicates it is a long-term brooder, with a gravid period from October through May. Produces conglutinates that resemble worms. Conglutinates are red when containing developing embryos, but pale with maturity. Hosts for glochidia include mottled and banded sculpins; greenside, snubnose, banded, tangerine, and Roanoke darters; blotchside logperch; and logperch (Jones and Neves 2002*b*). However, Roanoke darter does not occur sympatrically with this species, so is not a natural host.

BASIS FOR STATUS CLASSIFICATION. Has suffered drastic habitat loss and population fragmentation with impoundment of rivers throughout its distribution. Vulnerable to extinction due to limited distribution, specialized habitat requirements, and susceptibility to habitat degradation. Was classified as endangered throughout its distribution (Williams *et al.* 1993) and in Alabama (Stansbery 1976*a*), but now listed as extirpated from the state (Lydeard *et al.* 1999). **Listed as *endangered* by the U.S. Fish and Wildlife Service in 1991**.

***Prepared by*: Jeffrey T. Garner**

ALABAMA SPIKE

Elliptio arca (Conrad)

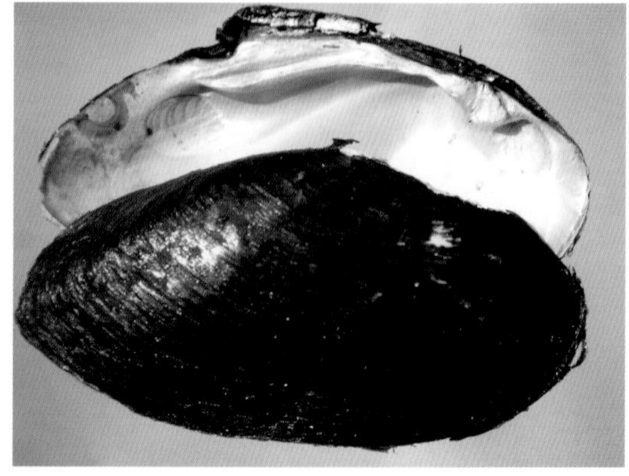

Photo—Jim Williams

OTHER NAMES. None.

DESCRIPTION. Has solid and compressed to moderately inflated shell (max. length = 75 mm [2 15/16 in.]) elliptical in outline, with a rounded anterior margin and bluntly pointed or biangulate posterior margin; dorsal and ventral margins more or less straight. Posterior ridge variable and varies from well formed and angled, to low and rounded or absent. Disk and slope of shell unsculptured. Umbos low and compressed, not elevated above the hinge line. Umbo sculpture in form of strong longitudinal bars that may be double looped. Periostracum rough and dull olive brown to very dark brown, becoming darker with age. Pseudocardinal teeth triangular, usually deeply serrated and widely divergent; lateral teeth long and straight. Interdentum very narrow or absent and there is no umbo cavity. Shell nacre highly variable, ranging from purple to salmon or white. (Modified from Simpson 1914, Parmalee and Bogan 1998)

DISTRIBUTION. Endemic to Mobile Basin in Alabama, Georgia, Mississippi, and Tennessee. Although widespread in Mobile Basin, population in Sipsey River is only one that appears healthy.

HABITAT. Lotic areas in medium to large streams both above and below the Fall Line. Highest densities occur in shallow, swift shoals in gravel/sand substrata, but individuals also can be found in deep gravel and sand-bottomed runs with slow but steady current. Individuals rarely found in pools, silty stream margins, or backwater areas (W. R. Haag and M. L. Warren, USDA Forest Service, unpubl. data).

LIFE HISTORY AND ECOLOGY. A short-term brooder, and females release glochidia in June and July. Method of host infestation unknown, but glochidia released in association with copious amounts of mucus, which may serve to entangle fishes. Primary hosts for glochidia are redspotted and blackbanded darters. Southern sand darter is a marginal host. (Summarized from Haag and Warren 2003)

BASIS FOR STATUS CLASSIFICATION. Vulnerable to extinction due to its restricted distribution, rarity, and declining population trend. Occurs primarily in small, isolated populations in Black Warrior, Coosa, and Tombigbee River systems, many of which could be easily lost with further habitat degradation. Was classified as threatened throughout its distribution (Williams *et al.* 1993), and in Alabama was listed as endangered (Stansbery 1976*a*) and imperiled by Lydeard *et al.* (1999).

Prepared by: **Wendell R. Haag**

DELICATE SPIKE

Elliptio arctata (Conrad)

OTHER NAMES. None.

DESCRIPTION. Shell elongate and somewhat compressed (max. length = 80 mm [3 1/8 in.]), elliptical and arcuate in outline, with a ventral margin straight to slightly concave, dorsal margin slightly convex, anterior margin narrowly rounded, and posterior margin narrowly rounded to somewhat truncate. Posterior ridge low and rounded. Shell disk and posterior ridge without sculpture. Umbos low and not elevated above the hinge line. Umbo sculpture consists of corrugated ridges parallel to growth lines. Periostracum varies from dull yellowish green in juveniles to dark brown or black in old specimens. Pseudocardinal teeth triangular, low, and stumpy; lateral teeth long, thin, slightly curved, and remote from pseudocardinals. No interdentum, and umbo cavity very shallow. Shell nacre dull bluish white, but may be salmon in umbo cavity. (Modified from Simpson 1914, Parmalee and Bogan 1998)

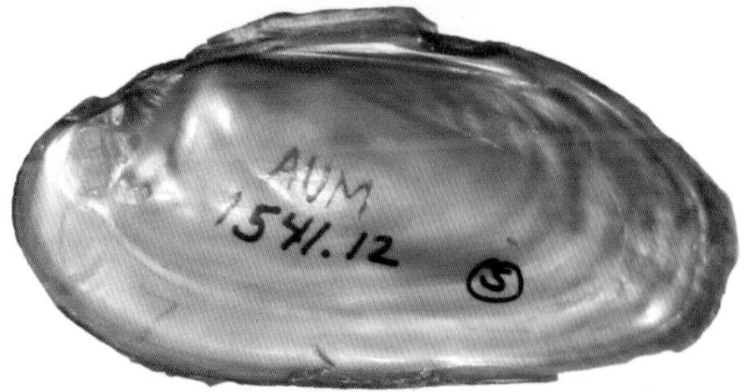

Photo—Mike Gangloff

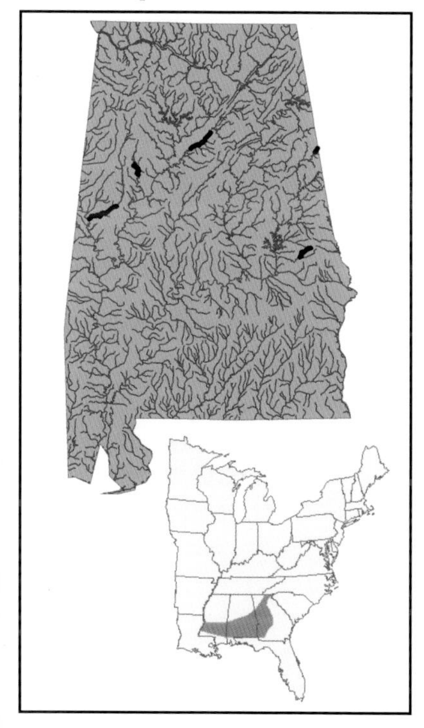

DISTRIBUTION. Problematic. Has been reported from throughout Mobile Basin. A form closely resembling *E. arctata* occurs in Alabama reaches of Escambia River system, but its true identity is uncertain at this time.

HABITAT. In coarse sand and gravel, often under and around large rocks, and usually in current (Parmalee and Bogan 1998). Several may be found packed vertically under large rocks (Hurd 1974).

LIFE HISTORY AND ECOLOGY. Unknown, but presumably a short-term brooder like its congeners.

BASIS FOR STATUS CLASSIFICATION. Limited distribution, rarity, and specialized habitat requirements make *E. arctata* vulnerable to extinction. Was listed as threatened throughout its distribution (Williams *et al.* 1993) and imperiled in Alabama (Lydeard *et al.* 1999).

Prepared by: **Jeffrey T. Garner**

SPIKE

Elliptio dilatata (Rafinesque)

OTHER NAMES. Lady Finger.

Photo—Mike Gangloff

DESCRIPTION. Has thick, compressed shell (max. length = 120 mm [4 3/4 in.]), elliptical in outline, with a rounded anterior margin, bluntly pointed posterior margin, slightly convex dorsal margin, and straight to slightly convex ventral margin. Posterior ridge strongly developed and rounded. Shell disk and posterior ridge unsculptured. Umbos depressed and flattened, not elevated above the hinge line. Umbo sculpture consists of moderate to heavy loops. Periostracum varies from light brown to yellowish green in young shells, but becomes dark greenish brown to black with age. Rays may be present in young shells, but often become obscure with age. Pseudocardinal teeth strong and rough, two in left valve, one in right; lateral teeth straight and rough. Interdentum moderately wide. Shell nacre variable, varying from purple to light pink, rarely white (Modified from Cummings and Mayer 1992, Howells *et al.* 1996, Parmalee and Bogan 1998, Strayer and Jirka 1997)

DISTRIBUTION. Widespread in eastern United States, occurring throughout much of the Mississippi River system and portions of the Great Lakes drainage. In Alabama, restricted to Tennessee River drainage (Cummings and Mayer 1992, Howells *et al.* 1996, Parmalee and Bogan 1998, Strayer and Jirka 1997). Extant in tailwaters of Wilson and Guntersville Dams, where it was listed as uncommon and rare, respectively (Garner and McGregor 2001).

HABITAT. Creeks and rivers, primarily in areas with current, but also may be found in lakes under some conditions. Preferred substrata appear to be sand and gravel (Cummings and Mayer 1992, Howells *et al.* 1996, Parmalee and Bogan 1998, Strayer and Jirka 1997).

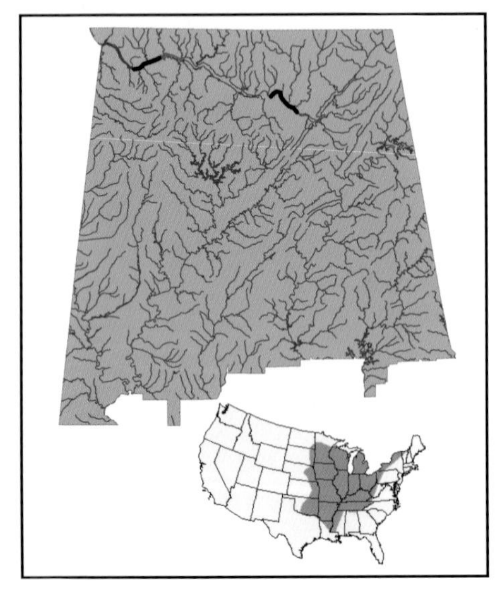

LIFE HISTORY AND ECOLOGY. A short-term brooder, with a gravid season from May to August. Reported glochidial hosts include rock bass, white and black crappie, rainbow darter, yellow perch, banded sculpin, gizzard shad, and flathead catfish. A dominant species in some mussel beds (Cummings and Mayer 1992, Howells *et al.* 1996, Parmalee and Bogan 1998, Strayer and Jirka 1997).

BASIS FOR STATUS CLASSIFICATION. Restricted distribution, rarity, and specialized habitat requirements make *E. dilatata* vulnerable to extirpation from Alabama. Currently listed as stable throughout its distribution (Williams *et al.* 1993) and in Alabama (Lydeard *et al.* 1999).

***Prepared by:* Jeffrey J. Herod**

FLUTED ELEPHANTEAR

Elliptio mcmichaeli Clench and Turner

OTHER NAMES. None.

Photo—Arthur Bogan

DESCRIPTION. Has thin and somewhat compressed shell (max. length = 100 mm [3 15/16 in.]), elliptical in outline and rounded anteriorly, and bluntly pointed posteriorly. Posterior ridge well defined and posterior slope slightly concave, often with a series of small, arcuate ridges extending from posterior ridge to dorsal margin. Umbos broad, but not elevated above the hinge line, and positioned well anterior of center. Periostracum very dark brown and may be faintly rayed in young specimens. One pseudocardinal tooth present in right valve, one large and one small pseudocardinal tooth present in left valve. Shell nacre white to pale pink or salmon. (Modified from Clench and Turner 1956)

DISTRIBUTION. Endemic to Choctawhatchee River system in Alabama and Florida (Clench and Turner 1956). Has suffered recent declines from some portions of its distribution (Blalock-Herod *et al.*, in review).

HABITAT. Areas with moderate current and sand, or a mixture of sand and gravel substrata. Appears somewhat tolerant of silty conditions (Blalock-Herod *et al.*, in review).

LIFE HISTORY AND ECOLOGY. A short-term brooder, found gravid with viable glochidia in April and May (J. J. Herod, USFWS, pers. comm., 2002). Glochidial hosts unknown, but its disappearance from Pea River upstream of Elba Dam suggests it uses a migratory fish (Blalock-Herod *et al.*, in review).

BASIS FOR STATUS CLASSIFICATION. Imperiled based on its restricted distribution, vulnerability to habitat degradation, and recent population declines. Minimal additional habitat degradation could result in extinction. Classified as a species of special concern throughout its distribution (Williams *et al.* 1993, Blalock-Herod *et al.*, in review). Listed as a species of special concern in Alabama (Lydeard *et al.* 1999).

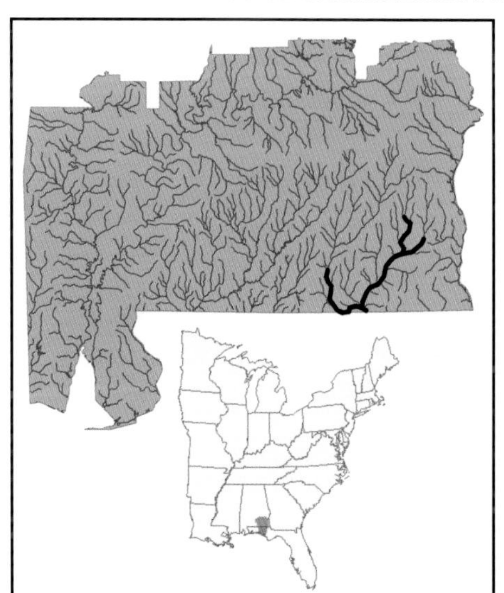

Prepared by: **Stuart W. McGregor**

INFLATED SPIKE

Elliptio purpurella (Lea)

Photo—Paul Johnson

OTHER NAMES. None.

DESCRIPTION. Has moderately thick, inflated shell (max. length = 65 mm [2 5/8 in.]), elliptical and slightly arcuate in outline, with a rounded anterior margin, posterior margin rounded to bluntly pointed, dorsal margin slightly convex, and ventral margin slightly concave. Posterior ridge well developed, but low and rounded. Shell disk and posterior ridge unsculptured. Umbos somewhat full, but not elevated above the hinge line. Periostracum brown with broad green rays that may be obscure. Two pointed, divergent pseudocardinal teeth in each valve; lateral teeth short and curved. No interdentum present, and umbo cavity shallow. Shell nacre purple. (Modified from Brim Box and Williams 2000)

DISTRIBUTION. Endemic to Apalachicola Basin. Historically, occurred in Chattahoochee and Chipola River systems in Alabama. Brim Box and Williams (2000) considered it extirpated from Alabama. However, specimens recently collected from Big Creek, Chipola River headwaters in Houston County, Alabama, appear to be *E. purpurella*, but identification unconfirmed (Jeff Garner, Ala. Div. Wildl. Freshwater Fish., unpubl. data).

HABITAT. Creeks and rivers, primarily in areas with current, but also may be found in lakes under some conditions. Preferred substrata appear to be sand and clay, often associated with limestone (Brim Box and Williams 2000).

LIFE HISTORY AND ECOLOGY. Nothing known; presumably a short-term brooder.

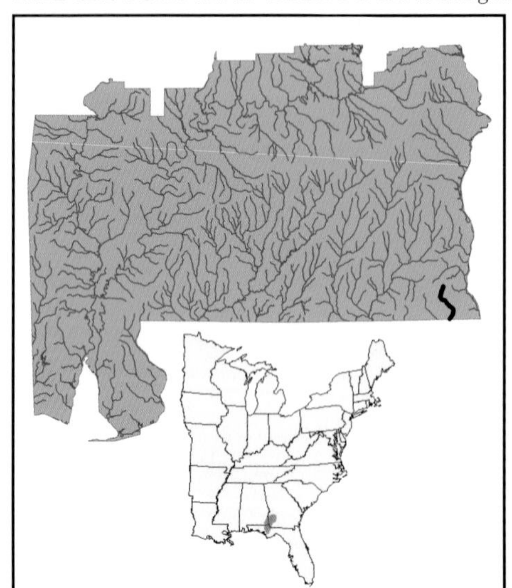

BASIS FOR STATUS CLASSIFICATION. Limited distribution and rarity make *E. purpurella* vulnerable to extinction. Only recently recognized as a valid species (Brim Box and Williams 2000), so was not addressed in Williams *et al.* (1993) or Lydeard *et al.* (1999).

Prepared by: **Jeffrey J. Herod**

PURPLE BANKCLIMBER

Elliptoideus sloatianus (Lea)

Photo—Paul Johnson

OTHER NAMES. None.

DESCRIPTION. Has thick shell (max. length ≥ 200 mm [7 7/8 in.]) rhomboidal in outline, with a broadly rounded anterior margin and obliquely truncate posterior margin that has a slight biangulation ventrally; dorsal and ventral margins more or less straight. Posterior ridge well defined, doubled on the distal half resulting in the biangulation at its ventral margin. Shell disk and posterior ridge heavily sculptured with somewhat radial plications that are less prominent anteriorly. Umbos full, but only slightly elevated above the hinge line. Periostracum chestnut to dark brown. Pseudocardinal teeth stumpy, but strong, and striate; lateral teeth fairly short, two in the left valve and one in the right valve. Interdentum very narrow and umbo cavity shallow. Shell nacre white with purple along margins. (Modified from Lea 1840; Simpson 1900, 1914; Brim Box and Williams 2000)

DISTRIBUTION. Endemic to Apalachicola Basin and Ochlockonee River system in Alabama, Florida, and Georgia (Clench and Turner 1956). In Alabama, known only from main channel of Chattahoochee River (Brim Box and Williams 2000). Was considered extirpated until a single specimen was found recently in Goat Rock Reservoir (C. Stringfellow, University of Columbus, Columbus, Georgia, pers. comm., 2002).

HABITAT. Main channels of large streams and rivers in a variety of substrata, ranging from firm clay to sand and fine gravel, or sand over limestone (Clench and Turner 1956, Brim Box and Williams 2000). Does not appear to tolerate reservoir conditions (Brim Box and Williams 2000).

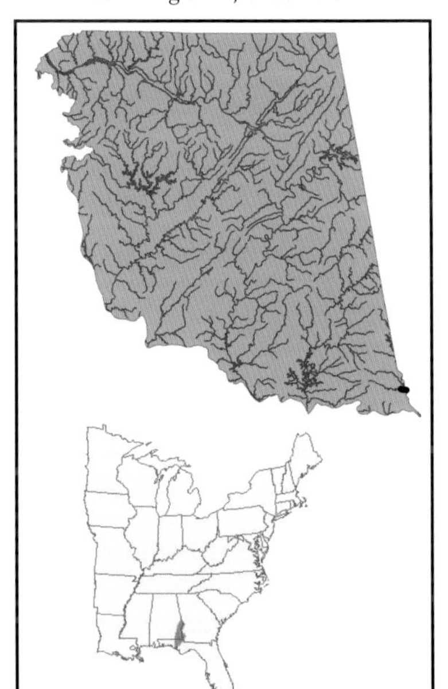

LIFE HISTORY AND ECOLOGY. A short-term brooder; found gravid in February and March (Brim Box and Williams 2000). Primary glochidial hosts have not been identified, but O'Brien (1997) reported secondary host to be eastern mosquitofish.

BASIS FOR STATUS CLASSIFICATION. Limited distribution, rarity, and specialized habitat requirements make species susceptible to extirpation from Alabama. Considered rare throughout its distribution (Clench and Turner 1956, Athearn 1970) and in Alabama (Stansbery 1971). More recently listed as threatened throughout its distribution (Williams *et al.* 1993) and in Alabama (Lydeard *et al.* 1999). **Listed as *threatened* by the U.S. Fish and Wildlife Service in 1998.**

***Prepared by:* Holly Blalock-Herod**

CUMBERLAND COMBSHELL

Epioblasma brevidens (Lea)

OTHER NAMES. None.

Female cumberland combshell.
Photo—From Parmalee and Bogan 1998

DESCRIPTION. Has solid, moderately thick shell (max. length = 80 mm [3 1/8 in.]), somewhat ovate to subtriangular in outline. Posterior margin of females rounded and those of males somewhat bluntly pointed. Shells of females moderately to greatly inflated, depending on age, but those of males hardly inflated. Posterior ridge rounded. Females of reproductive age usually have a narrow marsupial swelling that is manifested as a ventral extension of the posterior ridge, with sulci anteriorly and posteriorly separating it from the rest of the shell. Ventral margin of marsupial swelling ornamented with small serrations. Shell disk and posterior slope without sculpture. Umbos elevated slightly above the hinge line. Umbo sculpture consists of weak double loops. Periostracum smooth or cloth-like in texture and yellowish to tawny brown, marked with narrow, broken, green rays radiating out from the umbo. Pseudocardinal teeth triangular and ragged, separated from short, heavy lateral teeth by a short, narrow interdentum. Umbo cavity shallow. Shell nacre white. (Modified from Lea 1831, Simpson 1914, Johnson 1978, Parmalee and Bogan 1998)

DISTRIBUTION. Historic distribution includes entire Tennessee and Cumberland Rivers and some of their tributaries (Johnson 1978, Parmalee and Bogan 1998). However, has not been reported from Tennessee River proper since it was impounded (Garner and McGregor 2001). Only known extant population in Alabama is in Bear Creek, Colbert County (McGregor and Garner, in review).

HABITAT. Lotic areas of small to large rivers and large creeks (Wilson and Clark 1914). Apparently not tolerant of silty conditions.

LIFE HISTORY AND ECOLOGY. A long-term brooder, with females brooding glochidia from autumn through June of the following year (TVA 1986). Glochidia discharged individually, rather than being discharged in host-attracting conglutinates (TVA 1986). Fishes reported to serve as hosts include greenside, fantail, spotted, redline, snubnose, wounded, and Roanoke darters; logperch; and mottled, black, and banded sculpins (TVA 1986, Yeager and Saylor 1995, Jones and Neves 2002*a*). The Roanoke darter does not occur sympatrically with *E. brevidens*, so not a natural host.

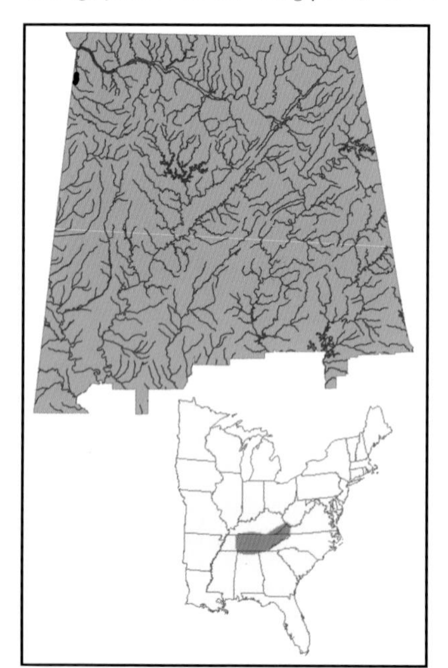

BASIS FOR STATUS CLASSIFICATION. Apparently, last extant population in Alabama decimated by construction of four dams in the Bear Creek system in Colbert and Franklin Counties, and it survives in only a 5.5-kilometer (3.4-mile) reach of Bear Creek in Colbert County (McGregor and Garner, in review). A restricted distribution, declining population trend, and specialized habitat requirements make *E. brevidens* very susceptible to extirpation from the state. Stansbery (1976*a*) originally listed species as threatened in Alabama. Now classified as endangered throughout its distribution (Williams *et al.* 1993) and in Alabama (Lydeard *et al.* 1999). **Listed as *endangered* by the U.S. Fish and Wildlife Service in 1997**.

***Prepared by:* Jeffrey T. Garner**

SOUTHERN COMBSHELL

Epioblasma penita (Conrad)

Male, left, and female, right, southern combshell. Photo—Paul Johnson

OTHER NAMES. None.

DESCRIPTION. Has solid shell (max. length = 55 mm [2 1/8 in.]) rhomboidal to subtriangular in outline, with a rounded anterior margin and truncate posterior margin. Females of reproductive age have a distinct marsupial swelling posteriorly and may be somewhat more rounded than males along that margin. Posterior ridge sharp dorsally, but becomes rounder with a ventral progression, and incorporated into marsupial swelling of females. Shell adorned with radial sculpturing posteriorly that is generally stronger in females. Umbos full and elevated above hinge line. Periostracum varies from brownish yellow to greenish yellow, sometimes with inconspicuous dots or triangles in a radial arrangement on posterior part of shell. Pseudocardinal teeth ragged and irregular; lateral teeth short and straight. Interdentum short and very narrow when present, and umbo cavity shallow. Shell nacre white or straw colored. (Modified from Simpson 1914, USFWS 2000)

DISTRIBUTION. Endemic to Mobile Basin in Alabama and Mississippi. Historic distribution within basin uncertain. Most records are from Tombigbee and lower Alabama Rivers, but a few historic records exist from the Cahaba and Coosa River systems (Stansbery 1976*a*, 1983*a*). Only known extant population is in lower Buttahatchee River, possibly ranging upstream into Alabama for a short distance.

HABITAT. Lotic areas of medium to large rivers, generally in gravel and sand substrata in moderate to swift current (Williams 1982).

LIFE HISTORY AND ECOLOGY. Unknown. Other species of *Epioblasma* are long-term brooders and use darters (Percidae) as glochidial hosts (Yeager and Saylor 1995, Rogers *et al.* 2001).

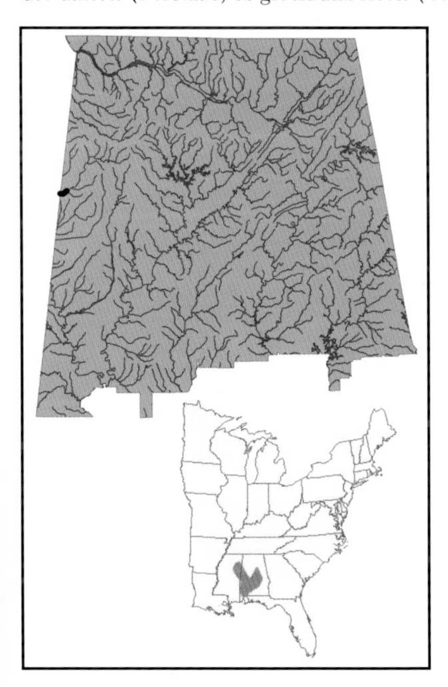

BASIS FOR STATUS CLASSIFICATION. Prior to 1980, the largest populations occurred in the Tombigbee River, but most of these populations were destroyed by construction of the Tennessee-Tombigbee Waterway. A small population persisted until mid-1980s in a short reach of the Tombigbee River in Mississippi not directly modified by the waterway (R. Jones, Mississippi Museum of Natural Science, pers. comm., 2002). Only known remaining population in Buttahatchee River, but McGregor (1999) reported habitat to be poor in Alabama reaches of river downstream of the city of Hamilton. Vulnerable to extinction due to its severely limited distribution, declining population trend, and specialized habitat requirements. Was classified as endangered throughout its distribution (Williams *et al.* 1993), in Alabama (Stansbery 1976*a*), and Lydeard *et al.* (1999) more recently considered it extirpated from Alabama. **Listed as *endangered* by the U.S. Fish and Wildlife Service in 1987.**

***Prepared by:* Wendell R. Haag**

SNUFFBOX

Epioblasma triquetra (Rafinesque)

OTHER NAMES. None.

Photo—From Parmalee and Bogan 1998

DESCRIPTION. Has solid, thick shell (max. length = approx. 70 mm [2 3/4 in.] in males; 50 mm [2 in.] in females) generally inflated and triangular, with a rounded anterior margin and truncate posterior margin. Ventral margin slightly convex in males and almost straight in females. Females have marsupial swelling posterior ventrally. Posterior ridge sharp, especially in females, with a widely flattened posterior slope. Shell disk and posterior slope without sculpture. Umbos inflated, elevated above hinge line, turned forward and inward, and located anterior to center of shell. Periostracum generally smooth, yellow to yellowish green, with irregular broken dark green rays that often appear as square, triangular, or chevron-shaped markings. Pseudocardinal teeth strong, elevated and roughened, or serrated; lateral teeth short, strong, serrated, and erect, with two in left valve and one in right valve. Interdentum narrow and short, or absent. Umbo cavity wide and deep. Shell nacre white, often with a silvery luster and a gray-blue tinge in umbo cavity. (Modified from Parmalee and Bogan 1998)

DISTRIBUTION. Distributed widely in Mississippi, lower Missouri, Ohio, Cumberland, and Tennessee River systems, and in middle Great Lakes Basin. In Alabama, once occurred in Tennessee River and several of its tributaries. However, now known to persist only in Paint Rock River system, Jackson County (Steve Fraley, North Carolina Wildlife Resources Commission, pers. comm., 2002).

HABITAT. Large creeks to large rivers, generally in gravel and sand substrata in shoal and riffle habitats. Individuals often completely buried, or with only their posterior slopes exposed (Parmalee and Bogan 1998).

LIFE HISTORY AND ECOLOGY. A long-term brooder. Gravid females observed from September to May, with glochidial discharge taking place in late May (Ortmann 1919). Reported hosts include logperch, Roanoke darter, and banded and black sculpins (Yeager and Saylor 1995; J.W. Jones, Virginia Polytechnic Institute and State University, pers. comm., 2002). However, the Roanoke darter does not occur sympatrically with *E. triquetra*, so not a natural host.

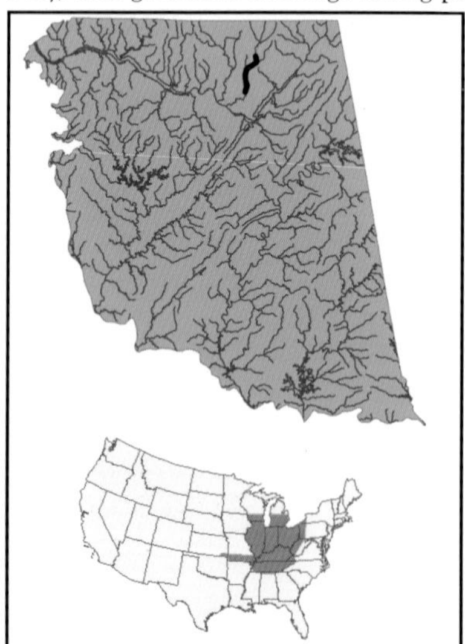

BASIS FOR STATUS CLASSIFICATION. Only extant population in Alabama located in Paint Rock River system, Jackson County. Densities very low and long-term viability questionable. Many populations distribution-wide declining or have become extirpated. Considered threatened throughout its distribution (Williams *et al.* 1993) and endangered (Stansbery 1976*a*) or imperiled in Alabama (Lydeard *et al.* 1999). **Considered a *species of concern* by U.S. Fish and Wildlife Service, Southeast Region**.

Prepared by: **Robert S. Butler**

SHINY PIGTOE

Fusconaia cor (Conrad)

Photo—From Parmalee and Bogan 1998

OTHER NAMES. None.

DESCRIPTION. Has solid and somewhat inflated shell (max. length = 80 mm [3 1/8 in.]) subtriangular in outline, with anterior margin broadly rounded and somewhat obliquely truncate above, and posterior margin nearly straight, but obliquely angled; dorsal and ventral margins nearly straight. Posterior ridge narrowly rounded or angular and may be slightly doubled, ending in a point or slight biangulation ventrally. A medial swelling is located anterior of posterior ridge, with a shallow sulcus separating the two. Shell disk and posterior ridge without sculpture. Umbo moderately high and full, oriented anteriorly and elevated only slightly above hinge line. Umbo sculpture consists of a few broken, subnodulous ridges. Periostracum greenish yellow to yellowish tan, or brown, in young individuals, often darkening with age to dark brown, or black. Both wide and narrow green rays present in younger specimens, but may become indistinct with age. One of most distinctive characteristics is shininess of periostracum that persists even in old individuals. Pseudocardinal teeth heavy and triangular, separated from stout, straight lateral teeth by a short, but wide, interdentum. Umbo cavity deep and compressed. Shell nacre white. (Modified from Simpson 1914, Parmalee and Bogan 1998)

DISTRIBUTION. Endemic to Tennessee River system, historically occurring from its headwaters downstream to Muscle Shoals and in some of its large tributaries (Parmalee and Bogan 1998). Extirpated from Tennessee River proper (Garner and McGregor 2001), but still occurs in several tributaries, including Paint Rock River in Alabama (Ahlstedt 1995).

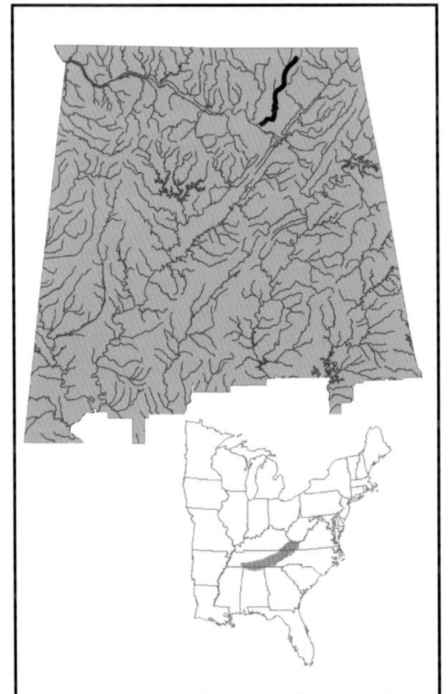

HABITAT. Shoal and riffle habitat of medium to large rivers (Neves 1991).

LIFE HISTORY AND ECOLOGY. A short-term brooder, spawning from late May to early June and gravid from mid-May to mid-July (Ortmann 1921, Kitchel 1985). Eggs and developing embryos usually crimson, but may be pink (Ortmann 1921). Fishes reported to serve as hosts for glochidia include telescope, warpaint, and common shiners (Kitchel 1985).

BASIS FOR STATUS CLASSIFICATION. Suffered severe distribution reductions during the twentieth century. Imperiled due to restricted distribution, specialized habitat requirements, and declining population trend. Classified as endangered throughout its distribution (Williams *et al.* 1993), and in Alabama (Stansbery 1976*a*, Lydeard *et al.* 1999). **Listed as *endangered* by the U.S. Fish and Wildlife Service in 1976.**

***Prepared by:* Jeffrey T. Garner**

FINERAYED PIGTOE

Fusconaia cuneolus (Lea)

OTHER NAMES. None.

Photo—From Parmalee and Bogan 1998

DESCRIPTION. Has solid, somewhat inflated, shell (max. length = 80 mm [3 1/8 in.]), subtriangular to rhomboidal in outline, with an anterior margin that is rounded, but often obliquely truncate on dorsal half, and posterior margin that is bluntly pointed to narrowly rounded. Posterior ridge well-developed, angular near umbo, but becoming more rounded ventrally, where it ends in a blunt point. Shell disk and posterior slope unsculptured. Umbos moderately full and elevated above hinge line. Periostracum has satiny texture and is dull yellowish green to olive brown, marked with green rays of varying length, which may be indistinct. Pseudocardinal teeth low, rough, and triangular; lateral teeth fairly long and usually straight, but may be slightly curved. Interdentum short and wide, and umbo cavity moderately deep. Shell nacre white. (Modified from Lea 1840, Simpson 1914, Parmalee and Bogan 1998)

DISTRIBUTION. Endemic to Tennessee River system, historically occurring from headwaters in Virginia, downstream to Muscle Shoals, and in some tributaries (Parmalee and Bogan 1998). However, has been extirpated from Tennessee River proper (Garner and McGregor 2001). A population in Paint Rock River (Ahlstedt 1995) appears to be only one extant in Alabama.

HABITAT. Shoal habitat of medium to large rivers. Typically lives in stable, mixed substrata, with particle sizes ranging from sand to cobble (Neves 1991). However, Ortmann (1925) reported finding it in a sand and mud mixture in a small creek.

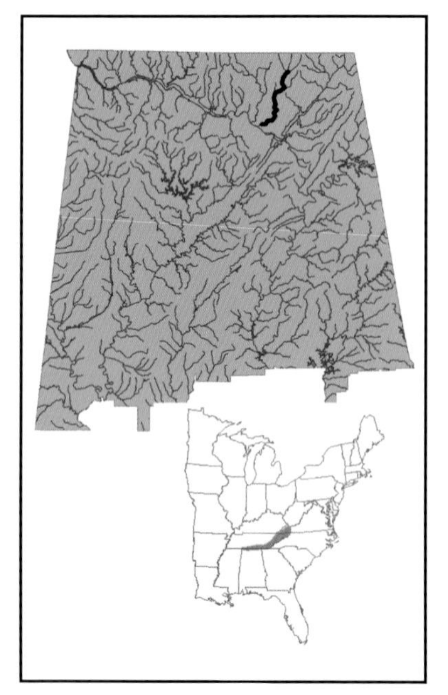

LIFE HISTORY AND ECOLOGY. A short-term brooder, spawning in May, with females being gravid until late July (Ortmann 1921, Bruenderman and Neves 1993). Embryos bright pink, but change to light orange or peach as they mature (Ortmann 1921, Bruenderman and Neves 1993). Fishes reported to serve as hosts include river chub; whitetail, white, telescope, and Tennessee shiners; central stoneroller; fathead minnow; and mottled sculpin (Bruenderman and Neves 1993).

BASIS FOR STATUS CLASSIFICATION. Limited distribution, specialized habitat requirements, and declining population trend make *F. cuneolus* vulnerable to extinction. Classified as endangered throughout its distribution (Williams *et al.* 1993) and in Alabama (Stansbery 1976*a*, Lydeard *et al.* 1999). **Listed as *endangered* by the U.S. Fish and Wildlife Service in 1976.**

Prepared by: **Jeffrey T. Garner**

NARROW PIGTOE

Fusconaia escambia Clench and Turner

Photo—Paul Johnson

OTHER NAMES. None.

DESCRIPTION. Has heavy, inflated shell (max. length ≥ 70 mm [2 3/4 in.]) subcircular in outline, with a somewhat pointed posterior margin. Posterior ridge well defined and posterior slope slightly concave. Shell disk and posterior slope without sculpture. Umbos broad and full, elevated above hinge line, and positioned slightly anterior of center. Periostracum smooth on upper disk, but slightly roughened on posterior slope and along ventral margin; reddish brown in young specimens, but darkens with age to very dark brown. Pseudocardinal teeth crenulate, one large one in right valve, two in left valve. Shell nacre usually deep salmon, but rarely white. (Modified from Clench and Turner 1956, Williams *et al.*, in review)

DISTRIBUTION. Endemic to Gulf Coast drainages, where known from Escambia and Yellow River systems in Alabama and Florida (Johnson 1969, Williams and Butler 1994). Apparently extirpated from Yellow River system (Williams *et al.*, in review).

HABITAT. Small to medium rivers with sand, gravel, or sandy gravel substrata and slow to moderate flow (Williams and Butler 1994).

LIFE HISTORY AND ECOLOGY. A short-term brooder. Gravid females with red eggs and red glochidia reported in June (Williams *et al.*, in review). Glochidial hosts unknown.

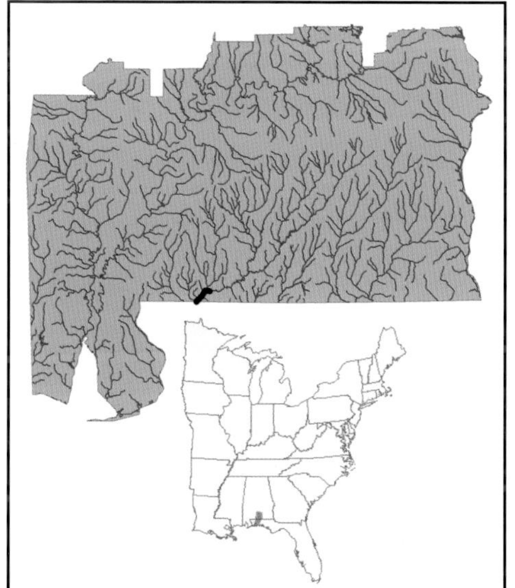

BASIS FOR STATUS CLASSIFICATION. Vulnerable to extinction because of limited distribution, rarity, and susceptibility to habitat degradation. Classified as threatened throughout its distribution (Williams *et al.* 1993), and as a species of special concern (Stansbery 1976*a*) and imperiled in Alabama (Lydeard *et al.* 1999).

Prepared by: **Stuart W. McGregor**

ROUND EBONYSHELL

Fusconaia rotulata (Wright)

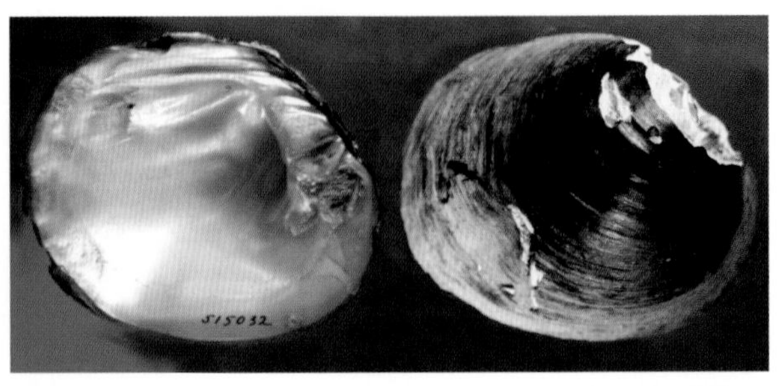

Photo—Jim Williams

OTHER NAMES. None.

DESCRIPTION. Has heavy, inflated shell (max. length = 70 mm [2 3/4 in.]) almost circular in outline, with rounded margins on all sides. Posterior ridge faint, but well rounded, and posterior slope very slightly concave. Shell disk and posterior slope without sculpture. Umbos inflated and elevated above hinge line. Periostracum smooth and very dark brown to black. Pseudocardinal teeth heavy, broad, and divergent; lateral teeth short and slightly curved. Interdentum short, but broad, and umbo cavity deep. Shell nacre white. (Modified from Simpson 1914, Williams and Butler 1994.) Recently considered to belong to *Obovaria* until genetic evaluation and comparative studies of soft anatomy suggested a closer relationship to *Fusconaia* group (Williams and Butler 1994, Lydeard *et al.* 2000).

DISTRIBUTION. Endemic to Escambia River drainage in Alabama and Florida (Williams and Butler 1994). In Alabama, appears confined to main channel of Conecuh River, where occurs as far upstream as Conecuh/Covington County line.

HABITAT. Known only from main channel of Escambia and Conecuh Rivers, in areas with moderate current and with sand or a mixture of sand and gravel substrata (Williams and Butler 1994).

LIFE HISTORY AND ECOLOGY. Although little known of life history, presumably a short-term brooder, as are its congeners. Hosts of glochidia unknown.

BASIS FOR STATUS CLASSIFICATION. Limited distribution and rarity make *F. rotulata* susceptible to extinction from habitat degradation within Escambia River watershed. Classified as endangered throughout its distribution (Williams *et al.* 1993), and deemed imperiled in Alabama (Lydeard *et al.* 1999).

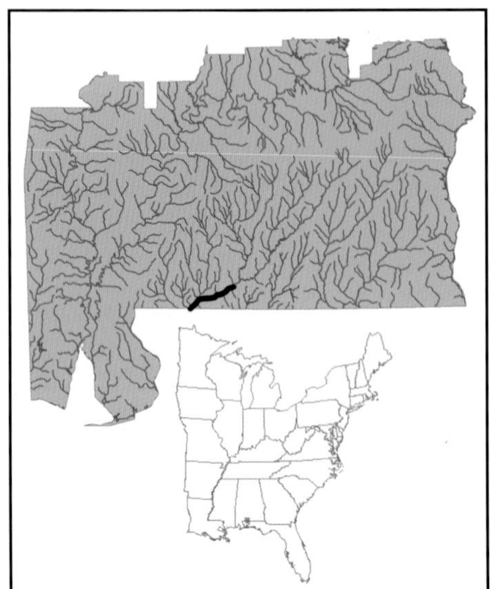

Prepared by: **Stuart W. McGregor**

LONGSOLID

Fusconaia subrotunda (Lea)

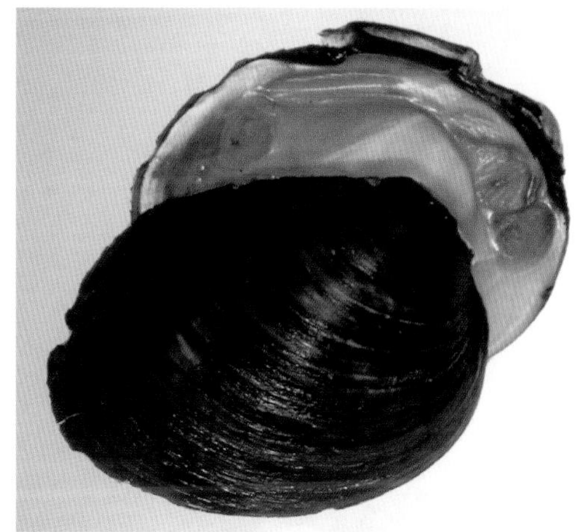

Photo—From Parmalee and Bogan 1998

OTHER NAMES. Round Solid, Pigtoe.

DESCRIPTION. Has solid shell (max. length = 100 mm [3 15/16 in.]) oval to broadly elliptical in outline, with dorsal and ventral margins gently convex, and anterior margin broadly rounded. Specimens from large rivers usually inflated, but those from smaller streams compressed to only moderately inflated. Posterior margin gently rounded to obliquely truncate. Posterior ridge weak to lacking. Shell disk and posterior slope without sculpture. Umbos high and full, elevated well above hinge line and turned anteriorly. Umbo sculpture weak, consisting of a few subnodulous ridges or wrinkles. Periostracum somewhat cloth-like in texture, tawny to greenish brown, darkening with age to almost black in some individuals. Pseudocardinal teeth low, heavy, and divergent; lateral teeth moderately long and straight. Interdentum somewhat short, but wide. Umbo cavity deep and somewhat compressed. Shell nacre white. (Modified from Simpson 1914, Parmalee and Bogan 1998)

DISTRIBUTION. Throughout much of the Tennessee, Cumberland, and Ohio River systems (Burch 1975). In Alabama, known to be extant only in Tennessee River tailwaters of Guntersville and Wilson Dams.

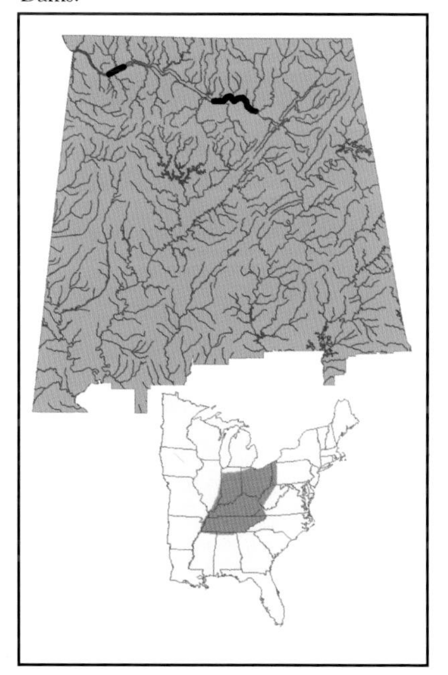

HABITAT. Lotic areas. In tailwaters of Wilson and Guntersville Dams, occurs in substrata composed of mixtures of sand, gravel and cobble, with minimal siltation, usually at depths of four to six meters (13 to 19 3/4 feet.).

LIFE HISTORY AND ECOLOGY. Little known; presumably a short-term brooder, being gravid from spring to mid-summer, like its congeners. Glochidial hosts unknown.

BASIS FOR STATUS CLASSIFICATION. Vulnerable to extirpation from Alabama because of limited distribution, rarity, and specialized habitat requirements. Rare in two remaining populations in state (Garner and McGregor 2001). Listed as a species of special concern throughout its distribution (Williams *et al.* 1993) and in Alabama (Lydeard *et al.* 1999), although Stansbery (1976*a*) considered it endangered in Alabama.

Prepared by: **Jeffrey T. Garner**

CRACKING PEARLYMUSSEL

Hemistena lata (Rafinesque)

OTHER NAMES. None.

Photo—From Parmalee and Bogan 1998

DESCRIPTION. Has thin, slightly inflated shell (max. length = 90 mm [3 1/2 in.]) elongate and elliptical to rhomboidal in outline, with rounded anterior margin and bluntly pointed to obliquely truncate posterior margin. Posterior ridge low and rounded, occasionally sculptured with a few strong ridges. Umbos low and compressed, turned slightly anteriorly. Periostracum dull yellow to brownish green or brown, usually with scattered, broken, dark green rays of varying width. Pseudocardinal teeth only a single raised knob or ridge; lateral teeth appear as a thickened hinge line with no interdentum. Umbo cavity very shallow or absent. Shell nacre pale bluish white, with a dark purple umbo cavity. (Modified from Simpson 1914, Parmalee and Bogan 1998)

DISTRIBUTION. Historically, in the Ohio, Cumberland, and Tennessee River systems. In Alabama, extant only in the Elk River, where densities very low and viability of population questionable.

HABITAT. Riverine reaches of large rivers, but more typically found in medium rivers, usually in less than one meter (3 1/4 feet) of water (Parmalee and Bogan 1998).

LIFE HISTORY AND ECOLOGY. A short-term brooder; although Ortmann (1915) observed specimens gravid with eggs and glochidia in mid-May, Jones and Neves (2002*a*) reported finding only eggs and embryos in three gravid specimens collected from Clinch River between late April and late June. Hosts of glochidia include whitetail shiner, central stoneroller, and streamline chub, as well as banded sculpin and fantail darter (J.W. Jones, Virginia Polytechnic Institute and State University, pers. comm., 2002). Often difficult to collect, due to propensity for burrowing deeply into substrata of mud, sand, and fine gravel using its proportionally large foot (Yokley 1972, Parmalee and Bogan 1998).

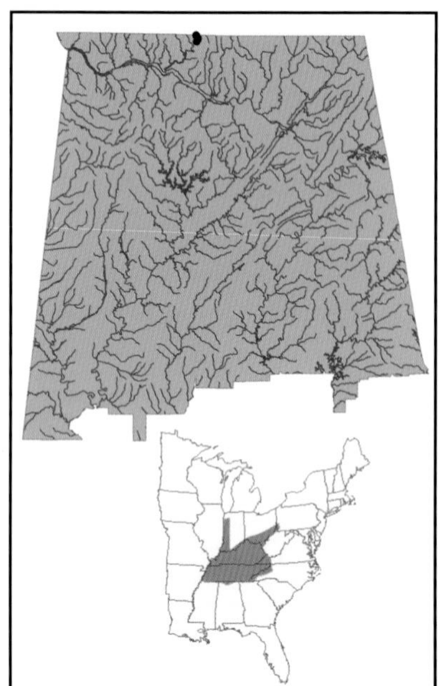

BASIS FOR STATUS CLASSIFICATION. Severely restricted distribution, rarity, and specialized habitat requirements make *H. lata* vulnerable to extinction. Classified as endangered throughout its distribution (Williams *et al.* 1993) and in Alabama (Stansbery 1976*a*), although most recently listed as extirpated from Alabama (Lydeard *et al.* 1999). **Listed as *endangered* by the U.S. Fish and Wildlife Service in 1989**.

***Prepared by*: Jeffrey T. Garner**

PINK MUCKET

Lampsilis abrupta (Say)

OTHER NAMES. Ohio Mucket, Tan Mucket, Square Mucket.

Female pink mucket.
Photo—From Parmalee and Bogan 1998

DESCRIPTION. Has very solid, somewhat inflated shell (max. length = 120 mm [4 3/4 in.]). Males generally attain greater size than females. Shell ovate to subquadrate in outline. Males rounded to very bluntly pointed posteriorly; females have a broadly rounded to truncate posterior margin. Valves often gape anteriorly, along ventral margin, especially in females. Posterior ridge well-defined in males and lies adjacent to dorsal margin, but indistinct in more inflated females. Shell disk and posterior slope without sculpture. Umbos inflated and raised above hinge line. Umbo sculpture consists of faint, scarcely looped ridges. Periostracum yellow to dark brown, darkening with age; variable dark green rays may be present. Pseudocardinal teeth large and triangular, separated from strong, slightly curved lateral teeth by a broad interdentum. Umbo cavity broad and deep. Shell nacre varies from white to pink or salmon. (Modified from Simpson 1914, Parmalee and Bogan 1998)

DISTRIBUTION. Encompasses Ohio, Tennessee, Cumberland, and lower Mississippi Rivers, as well as some of their larger tributaries. Although historically found in entire reach of Tennessee River across northern Alabama, it currently occurs only in riverine reaches downstream of Wilson and Guntersville Dams, where reportedly uncommon to rare (Garner and McGregor 2001). A single gravid female recently found in lower, unimpounded reaches of Bear Creek, Colbert County (McGregor and Garner, in review).

HABITAT. Free-flowing reaches of large rivers, typically in silt-free, gravel substrata.

LIFE HISTORY AND ECOLOGY. A long-term brooder, gravid from August to the following June (Ortmann 1912, 1919). Fishes reportedly serving as hosts for glochidia include smallmouth, spotted, and largemouth bass, as well as freshwater drum and possibly sauger (Parmalee and Bogan 1998, Madison and Layzer 2000).

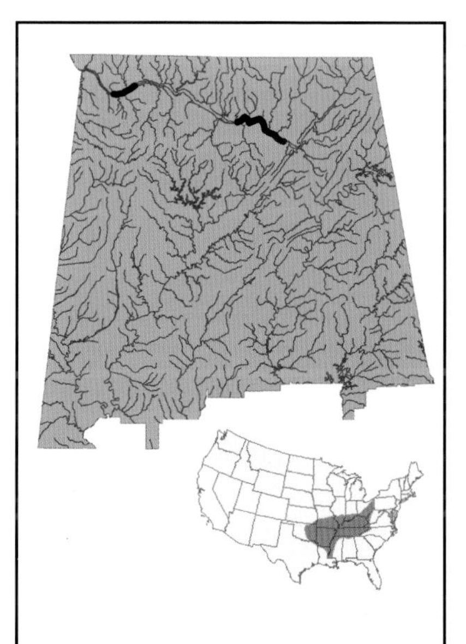

BASIS FOR STATUS CLASSIFICATION. Although widespread, has suffered severe habitat loss and population fragmentation during the twentieth century. Rarity, specialized habitat requirements, and susceptibility to habitat degradation make it vulnerable to extinction. Was classified as a species of special concern in Alabama (Stansbery 1976*a*), and, more recently, as endangered throughout its distribution (Williams *et al.* 1993) and in Alabama (Lydeard *et al.* 1999). Specimens younger than 10 years of age reportedly rare in Wilson and Guntersville Dam tailwaters (Garner and McGregor 2001). **Listed as *endangered* by the U.S. Fish and Wildlife Service in 1976.**

Prepared by: **Jeffrey T. Garner**

SOUTHERN SANDSHELL

Lampsilis australis Simpson

Photo—Arthur Bogan

OTHER NAMES. None.

DESCRIPTION. Has a moderately thick, elliptical, and moderately inflated shell (max. length ≥ 80 mm [3 1/8 in.]), rounded anteriorly and somewhat pointed posteriorly; dorsal and ventral margins slightly convex. Females slightly more inflated and rounded posterior ventrally than males. Posterior ridge low and rounded. Shell disk and posterior slope without sculpture. Umbos moderately inflated, elevated slightly above hinge line. Periostracum smooth and shiny, varying from greenish yellow with green rays in younger specimens to almost black, with obscure rays, in older specimens. Pseudocardinal teeth delicate and somewhat compressed; lateral teeth straight and interdentum narrow. Umbo cavity moderately deep. Shell nacre white to bluish white and iridescent posteriorly. (Modified from Simpson 1900, Athearn 1964, Williams *et al.*, in review).

DISTRIBUTION. Endemic to Gulf Coast drainages, occurring in Escambia, Yellow, and Choctawhatchee River systems in southern Alabama and western Florida (Clench and Turner 1956, Blalock-Herod *et al.* 2002).

HABITAT. Usually clear, medium-sized creeks to rivers, with slow to moderate current and sandy substrata (Williams and Butler 1994).

LIFE HISTORY AND ECOLOGY. A long-term brooder; one of four species that produce superconglutinates to facilitate larval dispersal. Superconglutinates pigmented to resemble small fish and display a darting motion in stream currents, eliciting attacks from potential host fishes (Haag and Warren 1999). Superconglutinates discharged between April and June (Blalock-Herod *et al.* 2002; Haag *et al.* 1995). Preliminary data suggest *L. australis* uses bass species (*Micropterus* spp.) as glochidial hosts, as do other superconglutinate-producing species (Blalock-Herod *et al.* 2002, Haag *et al.* 1995, O'Brien and Brim Box 1999).

BASIS FOR STATUS CLASSIFICATION. Has very restricted distribution, somewhat rare, and has experienced recent declines in habitat. Considered endangered in Alabama for 30-plus years (Athearn 1970, Stansbery 1971). Was classified as threatened throughout its distribution (Williams *et al.* 1993), but now considered endangered throughout (Williams *et al.*, in review; Blalock-Herod *et al.*, in review; Blalock-Herod 2002). Currently listed as imperiled in Alabama (Lydeard *et al.* 1999).

***Prepared by:* Holly Blalock-Herod**

SHINYRAYED POCKETBOOK

Lampsilis subangulata (Lea)

OTHER NAMES. None.

Photo—Arthur Bogan

DESCRIPTION. Has a thin, but solid shell (max. length = 85 mm [3 3/8 in.]), subelliptical in outline. Females of reproductive age generally more inflated and rounded posteriorly than males, which are bluntly pointed. Posterior ridge rounded and posterior slope concave. Shell disk and posterior slope without sculpturing. Umbos broad and somewhat inflated, but elevated little above hinge line. Periostracum smooth and shiny, light yellowish brown, with variable bright green rays in young specimens; darkening and obscuring of rays occurs with age. Pseudocardinal teeth double, large, and erect in left valve, and single, large, and spatulate in right valve; lateral teeth straight to very slightly curved. Interdentum narrow. Shell nacre white or bluish, occasionally with salmon tint in umbo cavity. (Modified from Clench and Turner 1956, Williams and Butler 1994)

DISTRIBUTION. Known only from Apalachicola and Ochlockonee River systems. In Alabama, confined to Chattahoochee and upper Chipola River systems (Williams and Butler 1994, Brim Box and Williams 2000). Extant in Big Creek, Houston County, and possibly Uchee Creek, Russell County.

HABITAT. Generally in medium-sized creeks to rivers in clean or silty sand substrata, and slow to moderate current.

LIFE HISTORY AND ECOLOGY. A long-term brooder; females hold superconglutinate packets of glochidia from autumn until late spring or early summer to attract potential fish hosts (O'Brien and Brim Box 1999, Brim Box and Williams 2000). These superconglutinates are pigmented to resemble small fish that display a darting motion in stream currents, and elicit attacks from potential hosts (Haag and Warren 1999). Largemouth and spotted bass serve as hosts (O'Brien and Brim Box 1999).

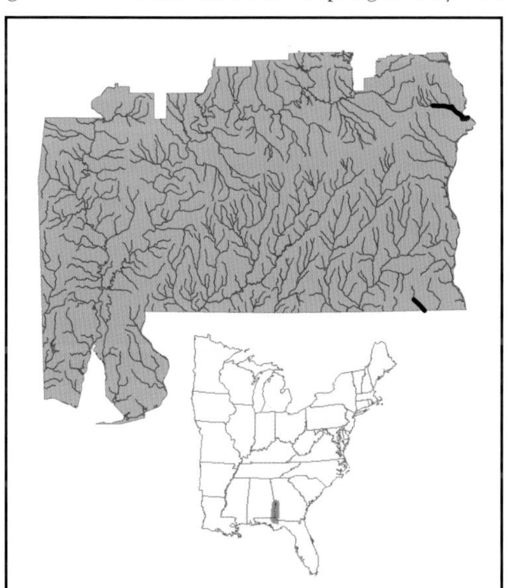

BASIS FOR STATUS CLASSIFICATION. Restricted distribution, rarity, and dwindling habitat quality make *L. subangulata* susceptible to extinction. Classified as threatened throughout its distribution (Williams *et al.* 1993) and in Alabama (Lydeard *et al.* 1999). **Listed as *endangered* by the U.S. Fish and Wildlife Service in 1998.**

***Prepared by:* Stuart W. McGregor**

ALABAMA LAMPMUSSEL

Lampsilis virescens (Lea)

Photo—Jeff Garner

OTHER NAMES. None.

DESCRIPTION. Has moderately thin shell (max. length = 70 mm [2 3/4 in.]), elliptical to long ovate in outline and somewhat inflated. Anterior margin rounded and posterior margin bluntly pointed in males, slightly more inflated and rounded in females. Dorsal margin slightly rounded and ventral margin straight, but curved upward posteriorly. Posterior ridge low and rounded. Shell disk and posterior slope unsculptured. Umbos moderately full and slightly elevated above hinge line. Umbo sculpture consists of numerous delicate ridges, looped up in the middle. Periostracum greenish to yellow and typically shiny. Thin green rays may be present, especially on posterior slope. Pseudocardinal teeth compressed and elevated and lateral teeth slightly curved and delicate. Interdentum narrow and curved and umbo cavity broad and deep. Shell nacre bluish white. (Modified from Simpson 1914, Parmalee and Bogan 1998)

DISTRIBUTION. Endemic to Tennessee River system and, historically, occurred from its headwaters downstream to Muscle Shoals (Ortmann 1925, Parmalee and Bogan 1998). Now only known to be extant in upper reaches of Paint Rock River system, Jackson County, Alabama (Ahlstedt 1995).

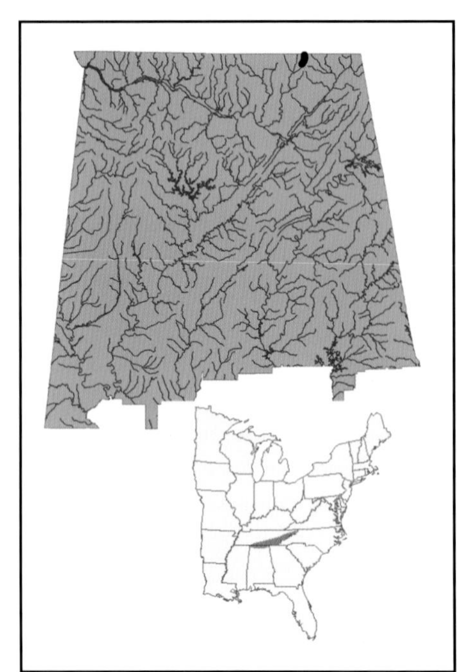

HABITAT. Shoals in small to medium rivers (Parmalee and Bogan 1998). However, its presence at Muscle Shoals, prior to impoundment, indicates ability to exist in larger rivers under some conditions.

LIFE HISTORY AND ECOLOGY. Unknown, but presumably a long-term brooder like its congeners.

BASIS FOR STATUS CLASSIFICATION. Apparently eliminated throughout its distribution, with exception of upper reaches of Paint Rock River system, where rare. Imperiled due to severely restricted distribution, rarity, and vulnerability to habitat degradation. Classified as endangered throughout its distribution (Williams *et al.* 1993) and in Alabama (Stansbery 1976*a*, Lydeard *et al.* 1999). **Listed as *endangered* by the U.S. Fish and Wildlife Service in 1976.**

***Prepared by:* Jeffrey T. Garner**

SLABSIDE PEARLYMUSSEL

Lexingtonia dolabelloides (Lea)

Photo—From Parmalee and Bogan 1998

OTHER NAMES. None.

DESCRIPTION. Has solid, moderately inflated, shell (max. length = 85 mm [3 3/8 in.]), generally subtriangular in outline, but exhibits considerable variability within and among populations. Anterior margin broadly rounded, posterior margin narrowly rounded to broadly pointed, dorsal margin slightly convex and ventral margin straight to slightly convex. Most shells have wide, flat disk, from umbo to ventral margin. Posterior ridge narrowly rounded and distinct, and dorsal slope strongly curved. Shell disk and posterior slope without sculpture. Umbos moderately inflated and turned slightly forward, elevated above hinge line and located toward anterior end. Periostracum greenish yellow in juveniles, darkening to tawny or brown with age, often with a few broken green rays or blotches, especially in young specimens. Pseudocardinal teeth triangular with a blade-like accessory tooth in left valve; lateral teeth large and curved. Interdentum well formed, but narrow, and umbo cavity shallow. Shell nacre usually white, but may be straw colored. In live specimens, foot always bright orange. (Modified from Lea 1840, Simpson 1914, Parmalee and Bogan 1998)

DISTRIBUTION. Historically, included the Tennessee River from headwaters downstream to at least area now comprising tailwaters of Pickwick Dam, with a disjunct population in Duck River of central Tennessee (Ortmann 1924*b*, Yokley 1972, Parmalee and Bogan 1998). In Alabama, appears to be extant in Paint Rock River system (Ahlstedt 1995) and a short reach of Bear Creek, Colbert County (McGregor and Garner, in review).

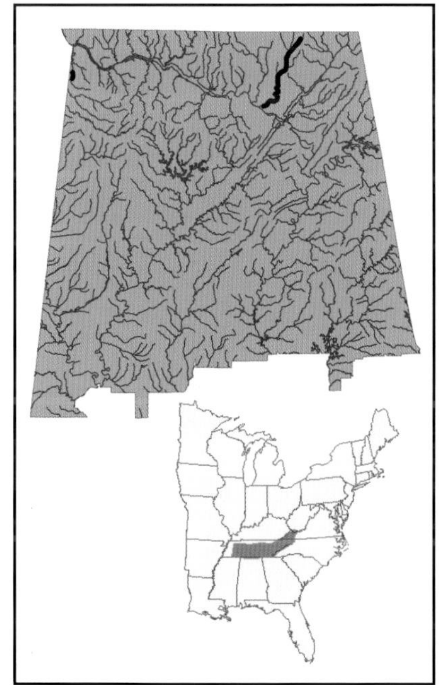

HABITAT. Flowing water of large creeks and medium to large rivers.

LIFE HISTORY AND ECOLOGY. A short-term brooder, gravid between mid-May and August (Ortmann 1921, Kitchel 1985). Eggs and developing embryos usually deep red, but may be orange (Ortmann 1921). Fishes reported to serve as hosts of glochidia are popeye, rosyface, saffron, silver, telescope, and Tennessee shiners (Kitchel 1985).

BASIS FOR STATUS CLASSIFICATION. Impoundment of Tennessee River resulted in extensive habitat loss and fragmentation of populations. Restricted distribution, specific habitat requirements, and declining population trend make it vulnerable to extirpation from state. Was classified as threatened throughout distribution (Williams *et al.* 1993) and endangered (Stansbery 1976*a*) and imperiled (Lydeard *et al.* 1999) in Alabama.

***Prepared by:* Jeffrey T. Garner**

CUMBERLAND MOCCASINSHELL

Medionidus conradicus (Lea)

Photo—From Parmalee and Bogan 1998

OTHER NAMES. None.

DESCRIPTION. Has moderately thin shell (max. length = 60 mm [2 3/8 in.]), elongate and elliptical in outline, with a rounded anterior margin and posterior margin that comes to a rounded point at distal end of posterior ridge. Old males often become arcuate. Posterior ridge low and indistinct. Females somewhat more inflated along middle of ventral margin, sometimes with faint radial grooves in that area. Posterior slope usually has weak wrinkles or corrugations that may extend anteriorly on to shell disk. Umbos slightly inflated and elevated above hinge line. Umbo sculpture consists of fine, irregular, corrugated ridges that may be double looped. Periostracum slightly shiny and yellowish green to tawny, with weak, broken, dark green rays. Pseudocardinal teeth stumpy and short and lateral teeth straight to slightly curved. No interdentum, and umbo cavity very shallow. Shell nacre bluish white to salmon. (Modified from Simpson 1914, Parmalee and Bogan 1998)

DISTRIBUTION. Endemic to Cumberland and Tennessee River systems, historically occurring in Tennessee River from its headwaters downstream to Muscle Shoals (Ortmann 1925, Parmalee and Bogan 1998). Known to be extant in Alabama only in Paint Rock River system (Ahlstedt 1995) and Foxtrap Creek, Colbert County.

HABITAT. Usually moderate to strong current, generally in small streams to medium rivers, and often under large flat rocks (Parmalee and Bogan 1998). However, its presence at Muscle Shoals, prior to impoundment of Tennessee River (Ortmann 1925) indicates ability to exist in large rivers under certain conditions.

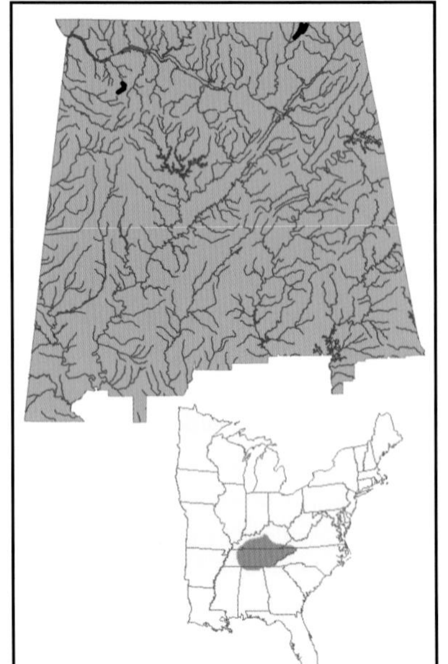

LIFE HISTORY AND ECOLOGY. A long-term brooder, spawning in July; females brood mature glochidia September through late May (Ortmann 1915, 1921; Zale and Neves 1982*a*). Glochidia found in stream drift every month of year except July and August, suggesting a slow, continual, long-term discharge (Zale and Neves 1982*b*). Fishes reportedly serving as hosts for glochidia are rainbow, fantail, redline, and striped darters (Zale and Neves 1982*a*, Luo 1993).

BASIS FOR STATUS CLASSIFICATION. Vulnerable to extirpation due to limited distribution, rarity, and susceptibility to habitat degradation. Long-term viability of Foxtrap Creek population questionable. Although was listed as endangered (Stansbery 1976*a*), more recently classified as a species of special concern throughout its distribution (Williams *et al.* 1993) and in Alabama (Lydeard *et al.* 1999).

Prepared by: **Jeffrey T. Garner**

GULF MOCCASINSHELL

Medionidus penicillatus (Lea)

OTHER NAMES. None.

Photo—Arthur Bogan

DESCRIPTION. Has small, moderately thin, fairly inflated shell (max. length = 55 mm [2 1/8 in.]), elongate elliptical to rhomboidal in outline, with a rounded anterior margin, bluntly pointed posterior margin, very slightly convex dorsal margin, and straight to slightly convex ventral margin. Females tend to have posterior point higher on shell and be somewhat more inflated than males. Posterior ridge well defined and typically has a series of small, thin, radial plications along posterior slope. Shell disk smooth. Umbos moderately inflated and elevated above hinge line. Periostracum yellowish with fine, interrupted green rays and chevrons. Left valve has two stubby pseudocardinal teeth and right valve has one; lateral teeth curved, double in left valve and single in right valve. Shell nacre purple or greenish and slightly iridescent posteriorly. (Modified from Williams and Butler 1994, Brim Box and Williams 2000)

DISTRIBUTION. Appears endemic to Apalachicola Basin in Alabama, Georgia, and Florida, with a disjunct population in Econfina Creek, Florida. In Alabama, known from main channel and tributaries of Chattahoochee River, as well as Chipola River headwater streams (Brim Box and Williams 2000). Currently restricted in Alabama to Big Creek, Houston County, a Chipola River headwater stream. Identity of *Medionidus* west of Apalachicola Basin (Choctawhatchee, Escambia, and Yellow River systems) uncertain, but have been assigned to *M. penicillatus* by some (*e.g.*, Johnson 1977, Butler 1989). However, until their true identity can be resolved with comparative anatomical and genetic studies, they are tentatively considered *M. acutissimus*. Apalachicola Basin shows little affinity to westward basins, whereas Mobile Basin, to which *M. acutissimus* belongs, shares a number of species with Gulf Coast systems west of Apalachicola Basin (J.D. Williams, U.S. Geological Survey, pers. comm., 2002).

HABITAT. Small creeks to large rivers, with slight to moderate current in a variety of substrata. Most commonly found in sand, sand and cobble mixtures, gravel, and clay sediments (Brim Box and Williams 2000).

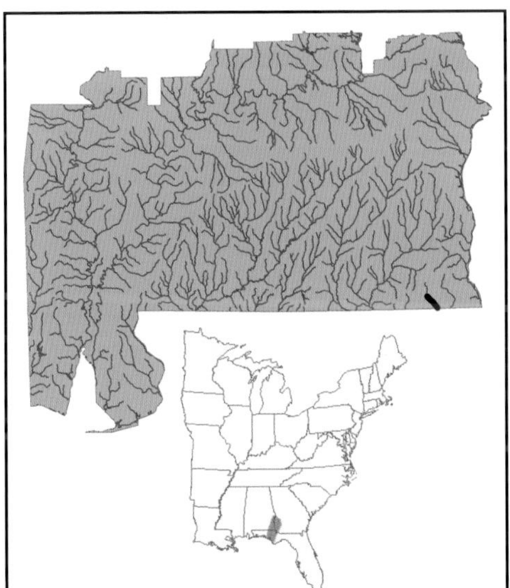

LIFE HISTORY AND ECOLOGY. A long-term brooder found with eggs in marsupia in September and mature glochidia in November, February, March, and April. In March, females detected lying on substratum surface displaying a mantle-flapping behavior, presumably to attract host fish. Primary glochidial hosts include brown and blackbanded darters. (Summarized from O'Brien 1997)

BASIS FOR STATUS CLASSIFICATION. Limited distribution, declining population trend, and reduction of quality habitat within distribution make it susceptible to extinction. Considered endangered throughout its distribution (Williams *et al.* 1993) and in Alabama (Lydeard *et al.* 1999). **Listed as *endangered* by the U.S. Fish and Wildlife Service in 1998.**

Prepared by: **Holly Blalock-Herod**

RING PINK

Obovaria retusa (Lamarck)

OTHER NAMES. Golf-stick.

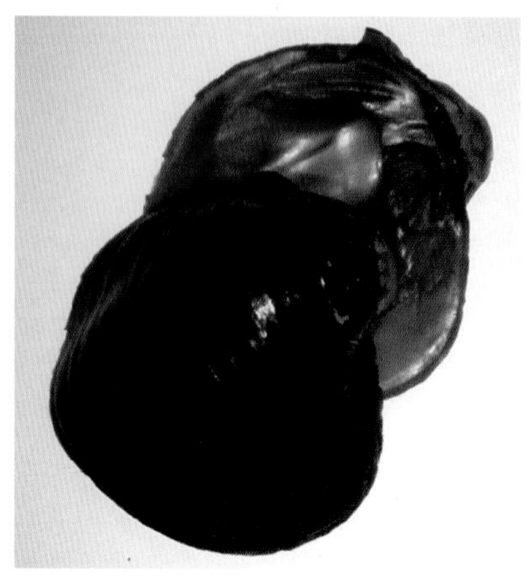

Photo—From Parmalee and Bogan 1998

DESCRIPTION. Has inflated, very solid shell (max. length = 95 mm [3 3/4 in.]). Males grow slightly larger than females. Shell usually ovate in outline, with all margins rounded, but may be irregularly quadrate. Posterior ridge low and rounded in males, but females have a distinct groove posterior to ridge, giving it better definition. Marsupial area of females, located posteriorly, slightly more inflated than same area of males. Shell disk and posterior slope without sculpture. Umbos swollen and turned anteriorly, elevated well above hinge line. Umbo sculpture consists of a few weak, double-looped ridges. Periostracum shiny and without rays, yellowish green to brown in young shells, becoming dark brown or black with age. Pseudocardinal teeth very heavy and triangular and lateral teeth short, heavy, and curved with a wide, but short, interdentum between. Umbo cavity deep and compressed. Shell nacre inside of pallial line varies from light pink to salmon or deep purple; outside pallial line nacre white. (Modified from Simpson 1914, Parmalee and Bogan 1998)

DISTRIBUTION. Historically, throughout the Ohio, Tennessee, and Cumberland Rivers, as well as many of their major tributaries (Parmalee and Bogan 1998). Only extant populations appear to be in Green River, Kentucky (R.S. Butler, U.S. Fish and Wildlife Service, Asheville, North Carolina, pers. comm., 2001), and possibly middle reaches of Cumberland River (Parmalee and Bogan 1998) and tailwaters of Wilson Dam (Garner and McGregor 2001).

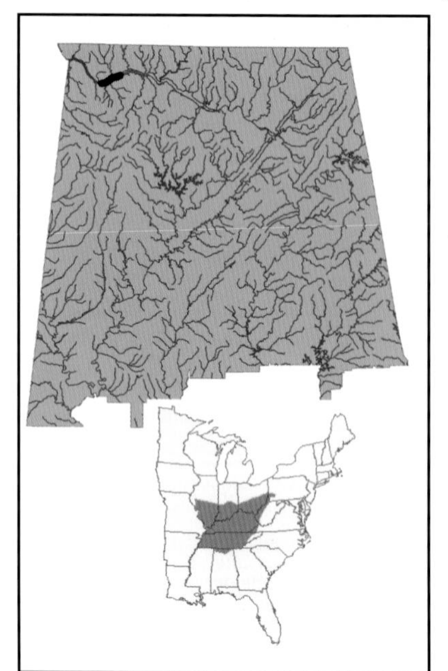

HABITAT. Primarily large rivers (Parmalee and Bogan 1998), but reported from Duck River, indicating it also could occur in small to medium rivers (Ortmann 1924*b*).

LIFE HISTORY AND ECOLOGY. A long-term brooder; females observed brooding eggs in late August and glochidia in September (Ortmann 1909, 1912). Hosts un-known.

BASIS FOR STATUS CLASSIFICATION. Suffered severe declines in distribution during the twentieth century. Has been classified as endangered throughout its distribution (Williams *et al.* 1993) and in Alabama (Stansbery 1976*a*, Lydeard *et al.* 1999). One credible report from Wilson Dam tailwaters during 1990s suggests it still occurs there, but in very small numbers (Garner and McGregor 2001). Based on its extremely limited distribution, rarity, and declining population trend, **listed as *endangered* by the U.S. Fish and Wildlife Service in 1989**.

***Prepared by:* Jeffrey T. Garner**

ROUND HICKORYNUT

Obovaria subrotunda (Rafinesque)

Photo—From Parmalee and Bogan 1998

OTHER NAMES. None.

DESCRIPTION. Has solid shell (max. length = approx. 60 mm [2 3/8 in.]), elliptical to circular in outline, with all margins rounded. Posterior margins of adult females may be more bluntly rounded than those of males. Shells become inflated with age. Shell surface evenly rounded, without a posterior ridge. Shell disk and posterior slope without sculpture. Umbos full and elevated well above hinge line, turned inward. Umbo sculpture consists of weak, slightly sinuous bars. Periostracum olive to dark brown, occasionally black in old specimens. Young shells may have green rays. Periostracum often has yellowish area on posterior dorsal surface. Pseudocardinal teeth heavy and triangular; lateral teeth short, thick, and curved. Interdentum narrow or absent and umbo cavity moderately deep. Shell nacre silvery white, sometimes with shades of pink or purple inside pallial line. (Modified from Parmalee and Bogan 1998)

DISTRIBUTION. Throughout the Tennessee and Cumberland River systems, and much of the Ohio River system. Also has been reported from some tributaries of Lakes Erie and St. Clair (Burch 1975, Parmalee and Bogan 1998). In Alabama, appears to be extant only in Paint Rock River (Ahlstedt 1995).

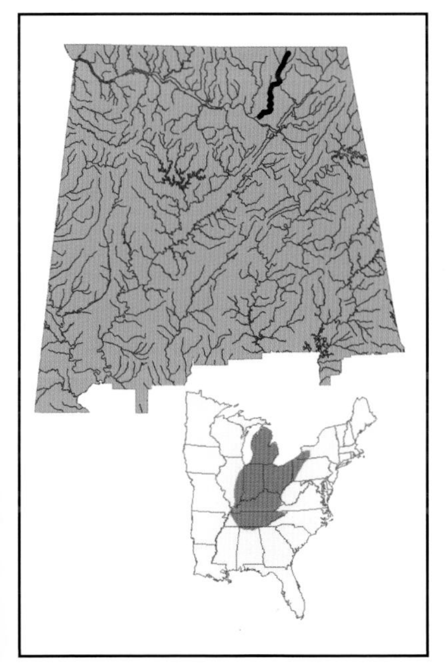

HABITAT. Generally, in medium to large rivers, where it occurs in sand and gravel substrata, usually at depths of less than two meters (6 1/2 feet) (Parmalee and Bogan 1998).

LIFE HISTORY AND ECOLOGY. A long-term brooder, gravid from September into June (Ortmann 1919). Hosts of glochidia unknown.

BASIS FOR STATUS CLASSIFICATION. Extant in Alabama only in Paint Rock River, where its restricted distribution, rarity, and susceptibility to habitat degradation make it vulnerable to extirpation. Although Stansbery (1976*a*) considered it endangered in Alabama, more recently listed as a species of special concern throughout its distribution (Williams *et al.* 1993) and in Alabama (Lydeard *et al.* 1999).

Prepared by: **Jeffrey T. Garner**

WHITE WARTYBACK

Plethobasus cicatricosus (Say)

OTHER NAMES. None.

Photo—From Parmalee and Bogan 1998

DESCRIPTION. Has thick, solid, moderately inflated shell (max. length = 100 mm [3 15/16 in.]), elongate oval to subtriangular in outline, with a broadly rounded anterior margin, narrowly rounded posterior margin, broadly rounded ventral margin, and slightly curved dorsal margin. Posterior ridge low and narrowly rounded to somewhat flattened. On disk of shell an irregular row of low knobs, beginning near umbo and extending obliquely to ventral margin. Umbos full and elevated well above hinge line, turned anteriorly. Periostracum has cloth-like texture and is rayless, yellow or greenish yellow in young specimens, darkening to yellowish brown with age. Pseudocardinal teeth triangular and lateral teeth short, thick, and slightly curved. Interdentum short and wide, and umbo cavity broad and shallow. Shell nacre white. (Modified from Simpson 1914, Parmalee and Bogan 1998)

DISTRIBUTION. Historically widespread in Ohio, Cumberland, Tennessee, and lower Wabash Rivers (Parmalee and Bogan 1998). Currently, only known extant population in tailwaters of Wilson Dam on Tennessee River, where it is rare (Garner and McGregor 2001).

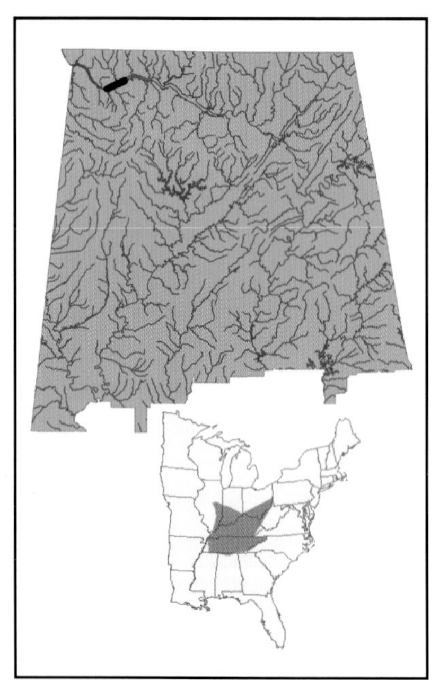

HABITAT. Lotic areas in large rivers. In Wilson Dam tailwaters it occurs in silt-free substrata composed of a mixture of gravel and sand.

LIFE HISTORY AND ECOLOGY. Unknown, but presumably a short-term brooder like its congeners.

BASIS FOR STATUS CLASSIFICATION. Suffered drastic declines with impoundment of almost all large rivers throughout distribution during twentieth century. Extremely limited distribution, rarity, and declining population trend make it vulnerable to extinction. Originally listed as endangered in Alabama, and possibly extinct, but more recently classified as endangered throughout its distribution (Williams *et al.* 1993) and in Alabama (Lydeard *et al.* 1999). **Listed as *endangered* by the U.S. Fish and Wildlife Service in 1976.**

Prepared by: **Jeffrey T. Garner**

ORANGEFOOT PIMPLEBACK

Plethobasus cooperianus (Lea)

OTHER NAMES. Cumberland Pigtoe.

DESCRIPTION. Has solid and moderately inflated shell (max. length = 90 mm [3 1/2 in.]), oval to subtriangular in outline, with a rounded anterior margin and posterior margin obliquely truncate dorsally and rounded ventrally. Posterior ridge low and rounded, or nonexistent. Posterior two-thirds of shell covered with numerous irregular pustules. Umbos full and elevated above hinge line, oriented anteriorly. Periostracum varies from yellowish brown to reddish brown, with juveniles usually having numerous dark green rays that are lost with age. Pseudocardinal teeth somewhat roughened and triangular; lateral teeth short and straight. Interdentum wide and umbo cavity deep and compressed. Shell nacre white, but may be pink inside of pallial line. Foot of live specimens bright orange. (Modified from Simpson 1914, Parmalee and Bogan 1998)

Photo—From Parmalee and Bogan 1998

DISTRIBUTION. Historically, throughout much of the Ohio, Cumberland, and Tennessee Rivers (Parmalee and Bogan 1998). Still survives in tailwaters of at least some Tennessee River dams (*e.g.*, Pickwick Dam, Hardin County, Tennessee). Although not reported from Alabama since 1979 (Garner and McGregor 2001), may still occur in Wilson or Guntersville Dam tailwaters in very low numbers.

HABITAT. Lotic areas in large rivers (Parmalee and Bogan 1998). In tailwaters of Pickwick Dam, found in silt-free areas in a mixture of sand and gravel.

LIFE HISTORY AND ECOLOGY. A short-term brooder; females gravid from early June through early August (Wilson and Clark 1914, Yokley 1972). Glochidial hosts unknown.

BASIS FOR STATUS CLASSIFICATION. Limited distribution, rarity, and susceptibility to habitat degradation make species vulnerable to extinction. Classified as endangered throughout its distribution (Williams *et al.* 1993) and in Alabama (Stansbery 1976*a*, Lydeard *et al.* 1999). May be extirpated from Alabama, but its presence in tailwaters of Pickwick Dam gives hope that it still occurs in very small numbers in tailwaters of Wilson and/or Guntersville Dams (Garner and McGregor 2001). **Listed as *endangered* by the U.S. Fish and Wildlife Service in 1976.**

***Prepared by:* Jeffrey T. Garner**

SHEEPNOSE

Plethobasus cyphyus (Rafinesque)

Photo—From Parmalee and Bogan 1998

OTHER NAMES. Bullhead.

DESCRIPTION. Has thick, moderately inflated shell (max. length = 120 mm [4 3/4 in.]), elongated and oval in outline, with a rounded anterior margin and bluntly pointed or slightly truncate posterior margin. Posterior ridge rounded and slightly curved ventrally. Most specimens have row of large, irregular, broad knobs on shell disk, extending obliquely from near umbo to ventral margin. A shallow depression lies between posterior ridge and row of knobs in most specimens. Umbos full and elevated above hinge line, located toward anterior margin. Umbo sculpture consists of irregular concentric ridges. Periostracum without rays and light yellow to yellowish brown, usually shiny in young specimens, but becoming dull with age. Pseudocardinal teeth erect, roughened and somewhat triangular and divergent; lateral teeth heavy, finely serrate, and slightly curved. Interdentum variable, ranging from very narrow to moderately wide, or may be absent. Umbo cavity shallow. Shell nacre white. (Modified from Simpson 1914, Parmalee and Bogan 1998)

DISTRIBUTION. Historically, throughout much of the Ohio, Cumberland, and Tennessee River systems, and north to upper Mississippi River (Parmalee and Bogan 1998). Now occurs in scattered populations within those systems. In Alabama, extant only in Tennessee River downstream of Wilson and Guntersville Dams, where it is rare (Garner and McGregor 2001).

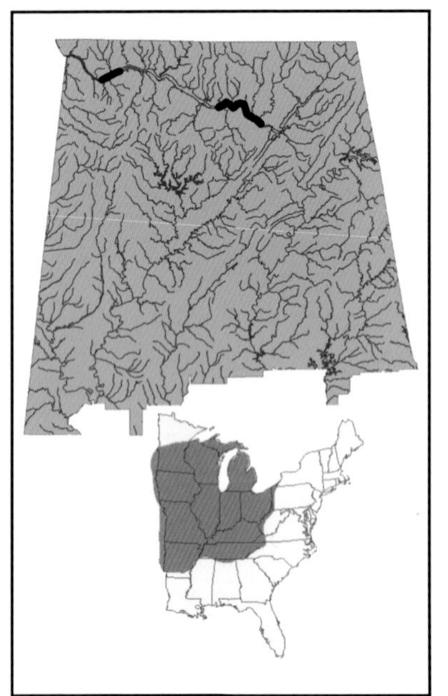

HABITAT. Large rivers (Neves 1991). May be found in clean sand and gravel substrata of shoals, as well as in deeper water, such as riverine reaches downstream of large river dams that often exceed six meters (19 3/4 feet) in depth.

LIFE HISTORY AND ECOLOGY. A short-term brooder, with gravid females reported from May through July (Surber 1912). Surber (1913) reported sauger as glochidial host.

BASIS FOR STATUS CLASSIFICATION. Limited distribution, rarity, and declining population trend make *P. cyphyus* vulnerable to extinction. Classified as threatened throughout its distribution (Williams *et al.* 1993). In Alabama, listed as a species of special concern (Stansbery 1976*a*) and as imperiled (Lydeard *et al.* 1999). ***Candidate for listing as endangered* by the U.S. Fish and Wildlife Service.**

***Prepared by:* Jeffrey T. Garner**

PAINTED CLUBSHELL

Pleurobema chattanoogaense (Lea)

OTHER NAMES. None.

Photo—From Parmalee and Bogan 1998

DESCRIPTION. Has thick, solid, moderately inflated shell (max. length = approx. 65 mm [2 5/8 in.]), ovate, broadly elliptical, or elongate in outline. Anterior margin broadly rounded and posterior margin narrowly rounded, bluntly pointed or biangulate; dorsal and ventral margins slightly convex. Posterior ridge low and rounded when present, but generally absent. Shell disk and posterior slope unsculptured. Umbos fairly high and inflated, extend only slightly above hinge line and located very near anterior end of shell. Periostracum cloth-like in texture and dull yellowish or greenish tan in juveniles, becoming darker with age. Very dark green, wide, and concentric rest lines also may appear. Juveniles also may have a few narrow green rays on umbo. Pseudocardinal teeth triangular, low, stumpy and double in left valve, moderately high, wedge shaped, often serrated, and single in right valve; lateral teeth long and straight. Interdentum short and fairly wide and umbo cavity shallow. Shell nacre a dull bluish white. (Modified from Parmalee and Bogan 1998)

DISTRIBUTION. Endemic to eastern reaches of Mobile Basin, where known from Coosa and lower Tallapoosa River systems. Thought to persist in Alabama only in Weiss bypass of Coosa River, downstream of Terrapin Creek, Cherokee County.

HABITAT. Large creeks to large rivers, generally in gravel and sand substrata of shoal and riffle habitats.

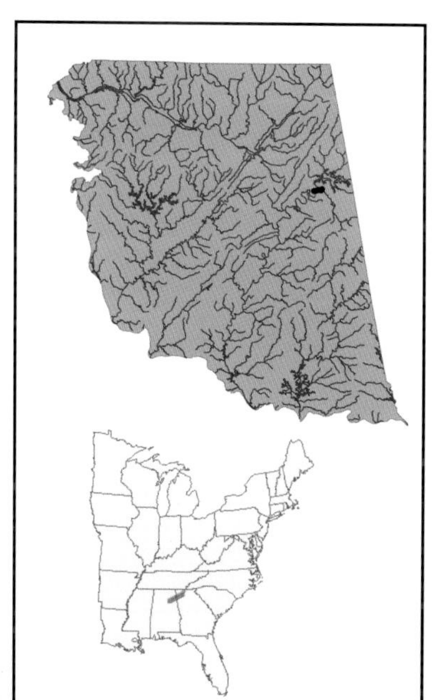

LIFE HISTORY AND ECOLOGY. A short-term brooder. Reported glochidial hosts include blacktail and Alabama shiners, with much higher transformation rates occurring on former (P.D. Johnson, Tennessee Aquarium Research Institute, pers. comm., 2002).

BASIS FOR STATUS CLASSIFICATION. Only extant population known in Alabama located in a short reach of Coosa River, Cherokee County. Many populations have been lost, and until recently was considered extinct. Limited distribution, specialized habitat requirements, and susceptibility to habitat degradation make it vulnerable to extinction. Considered endangered throughout its distribution (Williams *et al.* 1993) and in Alabama (Lydeard *et al.* 1999). ***Candidate for protection*** **by the U.S. Fish and Wildlife Service.**

Prepared by: **Robert S. Butler**

DARK PIGTOE

Pleurobema furvum (Conrad)

OTHER NAMES. None.

Photo—Arthur Bogan

DESCRIPTION. Has solid and somewhat inflated shell (max. length = 60 mm [2 3/8 in.]), ovate in outline, with broadly rounded anterior and bluntly pointed posterior margins. Dorsal and ventral margins convex. Posterior ridge rounded, steeper posteriorly. Shell disk and posterior slope without sculpture. Umbos full and elevated above hinge line, positioned anterior of center. Periostracum dull reddish brown to dark brown, sometimes nearly black. Pseudocardinal teeth heavy and angular; lateral teeth very slightly curved. Interdentum short, but well developed, and shell nacre white, sometimes reddish. (Modified from Simpson 1914, USFWS 2000)

DISTRIBUTION. Endemic to upper Black Warrior River system in Alabama (USFWS 2000). Extirpated from most of its former distribution, and currently known to occur only in Sipsey Fork headwaters of Bankhead National Forest and in North River (McGregor and Pierson 1999).

HABITAT. Lotic areas in medium to large streams. Individuals most frequently encountered in riffle habitats with gravel and sand substrata.

LIFE HISTORY AND ECOLOGY. A short-term brooder; females release mature glochidia in June. Transmission of glochidia to host fishes facilitated by females releasing glochidia in conglutinates that resemble food items of small fishes. Conglutinates flattened, oval shaped, and pink to peach colored. Primary glochidial hosts include blacktail shiner, Alabama shiner, and creek chub. The largescale stoneroller and blackspotted topminnow are marginal hosts. (Summarized from Haag and Warren 1997)

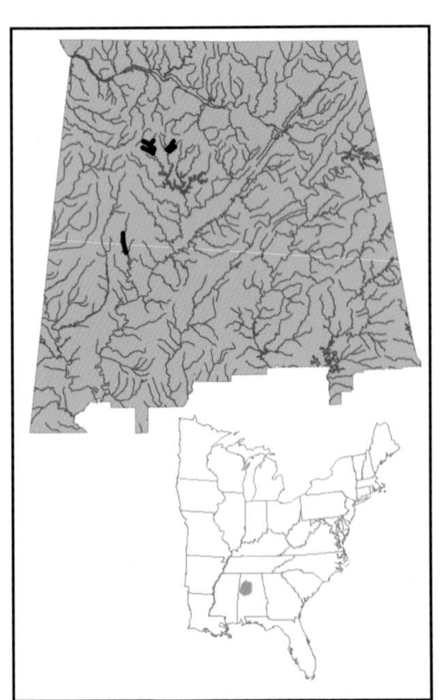

BASIS FOR STATUS CLASSIFICATION. Limited distribution, rarity, and declining population trend make *P. furvum* vulnerable to extinction. Classified as endangered throughout distribution (Williams *et al.* 1993) and in Alabama (Lydeard *et al.* 1999). **Listed as *endangered* by the U.S. Fish and Wildlife Service in 1993.**

***Prepared by:* Wendell R. Haag**

SOUTHERN PIGTOE

Pleurobema georgianum (Lea)

OTHER NAMES. None.

DESCRIPTION. Has solid shell (max. length = 65 mm [2 5/8 in.]), moderately inflated in juveniles, but becoming more compressed with age. Shell outline subrhomboid to ovate, with a broadly rounded anterior margin and bluntly pointed or obliquely truncate posterior margin. Posterior ridge low and rounded, often indistinct. Shell disk and posterior slope unsculptured. Umbos somewhat compressed and elevated only slightly above hinge line. Periostracum dull light yellow or greenish brown in young specimens, darkening to brown with age. Typically one wide, dark green ray on umbo and disk, usually broken and becoming indistinct ventrally. Pseudocardinal teeth low and triangular or linear; lateral teeth low, thick, and slightly curved. Interdentum wide in old specimens, but umbo cavity very shallow. Shell nacre bluish white. (Modified from Simpson 1914, Parmalee and Bogan 1998)

Photo—Mike Gangloff

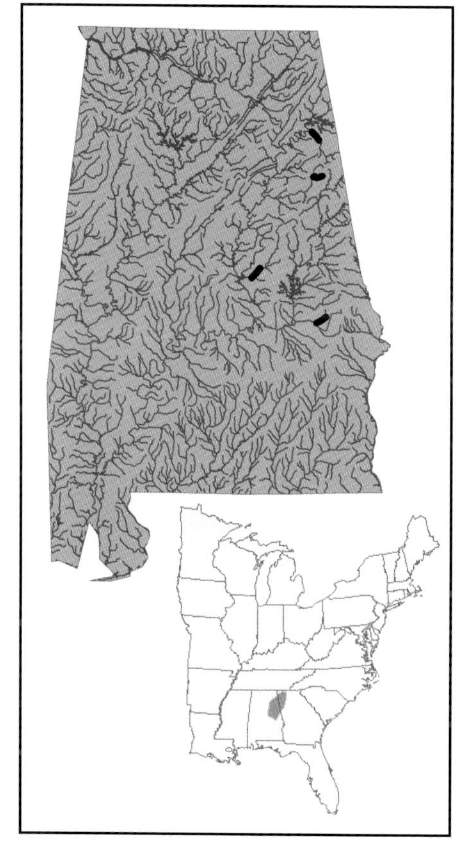

DISTRIBUTION. Endemic to Coosa River system in Alabama, Georgia, and Tennessee (USFWS 2000). Extant in a few widely scattered tributaries, including some streams in Talladega National Forest.

HABITAT. Lotic areas of small to medium streams. Individuals most frequently encountered in riffle habitats with gravel and sand substrata.

LIFE HISTORY AND ECOLOGY. Unknown. Other species of *Pleurobema* are short-term brooders and use cyprinids as glochidial hosts (Haag and Warren 2003).

BASIS FOR STATUS CLASSIFICATION. Extant in only a few, very small, isolated populations. Limited distribution, rarity, and declining population trend make it vulnerable to extinction. Has been classified as endangered throughout its distribution (Williams *et al.* 1993) and in Alabama (Lydeard *et al.* 1999). **Listed as *endangered* by the U.S. Fish and Wildlife Service in 1993**.

***Prepared by:* Wendell R. Haag**

TENNESSEE CLUBSHELL

Pleurobema oviforme (Conrad)

Photo—Arthur Bogan

OTHER NAMES. None.

DESCRIPTION. Has solid, fairly heavy shell (max. length = approx. 90 mm [3 1/2 in.]), obovate, elliptical, or rhomboidal in outline, and slightly to moderately inflated. Anterior margin broadly rounded and posterior margin bluntly pointed or somewhat truncate; dorsal and ventral margins slightly convex. Posterior ridge low and rounded, often indistinct. Shell disk and posterior slope without sculpture. Umbos moderately full, elevated slightly above hinge line, turned forward, and located very near anterior end of shell. Periostracum generally smooth and dull straw yellow, greenish yellow, or gray brown. Broken green rays of variable width may be associated with umbo and disk. Pseudocardinal teeth stout, triangular, deeply serrate and erect, projecting ventrally from hinge line at a nearly right angle; lateral teeth erect, heavy, long and straight, with two in left valve and one in right valve. Interdentum wide, but umbo cavity shallow. Shell nacre silvery or bluish white. (Modified from Parmalee and Bogan 1998)

DISTRIBUTION. Endemic to Cumberland and Tennessee River systems. Historically, in the Tennessee River downstream to Muscle Shoals and in several of its tributaries. In Alabama, now restricted to Paint Rock River system, Jackson County (Ahlstedt 1995).

HABITAT. Creeks and small to large rivers in shoals and riffles with substrata of coarse gravel and sand (Parmalee and Bogan 1998).

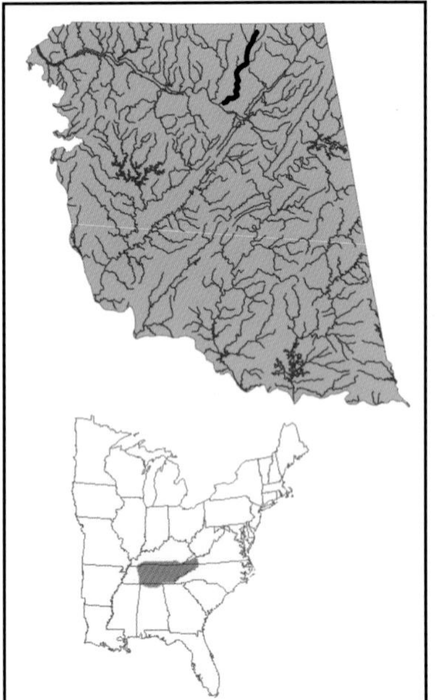

LIFE HISTORY AND ECOLOGY. A short-term brooder (Parmalee and Bogan 1998). Reported hosts include whitetail and common shiners, central stoneroller, and fantail darter (Weaver *et al.* 1991). Common shiners and central stonerollers do not occur in Tennessee River system of Alabama, but related native fishes such as striped shiners and largescale stonerollers do occur in state and may serve as hosts.

BASIS FOR STATUS CLASSIFICATION. Limited distribution, rarity, and declining population trend make *P. oviforme* vulnerable to extirpation from Alabama. Appears to be declining throughout most of its distribution and **considered a *species of concern* by the U.S. Fish and Wildlife Service** and Williams *et al.* (1993). Once listed as endangered in Alabama (Stansbery 1976*a*), but more recently listed as a species of concern in state (Lydeard *et al.* 1999).

***Prepared by:* Robert S. Butler**

OVATE CLUBSHELL

Pleurobema perovatum (Conrad)

OTHER NAMES. None.

Photo—Mike Gangloff

DESCRIPTION. Has moderately thick shell (max. length = 45 mm [1 3/4 in.]), oval in outline, with a rounded anterior margin and bluntly pointed posterior margin. Posterior ridge narrowly rounded. Shell disk and posterior slope without sculpture. Umbos inflated and rounded, elevated well above hinge line. Perio-stracum dull yellow to olive in young specimens, darkening to brown with age. Some specimens have a few weak green rays, variable in width. Pseudocar-dinal teeth erect and triangular; lateral teeth long and straight. When present, interdentum short and narrow. Umbo cavity shallow and open. Shell nacre white. (Modified from Simpson 1914, Parmalee and Bogan 1998)

DISTRIBUTION. Endemic to Mobile Basin (Parmalee and Bogan 1998). Populations remain only in widely scattered localities in Tombigbee, Coosa, and Tallapoosa River systems (McGregor *et al.* 1999, McCullagh *et al.*, in review). Populations in Coosa and Tallapoosa River systems small and isolated. Appears to be extirpated from Black Warrior and Cahaba River systems.

HABITAT. Medium to large streams. In coastal plain streams, found most frequently in silt or silty sand along stream margins or in side channels, and rarely found in main channel riffle or run habitats (W. R. Haag and M. L. Warren, USDA Forest Service, unpubl. data).

LIFE HISTORY AND ECOLOGY. A short-term brooder, gravid from May to July (W. R. Haag and M. L. Warren, USDA Forest Service, unpubl. data). Hosts of glochidia unknown, but other species of *Pleurobema* use cyprinids (Haag and Warren 2003).

BASIS FOR STATUS CLASSIFICATION. Limited distribution, rarity, and declining population trend make *P. perovatum* vulnerable to extinction. Also classified as endangered throughout its distribution (Williams *et al.* 1993) and in Alabama (Stansbery 1976*a*, Lydeard *et al.* 1999). **Listed as *endangered* by the U.S. Fish and Wildlife Service in 1993.**

***Prepared by:* Wendell R. Haag**

ROUGH PIGTOE

Pleurobema plenum (Lea)

OTHER NAMES. None.

DESCRIPTION. Has solid, inflated shell (max. length = 80 mm [3 1/8 in.]), subtriangular in outline, with a truncate anterior margin, straight posterior margin, slightly convex dorsal margin, and rounded ventral margin. Posterior ridge narrowly rounded and slightly curved, ending in a blunt point ventrally on posterior margin. A shallow sulcus often present just anterior to posterior ridge, with a median ridge located centrally. Shell disk and posterior slope without sculpture. Umbos elevated well above hinge line and turned anteriorly. Umbo sculpture consists of a few irregular, nodulous ridges. Periostracum has satiny texture, yellowish brown to reddish brown, sometimes with a series of fine, dark green rays on posterior half of shell. Pseudocardinal teeth large and radial; lateral teeth thick and short. Interdentum broad and umbo cavity moderately deep, wide, and open. Shell nacre usually white, but may be pink. (Modified from Simpson 1914, Parmalee and Bogan 1998)

Photo—From Parmalee and Bogan 1998

DISTRIBUTION. Historically, from the Ohio, Cumberland, and Tennessee Rivers, southwest to Kansas and Arkansas (Parmalee and Bogan 1998). In Alabama, extant populations exist in Tennessee River tailwaters of Wilson Dam, where very rare, and possibly Guntersville Dam (Garner and McGregor 2001).

HABITAT. Can occur in lotic areas of small to medium-sized rivers, but primarily found in large rivers (Neves 1991, Parmalee and Bogan 1998). In tailwaters of Wilson Dam, found in a mixture of sand and gravel in areas kept free of silt by moderate to strong current.

LIFE HISTORY AND ECOLOGY. A short-term brooder, with gravid females in May (Ortmann 1919). Glochidial hosts unknown.

BASIS FOR STATUS CLASSIFICATION. Limited distribution, rarity, and declining population trend make *P. plenum* vulnerable to extinction. Classified as endangered throughout its distribution (Williams *et al.* 1993) and in Alabama (Stansbery 1976*a*, Lydeard *et al.* 1999). **Listed as *endangered* by the U.S. Fish and Wildlife Service in 1976.**

***Prepared by:* Jeffrey T. Garner**

OVAL PIGTOE

Pleurobema pyriforme (Lea)

OTHER NAMES. None.

Photo—Arthur Bogan

DESCRIPTION. Has moderately thick, compressed shell (max. length = 60 mm [2 3/8 in.]), subovate to elliptical in outline, with a rounded anterior margin, bluntly pointed posterior margin, slightly convex ventral margin, and almost straight dorsal margin. Posterior ridge biangulate, ending in a blunt point posterior ventrally. Shell disk and posterior slope unsculptured. Umbos elevated slightly above hinge line and positioned anterior of center. Periostracum smooth and shiny, yellowish to chestnut brown, often shaded with green. Pseudocardinal teeth large, crenulate, and doubled in both valves; lateral teeth short and slightly curved. Interdentum very narrow and umbo cavity shallow. Shell nacre salmon to bluish white. (Modified from Simpson 1914, Williams and Butler 1994)

DISTRIBUTION. From Econfina Creek system, east to Suwannee River system (Brim Box and Williams 2000). In Alabama, confined to headwaters of Chipola River and lower Chattahoochee River system (Brim Box and Williams 2000). Only known extant population in Big Creek, a headwater tributary of Chipola River, Houston County.

HABITAT. Medium-sized creeks to small rivers, usually in slow to moderate current, especially in stream channels with clean sand or gravel substrata. Tolerates some silt (Williams and Butler 1994, Blalock-Herod 2000).

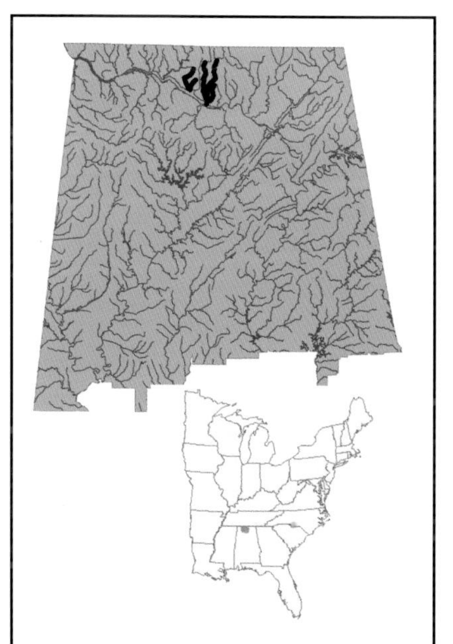

LIFE HISTORY AND ECOLOGY. A short-term brooder. Gravid individuals observed from March to July (Brim Box and Williams 2000). Fishes reportedly serving as hosts for glochidia include sailfin shiner and eastern mosquitofish (O'Brien 1997).

BASIS FOR STATUS CLASSIFICATION. Limited distribution, dwindling habitat quality within its distribution, and declining population trend make *P. pyriforme* vulnerable to extinction. Although once listed as threatened in Alabama (Stansbery 1976*a*), more recently classified as endangered throughout its distribution (Williams *et al.* 1993) and within Alabama (Lydeard *et al.* 1999). **Listed as *endangered* by the U.S. Fish and Wildlife Service in 1998.**

***Prepared by:* Stuart W. McGregor**

PYRAMID PIGTOE

Pleurobema rubrum (Rafinesque)

OTHER NAMES. Pink Pigtoe.

Photo—From Parmalee and Bogan 1998

DESCRIPTION. Has thick and solid shell (max. length = 90 mm [3 1/2 in.]), subtriangular in outline, with a somewhat truncate anterior margin, bluntly pointed posterior margin, and straight ventral margin. Posterior ridge low and rounded, but distinct. A very shallow and wide sulcus lies just anterior to posterior ridge, but generally does not extend dorsally to umbo. Shell disk and posterior slope without sculpture. Umbos greatly inflated and elevated well above hinge line, turned anteriorly and located at anterior end of shell. Periostracum light tan to brownish green, usually with green rays in juveniles, but darkens with age to almost black and rays become obscure. Pseudocardinal teeth low, triangular, and rough; lateral teeth straight and finely serrate. Interdentum wide, but umbo cavity very shallow and open. Shell nacre variable, ranging from white to pink or salmon. (Modified from Simpson 1914, Parmalee and Bogan 1998)

DISTRIBUTION. Historically, from the Ohio, Cumberland, and Tennessee Rivers, west to Arkansas and possibly Nebraska and upstream in Mississippi River to Wisconsin (Parmalee and Bogan 1998). In Alabama, extant, but uncommon, in tailwaters of Guntersville and Wilson Dams on Tennessee River (Garner and McGregor 2001).

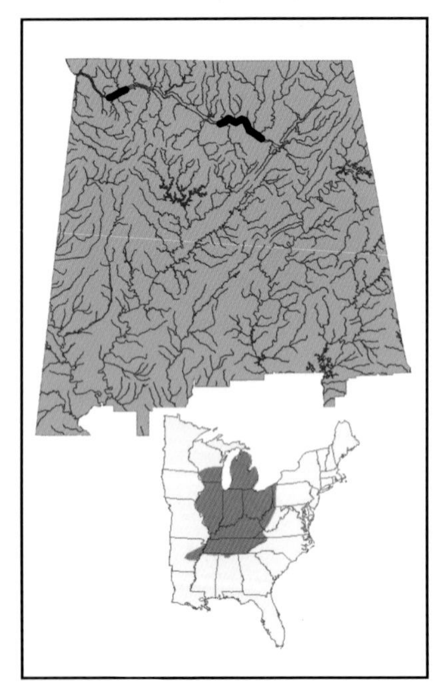

HABITAT. Lotic areas of medium to large rivers (Neves 1991). In unimpounded rivers, often found in shoals at depths of less than one meter (3 1/4 feet), but in tailwaters of dams on large rivers, may be found at depths of more than six meters (19 3/4 feet) (Parmalee and Bogan 1998).

LIFE HISTORY AND ECOLOGY. A short-term brooder, with gravid females in May and July (Ortmann 1919). Hosts of its glochidia unknown.

BASIS FOR STATUS CLASSIFICATION. Noted as uncommon with no specimens less than 10 years old encountered in two extant populations in Alabama (Garner and McGregor 2001). Limited distribution, declining population trend, and specific habitat requirements make species vulnerable to extirpation from state. Classified as threatened throughout its distribution (Williams *et al.* 1993) and imperiled in Alabama (Lydeard *et al.* 1999).

Prepared by: **Jeffrey T. Garner**

ROUND PIGTOE

Pleurobema sintoxia (Rafinesque)

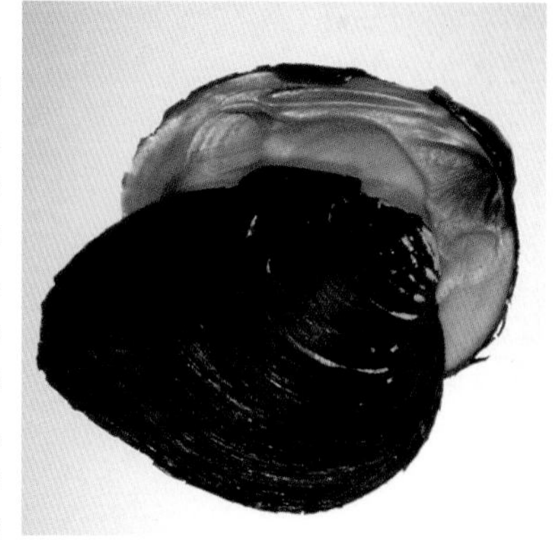

Photo—From Parmalee and Bogan 1998

OTHER NAMES. Solid Pigtoe, Pink Pigtoe, Flat Pigtoe.

DESCRIPTION. Has thick, solid shell (max. length = 120 mm [4 3/4 in.]), usually compressed and highly variable in shape. Most shells subtriangular, but may be elongate oval or subquadrate in outline. Ventral margin may be straight or slightly convex and dorsal margin generally slightly convex. Anterior margin often rounded below and obliquely truncate above. Posteriorly, shell obliquely truncate, with blunt point of posterior ridge located ventrally. Posterior ridge rounded, narrow toward umbo, becoming wider ventrally. Umbos compressed and only slightly elevated above hinge line in specimens from smaller streams, but large river specimens typically have umbos that are full and elevated well above hinge line. In both cases, umbos oriented anteriorly. Umbo sculpture consists of a few coarse, irregular ridges, curving upward posteriorly. Periostracum of juveniles dull tan, with distinct green rays, but darkening with age, becoming deep reddish brown to black, with obscure rays. Pseudocardinal teeth stout, triangular, and serrate; lateral teeth straight, moderately high and finely serrate. Interdentum wide, but umbo cavity very shallow. Shell nacre usually white, but may be various shades of pink. (Modified from Simpson 1914, Parmalee and Bogan 1998)

DISTRIBUTION. Historically, throughout the upper Mississippi River system, south to Alabama and Arkansas (Parmalee and Bogan 1998). In Alabama, extant, but rare, in the Tennessee River in the tailwaters of Wilson and Guntersville Dams.

HABITAT. Primarily medium to large rivers, occurring in lotic areas with firm, coarse sand and gravel substrata (Parmalee and Bogan 1998).

LIFE HISTORY AND ECOLOGY. A short-term brooder. Gravid from early May to late July in Wisconsin (Baker 1928). Fish hosts of glochidia include spotfin shiner, bluntnose minnow, and northern redbelly dace, as well as bluegill (Surber 1912, Hove *in* Parmalee and Bogan 1998).

BASIS FOR STATUS CLASSIFICATION. Vulnerable to extirpation from state due to its limited distribution, rarity, and specific habitat requirements. Classified as a species of special concern throughout its distribution (Williams *et al.* 1993) and imperiled in Alabama (Lydeard *et al.* 1999).

Prepared by: **Jeffrey T. Garner**

HEAVY PIGTOE

Pleurobema taitianum (Lea)

Photo—Jim Williams

OTHER NAMES. None.

DESCRIPTION. Has thick, solid, inflated shell (max. length = 50 mm [2 in.]), obliquely triangular in outline, with a broadly rounded to obliquely truncate anterior margin, slightly convex to obliquely truncate posterior margin, straight ventral margin, and slightly convex dorsal margin. Posterior ridge low, steeper dorsally. Shell disk and posterior slope unsculptured. A wide, radial swelling located just anterior of posterior slope, sometimes with a weak sulcus separating them. Umbos full and elevated well above hinge line, situated at, or exceeding anterior margin. Periostracum dull, varying from tawny to brown, usually darkening with age. Pseudocardinal teeth solid and ragged; lateral teeth moderately long and slightly curved. Interdentum short and well developed, but umbo cavity shallow. Shell nacre white or pink. (Modified from Simpson 1914, USFWS 2000)

DISTRIBUTION. Endemic to Mobile Basin in Alabama and Mississippi, including Cahaba, Coosa, and Tombigbee River systems (Williams 1982, Stansbery 1983c). However, known to be extant only in short reaches of Alabama and Tombigbee Rivers.

HABITAT. Main channel lotic habitat of large rivers, in gravel and sand substrata in moderate to swift current (Williams 1982).

LIFE HISTORY AND ECOLOGY. Unknown. Other species of *Pleurobema* are short-term brooders and use cyprinids as host fishes (Haag and Warren 2003).

BASIS FOR STATUS CLASSIFICATION. Prior to 1980, largest populations occurred in Tombigbee River in Alabama and Mississippi. These populations were destroyed by construction of Tennessee-Tombigbee Waterway. Small populations existed until mid-1980s in lower reaches of Buttahatchee and Sipsey Rivers, in Mississippi and Alabama, respectively. These populations were likely maintained by immigration from Tombigbee River and live individuals have not been found in these two rivers in almost 20 years. Populations in Cahaba and Coosa Rivers likely extirpated and only known remaining population is in a short reach of the Alabama River in Dallas County, and a short reach of the Tombigbee River in Choctaw County. Extremely limited distribution, declining population trend, and specific habitat requirements make *P. taitianum* vulnerable to extinction. Classified as endangered throughout its distribution (Williams *et al.* 1993) and in Alabama (Stansbery 1976*a*, Lydeard *et al.* 1999). **Listed as *endangered* by the U.S. Fish and Wildlife Service in 1987**.

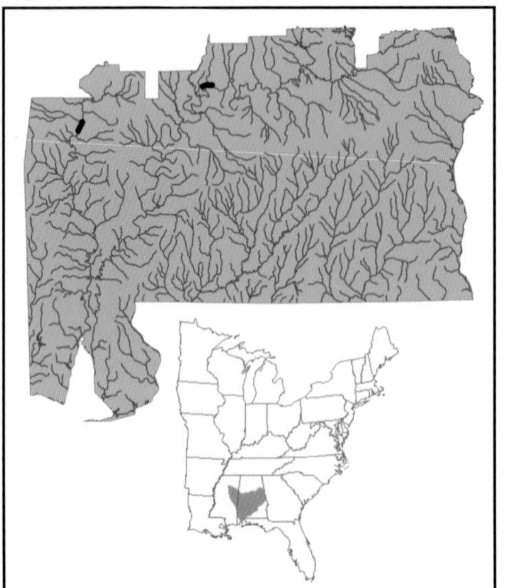

Prepared by: **Wendell R. Haag**

KIDNEYSHELL

Ptychobranchus fasciolaris (Rafinesque)

Photo—From Parmalee and Bogan 1998

OTHER NAMES. None.

DESCRIPTION. Has solid, but compressed shell (max. length = 150 mm [5 7/8 in.]), elongate and elliptical in outline, with a rounded anterior margin, bluntly pointed posterior margin, slightly convex dorsal margin, and almost straight ventral margin. Posterior ridge located more or less dorsally and rounded, but prominent. Shell disk and posterior ridge without sculpture. Umbos flattened and compressed, elevated little, if any, above hinge line. Umbo sculpture consists of fine, indistinct, wavy ridges. Periostracum yellow to yellowish green, darkening with age to dark chestnut brown. Most specimens adorned with variable green rays that are usually wide and broken and may be wavy. Pseudocardinal teeth low, thick, serrate, and triangular; lateral teeth heavy and usually widely separated, short, and straight. Interdentum long and moderately wide and umbo cavity shallow. Shell nacre white. (Modified from Simpson 1914, Parmalee and Bogan 1998)

DISTRIBUTION. Throughout Mississippi River system, including the Ohio, Tennessee, and Cumberland Rivers (Parmalee and Bogan 1998). In Alabama, extant populations exist only in Paint Rock River system (Ahlstedt 1995), a short reach of Bear Creek in Colbert County (McGregor and Garner, in review), and the tailwaters of Guntersville and Wilson Dams on the Tennessee River (Garner and McGregor 2001). Very rare in Tennessee River, and populations there may not be viable.

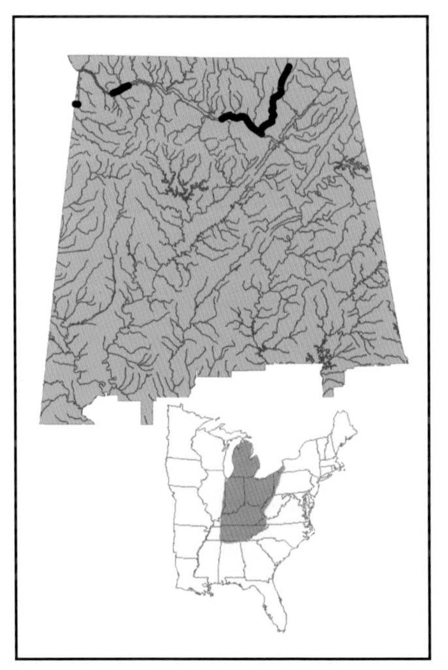

HABITAT. A variety of river types, from large creeks to large rivers, but always found associated with lotic conditions (Parmalee and Bogan 1998). Usually occurs in gravel and cobble substrata kept free of silt by moderate to swift current.

LIFE HISTORY AND ECOLOGY. A long-term brooder. Ortmann (1919) reported species to be gravid in August, and to brood glochidia until following summer, with glochidial discharge taking place from June through August. Glochidial hosts unknown.

BASIS FOR STATUS CLASSIFICATION. Although widespread, species has suffered severe distribution reductions. Limited distribution, rarity, and declining population trend make it vulnerable to extirpation from state. Classified as a species of special concern throughout its distribution (Williams *et al.* 1993) and in Alabama (Stansbery 1976*a*, Lydeard *et al.* 1999).

Prepared by: **Jeffrey T. Garner**

TRIANGULAR KIDNEYSHELL

Ptychobranchus greenii (Conrad)

Photo—From Parmalee and Bogan 1998

OTHER NAMES. None.

DESCRIPTION. Has solid and moderately inflated shell (max. length = 70 mm [2 3/4 in.]), elongate and subtriangular in outline, being broadly rounded anteriorly and broadly pointed posteriorly, with slightly convex dorsal and ventral margins. Posterior ridge broadly rounded and more or less doubled, usually distinct. Shell disk and posterior slope unsculptured. Umbos full and high, but not elevated above hinge line. Periostracum dull yellow, darkening to light brown with age. Pseudocardinal teeth low and triangular; lateral teeth high, thin, and straight. Interdentum narrow and umbo cavity moderately shallow. Shell nacre creamy or bluish white. (Modified from Simpson 1914, Parmalee and Bogan 1998)

DISTRIBUTION. Endemic to Mobile Basin, and has been reported from Black Warrior, Cahaba, and Coosa River systems in Alabama, Georgia, and Tennessee. Healthy populations appear to remain in only two streams in Bankhead National Forest. Very small, isolated populations exist in Locust Fork, Cahaba River, and upper Coosa River system (McGregor *et al.* 2000).

HABITAT. Lotic areas in medium to large streams above the Fall Line. Individuals most frequently encountered in riffle habitats with gravel and sand substrata.

LIFE HISTORY AND ECOLOGY. A long-term brooder that releases glochidia in April. Transmission of glochidia to hosts facilitated by females releasing glochidia in conglutinates that resemble food items of hosts. Conglutinates vary among individual mussels and may resemble dipteran larvae or fish eggs. Primary hosts include Warrior, Tuskaloosa, and blackbanded darters, and Mobile logperch. (Summarized from Haag and Warren 1997)

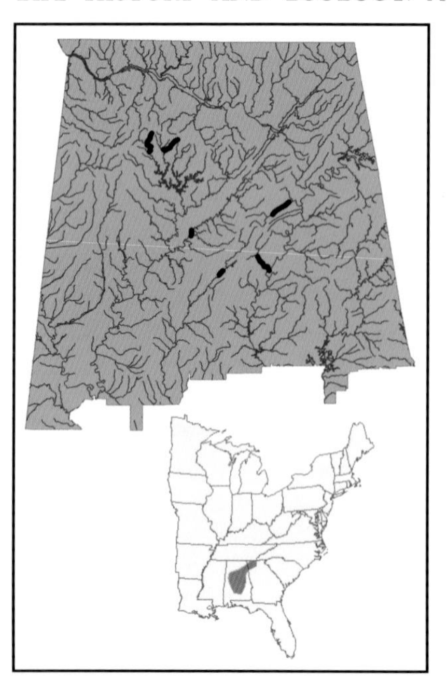

BASIS FOR STATUS CLASSIFICATION. Has experienced dramatic reduction in distribution. Limited distribution, rarity, and declining population trend make *P. greenii* vulnerable to extinction. Was listed as threatened in Alabama (Stansbery 1976*a*), but more recently classified as endangered throughout its distribution (Williams *et al.* 1993) and in Alabama (Lydeard *et al.* 1999). **Listed as *endangered* by the U.S. Fish and Wildlife Service in 1993.**

Prepared by: **Wendell R. Haag**

SOUTHERN KIDNEYSHELL

Ptychobranchus jonesi (van der Schalie)

Photo—Jeff Garner

OTHER NAMES. None.

DESCRIPTION. Has moderately thin, but inflated shell (max. length = 65 mm [2 5/8 in.]), elliptical in outline, with a rounded anterior margin and biangulate posterior margin. Females have a slight marsupial swelling posterior ventrally. Posterior ridge doubled and posterior slope moderately steep. Shell disk and posterior slope without sculpturing. Umbos full, but barely elevated above hinge line, positioned well anterior of center. Periostracum smooth, olivaceous to dark brown, with irregular, often obscure green rays. Two solid, compressed pseudocardinal teeth in left valve, one well developed and one rudimentary pseudocardinal tooth in right valve; lateral teeth thin and slightly curved. Interdentum narrow. Shell nacre bluish white. Gravid females may be easily distinguished from any other sympatric species by folded outer gill demibranchs. (Modified from van der Schalie 1934, Athearn 1964, Fuller and Bereza 1973, Williams and Butler 1994)

DISTRIBUTION. Includes the Choctawhatchee, Yellow, and Escambia River systems in Alabama and Florida (Burch 1975, Butler 1989). However, only recent records from West Fork Choctawhatchee River (Blalock-Herod *et al.*, in review; Williams *et al.*, in review).

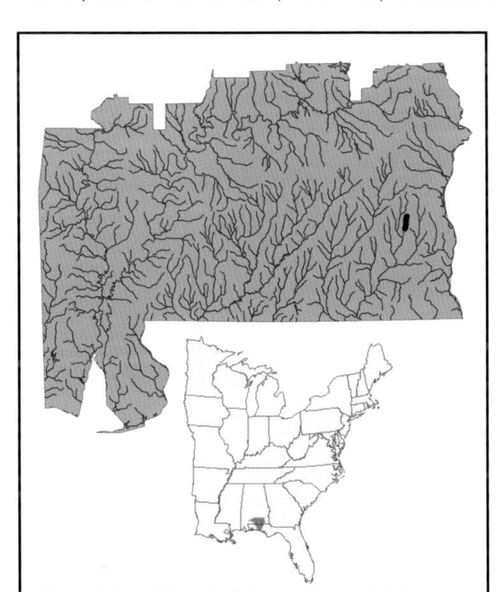

HABITAT. Medium-sized creeks to small rivers, usually in silty sand substrata and slow current (Williams and Butler 1994). Also found in small depressions in clay substrata (Blalock-Herod *et al.*, in review).

LIFE HISTORY AND ECOLOGY. Unknown. Presumably a long-term brooder.

BASIS FOR STATUS CLASSIFICATION. Suffered severe declines during recent past. Vulnerable to extinction due to limited distribution and rarity, along with dwindling habitat quality within its distribution. Classified as threatened throughout its distribution (Williams *et al.* 1993, Blalock-Herod *et al.*, in review) and imperiled in Alabama (Lydeard *et al.* 1999).

***Prepared by:* Stuart W. McGregor**

RABBITSFOOT

Quadrula cylindrica cylindrica (Say)

OTHER NAMES. Cylinder, Smooth Cob Shell, Spectaclecase Mussel.

Photo—From Parmalee and Bogan 1998

DESCRIPTION. Has solid shell (max. length = 120 mm [4 3/4 in.]), elongate and rhomboidal to rectangular in outline. Young specimens somewhat compressed, but shells become very inflated with age, becoming almost cylindrical. Anterior margin rounded; posterior margin truncate, sometimes obliquely; dorsal margin straight or slightly convex; and ventral margin straight or very slightly concave. Posterior ridge full and rounded, extending obliquely from umbo to posterior ventral margin. Posterior ridge generally adorned with a row of knobs that may be rudimentary in specimens from large rivers, taking the form of broad undulations. Specimens from smaller rivers may have additional, smaller, lachrymous nodules or plications (grading to a very tuberculate form of Tennessee River headwater streams, *Q. c. strigillata*.) Above posterior ridge a wide, shallow radial depression that sometimes ends in a slight biangulation on posterior margin. Posterior slope usually marked with a series of broadly rounded ridges that curve up toward dorsal margin. Posterior slope of specimens from large rivers may be smooth or only weakly sculptured. Umbos moderately elevated above hinge line. Umbo sculpture consists of a few irregular strong ridges or folds. Periostracum varies from light yellow to greenish, darkening to yellowish brown with age. Numerous dark green markings, in form of streaks, chevrons or ventrally pointed triangular spots, cover most of shell surface. Pseudocardinal teeth low and triangular; lateral teeth long and straight. Interdentum, when present, narrow. Umbo cavity moderately deep. Shell nacre usually white, often bluish gray within pallial line, but also may have a pinkish tinge. (Modified from Simpson 1914, Parmalee and Bogan 1998)

DISTRIBUTION. Throughout much of lower Mississippi River system, including the Tennessee River. In headwaters of Tennessee River, replaced by *Q. c. strigillata*. In Alabama, extant populations known to exist only in Paint Rock River system (Ahlstedt 1995) and a short reach of Bear Creek, Colbert County (McGregor and Garner, in review). A viable population still exists in tailwaters of Pickwick Dam on Tennessee River, so it is conceivable that it is present, but overlooked, in Wilson Dam tailwaters (Garner and McGregor 2001).

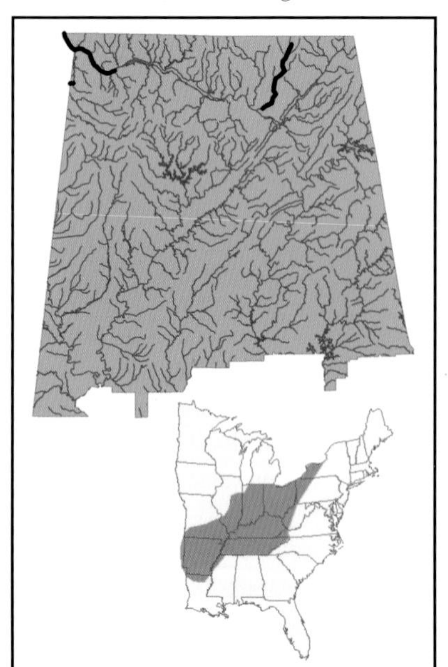

HABITAT. Large creeks to large rivers. In creeks and small rivers often occurs along margins of riffles and runs. In lotic reaches of larger rivers, may be found at depths of more than six meters (19 3/4 feet), as well as upon marginal shelf in shallower water. Majority of specimens in tailwaters of Pickwick Dam found in sand and clay substrata on channel slope and upon marginal shelf.

LIFE HISTORY AND ECOLOGY. A short-term brooder. Hosts of glochidia unknown. However, Yeager and Neves (1986) reported whitetail and spotfin shiners and bigeye chub serve as hosts for *Q. c. strigillata* glochidia.

BASIS FOR STATUS CLASSIFICATION. Although widespread, has suffered severe distribution reductions in Alabama. Limited distribution, rarity, and declining population trend make it vulnerable to extirpation. Classified as threatened throughout its distribution (Williams *et al.* 1993). Stansbery (1976) listed it as endangered and Lydeard *et al.* (1999) listed it as imperiled in Alabama.

***Prepared by:* Jeffrey T. Garner**

SCULPTURED PIGTOE
Quincuncina infucata (Conrad)

OTHER NAMES. None.

Photo—Paul Johnson

DESCRIPTION. Has small, heavy shell (max. length =50 mm [2 in.]), inflated and subcircular in outline. Anterior margin broadly rounded and posterior margin tapered to a rounded point posterior ventrally. Dorsal and ventral margins slightly convex. Posterior ridge poorly defined and posterior slope flat to slightly concave. Sculpturing highly variable, with inconspicuous to distinct chevron shaped plications across most of shell disk and arcuate ridges on posterior slope. Umbos only slightly inflated and not elevated above hinge line. Periostracum dark brown to black. Pseudocardinal teeth triangular and divergent, with two in each valve; lateral teeth short and straight. Shell nacre white to bluish white. (Modified from Clench and Turner 1956, Brim Box and Williams 2000)

DISTRIBUTION. Endemic to Apalachicola and Ochlockonee River basins in Alabama, Georgia, and Florida. In Alabama, historically occurred throughout Chattahoochee River system and headwaters of Chipola River. Only known extant populations in Alabama in Uchee and Little Uchee Creeks, Russell and Lee Counties, respectively (Brim Box and Williams 2000, Lydeard *et al.* 2000), and Big Creek, Houston County (J.T. Garner, Ala. Div. Wildl. Freshwater Fish., unpubl. data).

HABITAT. Small to large streams and rivers with moderate to swift current in substrata ranging from muddy sand to rocky areas (Brim Box and Williams 2000). Does not appear to tolerate impoundments (Brim Box and Williams 2000), but this species and a closely related species (*Q. kleiniana*) have been reported to be occasionally common in shallow pools and deeper reaches of rivers, often under detritus (Clench and Turner 1956, Jenkinson 1973, Blalock-Herod 2000).

LIFE HISTORY AND ECOLOGY. Little known. Presumably a short-term brooder, since closely related species found gravid in June (Brim Box and Williams 2000). All four gills used to brood glochidia (Ortmann and Walker 1922). Hosts of glochidia unknown.

BASIS FOR STATUS CLASSIFICATION. Limited distribution, rarity, and reduction of quality habitat within distribution make it susceptible to extirpation from Alabama. Was listed as a species of special concern throughout its distribution (Williams *et al.* 1993), in Alabama (Lydeard *et al.* 1999), and in Apalachicola Basin (Brim Box and Williams 2000).

Prepared by: **Holly Blalock-Herod**

CREEPER

Strophitus undulatus (Say)

OTHER NAMES. Squawfoot, Strange Floater, Sloughfoot.

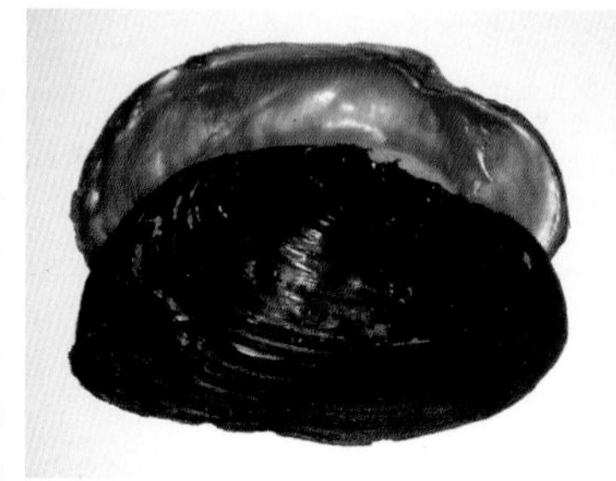

Photo—From Parmalee and Bogan 1998

DESCRIPTION. Has compressed and thin shell when young, becoming somewhat inflated and thick with age (max. length = 115 mm [4 1/2 in.]). Elliptical and somewhat rhomboid in outline, with a rounded anterior margin and bluntly pointed posterior margin; point sometimes positioned posterior ventrally. Posterior ridge broadly rounded and usually pronounced in older shells. Shell disk and posterior slope without sculpture. Umbos compressed and elevated only slightly above hinge line. Umbo sculpture consists of heavy bars, somewhat oblique to hinge line. Periostracum yellowish to greenish, marked by dark green rays, sometimes wavy. Shells darken with age to black or dark brown and rays become obscure. Pseudocardinal teeth rudimentary, consisting of thickened areas along hinge line; lateral teeth merely suggested by a thickened hinge line, or absent altogether. Shell without interdentum and umbo cavity shallow. Shell nacre white or bluish white, sometimes salmon or cream within pallial line. (Modified from Simpson 1914, Parmalee and Bogan 1998)

DISTRIBUTION. Throughout Mississippi River and Great Lakes systems, northern Atlantic Coast drainages, and parts of Canadian Interior Basin (Burch 1975, Parmalee and Bogan 1998). Presumably throughout Tennessee River system historically. In Alabama, current distribution appears restricted to short reach of Bear Creek, Colbert County, where population may not be viable (McGregor and Garner, in review).

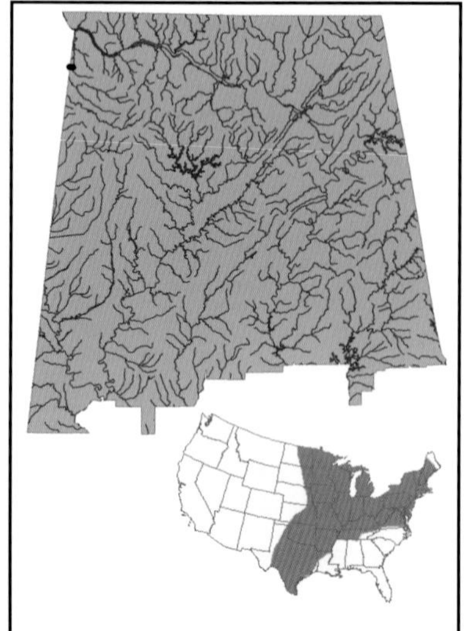

HABITAT. A generalist found in waters varying from high gradient streams to slow, meandering rivers, usually in substrata of fine sand and mud (Parmalee and Bogan 1998).

LIFE HISTORY AND ECOLOGY. A long-term brooder. Gravid from July through April or May of following year (Baker 1928). One of a few species in which glochidia have been reported to develop directly to juvenile stage in gills of female parent without a parasitic stage on a fish (Lefevre and Curtis 1910). However, glochidia also have been shown to undergo a parasitic stage in some cases, utilizing a variety of fishes as hosts, including green sunfish, bluegill, largemouth bass, spotfin shiner, fathead minnow, creek chub, black and yellow bullheads, Plains killifish, and walleye (Hove *in* Parmalee and Bogan 1998).

BASIS FOR STATUS CLASSIFICATION. Limited distribution, rarity, and susceptibility to habitat degradation make *S. undulatus* vulnerable to extirpation from Alabama. Listed as currently stable throughout most of its distribution (Williams *et al.* 1993), but Lydeard *et al.* (1999) considered it possibly extirpated from Alabama.

Prepared by: **Jeffrey T. Garner**

SOUTHERN PURPLE LILLIPUT
Toxolasma corvunculus (Lea)

OTHER NAMES. Southern Lilliput.

Photo—Paul Johnson

DESCRIPTION. Has moderately thick, somewhat inflated shell (max. length ≥ 30 mm [1 1/4 in.]), inequilateral and elliptical to obovate in outline. Anterior margin rounded. Males rounded posteriorly, but mature females have a small, angular marsupial swelling near the posterior end, giving the shell an obliquely truncate posterior margin. Posterior slope rudimentary, narrow, and elliptical. Shell disk without sculpture. Umbos low, sculptured, with undulations. Periostracum dark brown to black, without rays. Pseudocardinal teeth small, erect, somewhat compressed and crenulate; lateral teeth moderately long and slightly curved. Umbo cavity shallow. Shell nacre purple. (Modified from Lea 1868, Simpson 1914)

DISTRIBUTION. Endemic to Mobile River Basin in Alabama. Current distribution poorly known, but seldom encountered and has not been reported for several years.

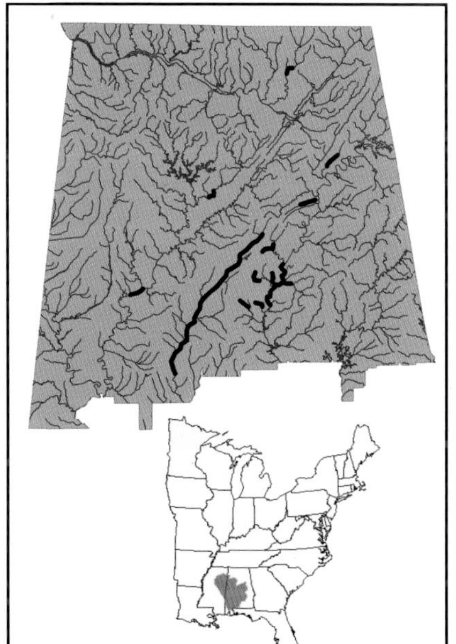

HABITAT. Creeks and rivers, usually found in sand or silt substrata in areas exposed to variable flow.

LIFE HISTORY AND ECOLOGY. Little known. Ortmann (1924*a*) reported it to be gravid. Glochidial hosts unknown.

BASIS FOR STATUS CLASSIFICATION. Limited distribution, rarity, and reduction of quality habitat make species susceptible to extinction. Listed as poorly known throughout its distribution (Williams *et al.* 1993) and imperiled in Alabama (Lydeard *et al.* 1999).

Prepared by: **Jeffrey J. Herod**

PALE LILLIPUT

Toxolasma cylindrellus (Lea)

OTHER NAMES. None.

Photo—From Parmalee and Bogan 1998

DESCRIPTION. Has moderately solid shell (max. length = 35 mm [1 3/8 in.]), elongate and elliptical in outline, and inflated in some older specimens. Anterior margin rounded and posterior margin obliquely angled dorsally, rounded ventrally. Dorsal and ventral margins usually straight. Females differ slightly from males in having a weak marsupial swelling posteriorly. Posterior ridge low or absent. Shell disk and posterior slope without sculpture. Umbos moderately inflated, but elevated only slightly above hinge line. Periostracum has cloth-like texture and is tawny or yellowish green, without rays. Pseudocardinal teeth short and stumpy; lateral teeth short and straight. Interdentum narrow and umbo cavity shallow. Shell nacre white outside of pallial line and coppery purple inside it. (Modified from Simpson 1914, Parmalee and Bogan 1998)

DISTRIBUTION. Endemic to Tennessee River drainage, where historically found in some tributaries from Sequatchie River system downstream to Duck River system (Parmalee and Bogan 1998). One possible record from a Mobile Basin stream in northwestern Georgia (Parmalee and Bogan 1998). Apparently, has been eliminated throughout distribution, except in Paint Rock River system, where rare (Ahlstedt 1995).

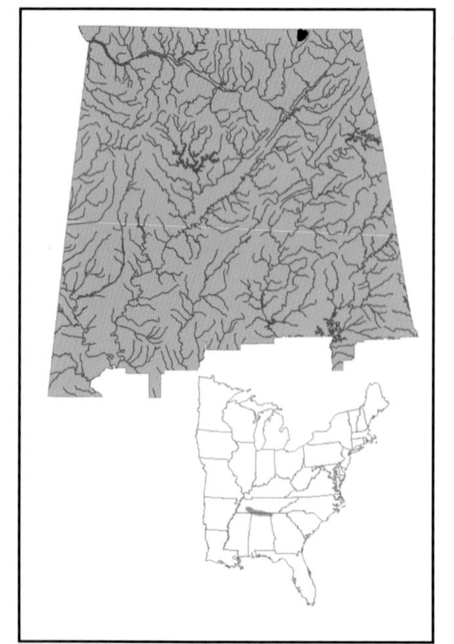

HABITAT. Large creeks and small rivers, typically found in gravel in moderate current (Parmalee and Bogan 1998).

LIFE HISTORY AND ECOLOGY. A long-term brooder. Glochidial hosts unknown.

BASIS FOR STATUS CLASSIFICATION. Vulnerable to extinction due to extremely limited distribution, rarity, and susceptibility to habitat degradation. Classified as endangered throughout its distribution (Williams *et al.* 1993) and in Alabama (Stansbery 1976*a*, Lydeard *et al.* 1999). **Listed as *endangered* by the U.S. Fish and Wildlife Service in 1976**.

***Prepared by:* Jeffrey T. Garner**

DEERTOE

Truncilla truncata Rafinesque

Photo—From Parmalee and Bogan 1998

OTHER NAMES. Deerhorn.

DESCRIPTION. Has solid and moderately inflated shell (max. length = 70 mm [2 3/4 in.], but usually less than 50 mm [2 in.]), somewhat triangular to subrhomboid in outline, with a rounded anterior margin and posterior margin that is pointed to obliquely truncate. Posterior ridge prominent and sharply angled, with a steep posterior slope. Often has a wide, shallow sulcus anterior to posterior ridge. Shell disk and posterior slope without sculpture. Umbos full and elevated well above hinge line, turned inward. Umbo sculpture consists of fine, concentric, and double looped ridges. Periostracum yellow, yellowish brown or greenish, rarely reddish, usually with numerous green rays of varying width and may have darker wavy or zigzag blotches. Pseudocardinal teeth strong, triangular and erect, somewhat compressed in left valve; lateral teeth high, compressed, and slightly curved. Interdentum narrow or absent and umbo cavity fairly shallow. Shell nacre usually white, but occasional specimens have salmon or pink nacre. (Modified from Simpson 1914, Parmalee and Bogan 1998)

DISTRIBUTION. Throughout Mississippi River system, as well as some tributaries of Lakes Erie and St. Clair (Burch 1975). In Alabama, extant in Tennessee River in tailwaters of Wilson Dam (Garner and McGregor 2001), Paint Rock River (Ahlstedt 1995), and Bear Creek in Colbert County (McGregor and Garner, in review).

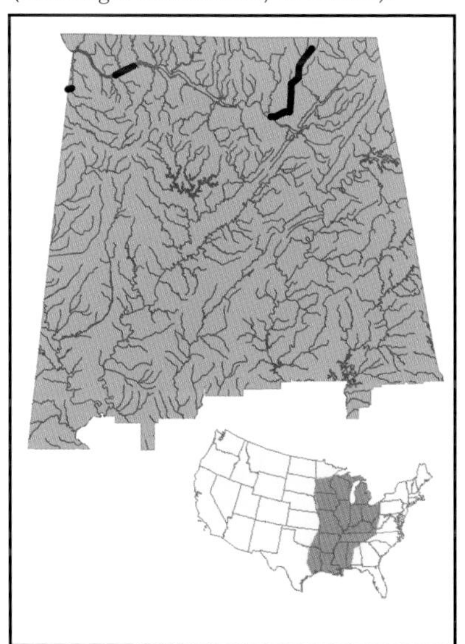

HABITAT. Most often medium-sized rivers, but occasionally in large rivers or lakes (Parmalee and Bogan 1998).

LIFE HISTORY AND ECOLOGY. A long-term brooder, remaining gravid from autumn until July of following year (Ortmann 1919). Sauger and freshwater drum reportedly serve as hosts for glochidia.

BASIS FOR STATUS CLASSIFICATION. Limited distribution, rarity, and susceptibility to habitat degradation make *T. truncata* vulnerable to extirpation from Alabama. Although Williams *et al.* (1993) considered it to be currently stable overall, Stansbery (1976*a*) classified it as threatened and Lydeard *et al.* (1999) listed it as a species of special concern in Alabama.

Prepared by: **Jeffrey T. Garner**

FRESHWATER MUSSELS

PRIORITY 2

HIGH CONSERVATION CONCERN

Taxa imperiled because of three of four of the following: rarity; very limited, disjunct, or peripheral distribution; decreasing population trend/population viability problems; specialized habitat needs/habitat vulnerability due to natural/human-caused factors. Timely research and/or conservation action needed.

Order Unionoida

Family Unionidae

RAYED CREEKSHELL *Anodontoides radiatus*
TENNESSEE PIGTOE *Fusconaia barnesiana*
FINERAYED POCKETBOOK *Lampsilis altilis*
ORANGENACRE MUCKET *Lampsilis perovalis*
FLUTEDSHELL *Lasmigona costata*
TENNESSEE HEELSPLITTER *Lasmigona holstonia*
BLACK SANDSHELL *Ligumia recta*
ALABAMA MOCCASINSHELL *Medionidus acutissimus*
ALABAMA HICKORYNUT *Obovaria unicolor*
SOUTHERN CLUBSHELL *Pleurobema decisum*
FUZZY PIGTOE *Pleurobema strodeanum*
ALABAMA HEELSPLITTER *Potamilus inflatus*
CHOCTAW BEAN *Villosa choctawensis*
COOSA CREEKSHELL *Villosa umbrans*
TAPERED PIGTOE *Quincuncina burkei*
ALABAMA CREEKMUSSEL *Strophitus connasaugaensis*
DOWNY RAINBOW *Villosa villosa*

RAYED CREEKSHELL

Anodontoides radiatus (Conrad)

OTHER NAMES. None.

Photo—Arthur Bogan

DESCRIPTION. Shell thin, moderately inflated (may be $\geq$ 70 mm [2 3/4 in.] long), oblong ovate in outline, with a rounded anterior margin, bluntly pointed posterior margin, straight dorsal margin, and gently convex ventral margin. Posterior ridge well formed dorsally, but becomes low and flattened near ventral margin. Shell disk and slope without sculpture. Umbos full and elevated slightly above the hinge line. Umbo sculpture in the form of profound undulations. Periostracum light olive with dark green rays, but tends to darken with age. Pseudocardinal teeth elongate and thin; lateral teeth rudimentary. Shell nacre yellowish. (Modified from Simpson 1914, Johnson 1967)

DISTRIBUTION. Eastern Gulf Coast drainages from Tickfaw River system in Louisiana, east to Apalachicola Basin (Williams and Butler 1994), including Mobile Basin. Also occurs in Mississippi River Basin in upper Yazoo River tributaries in northern Mississippi (Haag *et al.* 2002).

HABITAT. Most commonly in small to medium-sized coastal plain streams, but historical records exist for larger rivers as well (Brim Box and Williams 2000). Typically occurs in sand or silt substrata in areas of low to moderate current (Brim Box and Williams 2000, Haag *et al.* 2002, Blalock-Herod *et al.*, in review).

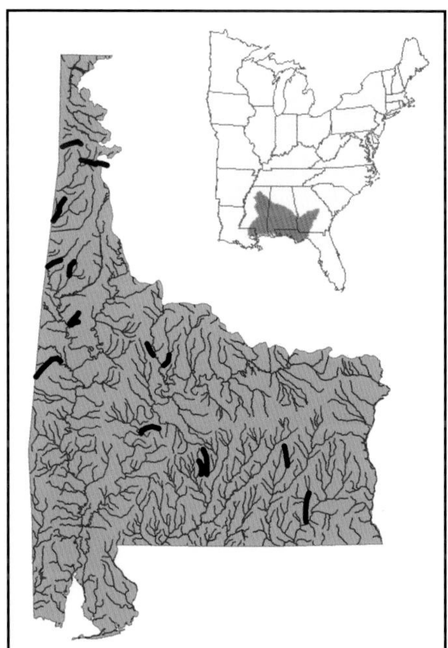

LIFE HISTORY AND ECOLOGY. Poorly known. Gravid females have been found in September and December (Brim Box and Williams 2000), suggesting it is a long-term brooder. Hosts of glochidia unknown, but many other related species are generalists and able to use a wide variety of fishes as hosts.

BASIS FOR STATUS CLASSIFICATION. Continues to survive throughout a large portion of its historical distribution, but remaining populations few in number, small, and widely scattered (Haag *et al.* 2002). Imperiled due to its vulnerability to habitat degradation and declining population trend. Was classified as a species of concern in Alabama (Lydeard *et al.* 1999) and throughout its distribution (Williams *et al.* 1993). Decline of this species appears to be more serious in Gulf Coast systems.

Prepared by: **Wendell R. Haag**

TENNESSEE PIGTOE

Fusconaia barnesiana (Lea)

Photo—Wendell Haag

OTHER NAMES. None.

DESCRIPTION. Has moderately thick shell (max. length = 95 mm [3 3/4 in.]) highly variable in outline, ranging from almost oval to triangular. Anterior margin well rounded and posterior margin obliquely truncate. Dorsal and ventral margins straight to slightly convex. Posterior ridge distinct, but usually low and rounded. Shell disk and posterior slope without sculpture. Umbos may be high and full or only slightly inflated, extending little above hinge line. Periostracum has satiny texture, and is dull yellowish olive to brown, often with variable dark green rays, usually darkening to dark brown with age. Pseudocardinal teeth erect and elongate, usually oriented perpendicular to lateral teeth, which are moderately long and straight. Interdentum short, but wide and umbo cavity shallow. Shell nacre white. (Modified from Simpson 1914, Parmalee and Bogan 1998)

DISTRIBUTION. Endemic to Tennessee and Cumberland River systems (Parmalee and Bogan 1998). In Alabama, extant in only a few Tennessee River tributaries, including Limestone and Round Island Creeks, Limestone County, and the Paint Rock River system.

HABITAT. Varies from small streams to medium-sized rivers. However, historically occurred in Tennessee River at Muscle Shoals, but was extirpated when river was impounded (Garner and McGregor 2001). Appears to prefer shallow water with moderate current and a substratum of coarse sand, silt, and gravel (Parmalee and Bogan 1998).

LIFE HISTORY AND ECOLOGY. Little known, but presumably a short-term brooder, being gravid from spring to mid-summer, like its congeners. Hosts of glochidia unknown (Parmalee and Bogan 1998).

BASIS FOR STATUS CLASSIFICATION. In Alabama, distribution has been reduced to a few tributaries of the Tennessee River. Limited distribution and susceptibility to habitat degradation make it vulnerable to extirpation from state. Was listed as endangered in Alabama (Stansbery 1976*a*), but more recently listed as a species of special concern throughout its distribution (Williams *et al.* 1993) and in Alabama (Lydeard *et al.* 1999).

Prepared by: **Jeffrey T. Garner**

FINERAYED POCKETBOOK

Lampsilis altilis (Conrad)

Photo—Malcolm Pierson

OTHER NAMES. None.

DESCRIPTION. Has moderately thin and inflated shell (max. length = 85 mm [3 3/8 in.]), subelliptical to ovate in outline, with a broadly rounded anterior margin and narrowly rounded or bluntly pointed posterior margin. Posterior ridge, when present, very low and broadly rounded, sometimes faintly doubled, giving a slight biangulation to posterior ventral margin. Shell disk and posterior slope without sculpture. Umbos moderately full and slightly elevated above hinge line. Periostracum dull yellowish brown, with most specimens having green, usually narrow, rays. Pseudocardinal teeth doubled in each valve, thin in left valve, triangular in right; lateral teeth thin, elevated, and straight or very slightly curved. A narrow interdentum separates pseudocardinal and lateral teeth. Umbo cavity moderately deep. Shell nacre white, but may have a pink or salmon tint. (Modified from Simpson 1914, Parmalee and Bogan 1998)

DISTRIBUTION. Endemic to eastern reaches of Mobile Basin in Alabama, Georgia, and Tennessee, including Coosa, Tallapoosa, and Cahaba River systems (Haag *et al.* 1999). Remains widespread, but populations isolated and often small.

HABITAT. Most frequently in small to medium-sized streams above the Fall Line, but historical records indicate also once occurred in larger rivers such as the Cahaba, Coosa, and Tallapoosa. Individuals occur in a wide variety of substrata from clean sand and gravel riffles to depositional areas along stream margins.

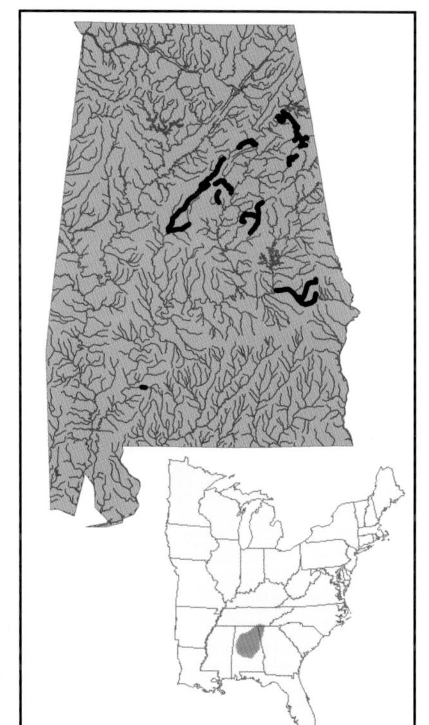

LIFE HISTORY AND ECOLOGY. A long-term brooder, with females reportedly releasing glochidia in March (Haag *et al.* 1999). Ortmann (1924*a*) reported a female from the Chattooga River that was gravid with eggs in May, which is atypical for a lampsiline. Only species known to attract host fishes using both mantle flap displays and superconglutinates (Haag *et al.* 1999). Superconglutinates pigmented to resemble small fish, and display a darting motion in stream currents eliciting attacks from potential host fish (Haag and Warren 1999). Superconglutinates fusiform in shape and dusky white, with a dark stripe dorsally. Mantle lure dusky gray, with numerous fine, black spots and distinct white bars, and with a black eyespot, surrounded by a white halo at posterior end. Margin of lure ornamented with long, finely branched papillae. Primary hosts include redeye, spotted, and largemouth bass. Green sunfish a marginal host (Haag *et al.* 1999).

BASIS FOR STATUS CLASSIFICATION. Has experienced a dramatic reduction in distribution, but still persists in low numbers at several sites in the Coosa and Tallapoosa River systems. Small populations vulnerable to habitat degradation. Currently extremely rare in Cahaba

River system (McGregor *et al.* 2000). Classified as threatened throughout its distribution (Williams *et al.* 1993) and in Alabama (Lydeard *et al.* 1999). **Listed as *threatened* by the U.S. Fish and Wildlife Service in 1993.**

***Prepared by:* Wendell R. Haag**

ORANGENACRE MUCKET

Lampsilis perovalis (Conrad)

OTHER NAMES. None.

Photo—Arthur Bogan

DESCRIPTION. Has moderately solid and somewhat inflated shell (max. length = 90 mm [3 1/2 in.]), ovate in outline, with a rounded anterior margin and bluntly pointed posterior margin; may be somewhat more truncate in females. Posterior ridge well developed and rounded, steeper dorsally. Shell disk and posterior slope without sculpture. Umbos moderately full and elevated above hinge line, positioned anterior of center. Umbo sculpture in form of strong, double-looped ridges. Periostracum greenish yellow to brown, darkening with age, with faint rays that usually are obscure in older specimens. Two strong, high pseudocardinal teeth in left valve and one strong pseudocardinal tooth in right valve accompanied by a rudimentary secondary tooth; lateral teeth short and strong, removed from pseudocardinal teeth. Interdentum absent and umbo cavity wide and moderately deep. Shell nacre salmon or pink, occasionally white. (Modified from Simpson 1914, USFWS 2000)

DISTRIBUTION. Endemic to Tombigbee and lower Alabama River systems in Alabama and Mississippi (Stansbery 1983*b*, Roe *et al.* 2001). Currently common only in upland streams of the Sipsey Fork drainage within Bankhead National Forest in Alabama.

HABITAT. Lotic areas in a wide variety of stream types from small, upland streams, to large Coastal Plain rivers. Individuals can be found in a wide variety of substrata, but most common in depositional areas along riffle margins or flowing pools.

LIFE HISTORY AND ECOLOGY. A long-term brooder; females attract host fish from approximately February to May by releasing glochidia in a pair of superconglutinate packets. Superconglutinates pigmented to resemble small fish display a darting motion in stream currents, eliciting attacks from potential hosts (Haag *et al.* 1995, Haag and Warren 1999) that include redeye, spotted, and largemouth bass (Haag and Warren 1997).

BASIS FOR STATUS CLASSIFICATION. Continues to survive throughout large portion of historical distribution, but remaining populations small and widely scattered, leaving them vulnerable to habitat degradation. Once listed as endangered in Alabama (Stansbery 1976*a*), more recently classified as threatened throughout its distribution (Williams *et al.* 1993) and in Alabama (Lydeard *et al.* 1999). **Listed as *threatened* by the U.S. Fish and Wildlife Service in 1993.**

***Prepared by:* Wendell R. Haag**

FLUTEDSHELL

Lasmigona costata (Rafinesque)

OTHER NAMES. Sand Mussel, Squawfoot.

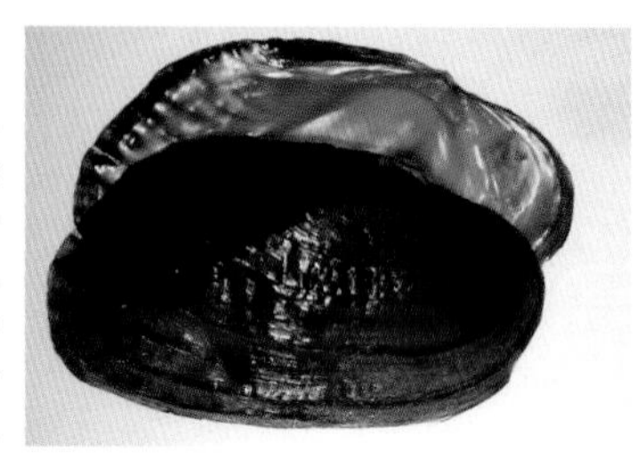

Photo—From Parmalee and Bogan 1998

DESCRIPTION. Has solid, compressed, and elongate shell (max. length = 200 mm [7 7/8 in.]), rhomboidal in outline, with straight dorsal and ventral margins and obliquely angled anterior and posterior margins. Posterior ridge usually well developed. Numerous, heavy costations, or rounded flutings, typically present on posterior slope and posterior end of shell, oriented more or less perpendicular to margin. Anterior end of shell usually unsculptured. Umbos compressed and flattened, scarcely elevated above hinge line. Umbo sculpture consists of heavy bars, parallel to hinge line, the first of which is curved, others double looped. Periostracum yellowish with numerous green rays in young shells, darkening with age to dark brown or black. Pseudocardinal teeth heavy and single, pyramidal and elevated in left valve, low and somewhat elongate in right valve; lateral teeth rudimentary, in form of thickened hinge lines. Interdentum, when present, narrow and umbo cavity very shallow. Shell nacre white, often tinged with salmon or cream. (Modified from Simpson 1914, Parmalee and Bogan 1998)

DISTRIBUTION. Throughout most of Mississippi River system, some of the southern and western tributaries of the Great Lakes, and some tributaries of Hudson Bay (Burch 1975, Parmalee and Bogan 1998). In Alabama, appears to be extant only in a short reach of Bear Creek, Colbert County, and in the Paint Rock and Elk Rivers. Viability of Bear Creek population questionable.

HABITAT. Generally in large creeks to medium-sized rivers (Parmalee and Bogan 1998). However, historically occurred in Tennessee River at Muscle Shoals prior to its impoundment, indicating that it occurred in large rivers under some conditions (Garner and McGregor, 2001).

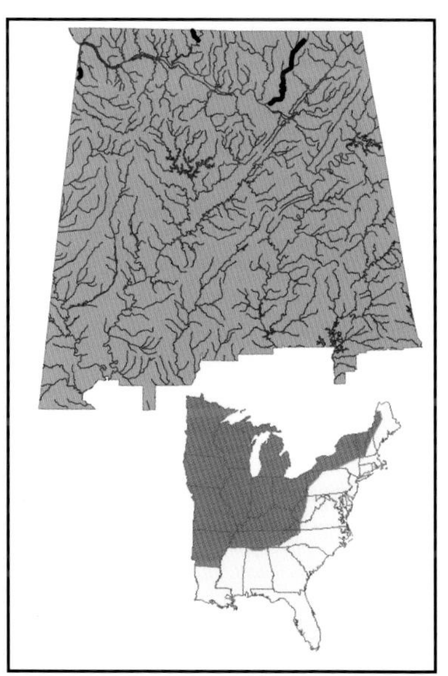

LIFE HISTORY AND ECOLOGY. A long-term brooder, gravid from August into May (Baker 1928). Glochidia use a wide variety of fishes as hosts, including green and longear sunfish; bluegill; rock, smallmouth, and largemouth bass; rainbow, fantail, and striped darters; yellow perch; walleye; banded sculpin; central stoneroller; brown bullhead; northern studfish; gizzard shad; river redhorse; bowfin; and northern pike (Luo 1993, Weiss and Layzer 1995, Hove *in* Parmalee and Bogan 1998). Also, glochidia can transform on common carp, an exotic species (Fuller 1974).

BASIS FOR STATUS CLASSIFICATION. Within Alabama, distribution has been reduced to three disjunct populations in tributaries of the Tennessee River. Restricted distribution and declining population trend make it vulnerable to extirpation from state. Listed as stable throughout distribution (Williams *et al.* 1993) and in Alabama (Lydeard *et al.* 1999).

***Prepared by:* Jeffrey T. Garner**

TENNESSEE HEELSPLITTER

Lasmigona holstonia (Lea)

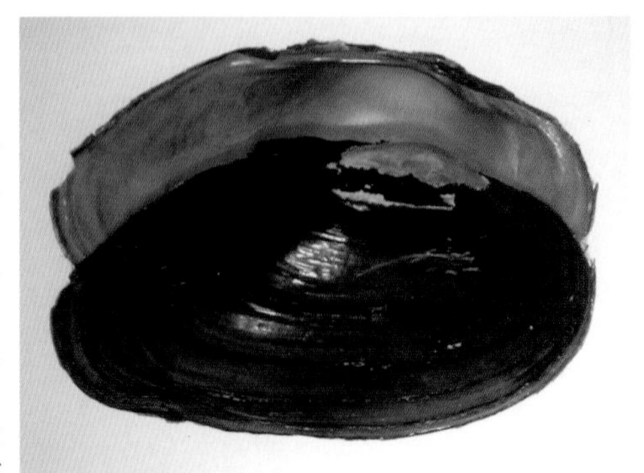
Photo—From Parmalee and Bogan 1998

OTHER NAMES. None.

DESCRIPTION. Has thin, but not fragile, moderately inflated shell (max. length = approx. 75 mm [2 15/16 in.]), somewhat elongate and rhomboidal in outline, with a broadly rounded anterior margin, broadly pointed or truncate posterior margin, and almost straight ventral margin. Posterior ridge pronounced, but broadly rounded, appearing double in some individuals. Shell disk and posterior ridge without sculpture. Umbos full, but not high, and project only slightly above hinge line, located anterior of center. Periostracum roughened and dull, greenish brown to yellowish brown, becoming dark brown with age. Pseudocardinal teeth compressed and nearly parallel with hinge line, double in left valve and single in right valve; lateral teeth rudimentary, represented by thickened hinge line. Interdentum absent and umbo cavity shallow. Shell nacre bluish white, often with a pale salmon wash in umbo cavity. (Modified from Parmalee and Bogan 1998)

DISTRIBUTION. In upper Tennessee and upper Coosa River systems. Extant in a few tributaries of Coosa River in northeastern Alabama, including Terrapin and Spring Creeks in Cherokee County. A single specimen reported from Hurricane Creek, Paint Rock River system, Jackson County (Ahlstedt 1995).

HABITAT. Creeks with flowing water over substrata of sand and mud (Parmalee and Bogan 1998). Sometimes found below riffles in shallow stream margins. May occur in very small creeks where often it is only mussel species present.

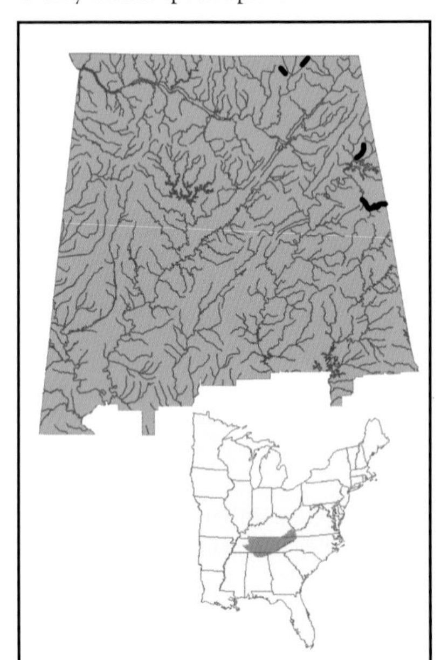

LIFE HISTORY AND ECOLOGY. A long-term brooder (Parmalee and Bogan, 1998). Hosts of glochidia include banded sculpin, rock bass, and possibly other headwater fish species, since it appears to be a host generalist like many other members of the subfamily Anodontinae (J. W. Jones, Virginia Polytechnic Institute and State University, pers. comm., 2002).

BASIS FOR STATUS CLASSIFICATION. Reduced to only a few populations in Alabama. **Considered a *species of concern* throughout its distribution by the U.S. Fish and Wildlife Service** and Williams *et al.* (1993), and in Alabama (Lydeard *et al.* 1999). Earlier listed as endangered in Alabama (Stansbery 1976*a*).

***Prepared by:* Robert S. Butler**

BLACK SANDSHELL

Ligumia recta (Lamarck)

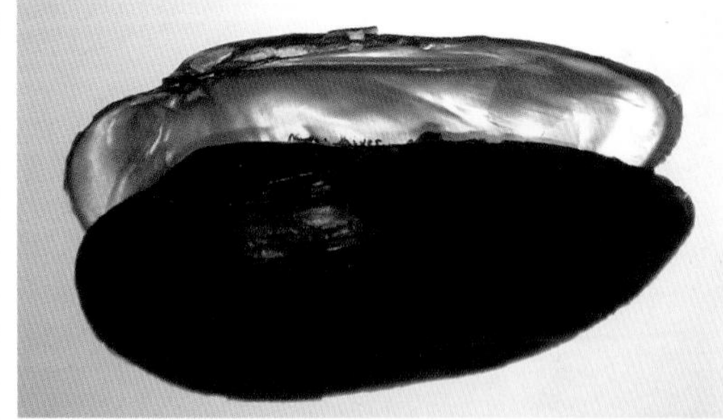
Photo—From Parmalee and Bogan 1998

OTHER NAMES. Black Sand Mussel, Long John, Sow's Ear, Lady's Slipper, Mule Ear.

DESCRIPTION. Has moderately thick shell (max. length = 160 mm [6 5/16 in.]), elliptical in outline, with a rounded anterior margin, and dorsal and ventral margins straight and parallel. Males pointed and compressed posteriorly, with point located toward the dorsal margin. Females more gently rounded and broadened posteriorly, with posterior margin obliquely truncate. Posterior ridge rounded, distinct near the umbo, but becoming flattened posterior ventrally. Shell disk and posterior ridge unsculptured. Umbos elevated slightly to moderately above the hinge line. Umbo sculpture consists of weak double-looped ridges. Periostracum smooth and shiny, dark green to brown in young specimens, darkening to dark brown or black with age. Green rays present on young shells. Pseudocardinal teeth strong and triangular, two in the left valve, one in the right; lateral teeth long, thin, and straight. Interdentum long and narrow, but may be absent. Umbo cavity shallow. Shell nacre usually white with pink or purple in umbo cavity, but may be entirely white and rarely entirely pink or purple. (Modified from Cummings and Mayer 1992, Parmalee and Bogan 1998, Strayer and Jirka 1997)

DISTRIBUTION. Widespread in eastern and central United States and Canada, occurring from the Great Lakes Basin south into the Mississippi River drainage to Louisiana and in some Gulf Coast drainages. In Alabama, occurs in Tennessee and Mobile River drainages (Cummings and Mayer 1992, Parmalee and Bogan 1998, Strayer and Jirka 1997). Extant in Tennessee River in tailwaters of Wilson and Guntersville Dams, where uncommon or rare (Garner and McGregor 2001). Extirpated from most reaches of Mobile Basin.

HABITAT. Coarse sand and gravel substrata in areas with current (Cummings and Mayer 1992, Parmalee and Bogan 1998, Strayer and Jirka 1997).

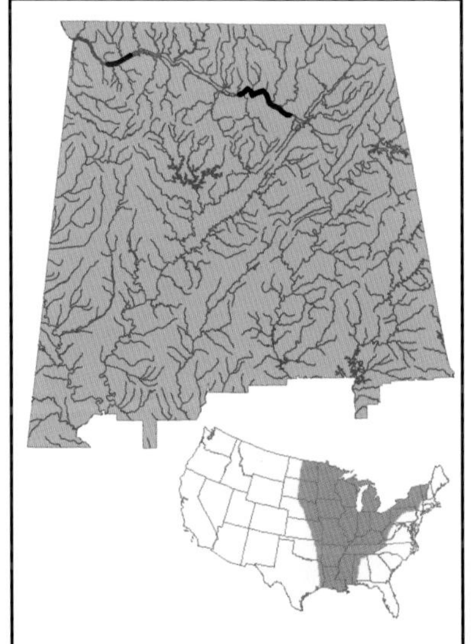

LIFE HISTORY AND ECOLOGY. A long-term brooder, with a gravid season from mid-August to July. Reported glochidial hosts are white crappie, bluegill, green and orangespotted sunfish, largemouth bass, walleye, sauger, and banded killifish (Cummings and Mayer 1992, Parmalee and Bogan 1998, Strayer and Jirka 1997).

BASIS FOR STATUS CLASSIFICATION. Has suffered drastic declines in Alabama from impoundment of major rivers. Although uncommon in tailwaters of Wilson Dam, it is reproducing there. Viability of Guntersville Dam tailwaters population is questionable. Almost extirpated from Mobile Basin. Its limited distribution and rarity make it vulnerable to extirpation from the state. Listed as a species of special concern throughout distribution (Williams *et al.* 1993) and in Alabama (Lydeard *et al.* 1999).

Prepared by: **Jeffrey J. Herod**

ALABAMA MOCCASINSHELL

Medionidus acutissimus (Lea)

OTHER NAMES. None.

Photo—Mike Gangloff

DESCRIPTION. Has small and moderately thin shell (max. length = 55 mm [2 1/8 in.]), rhomboidal to elliptical in outline, with a rounded anterior margin, pointed posterior margin, straight to slightly convex dorsal margin, and very slightly convex ventral margin. Males may be somewhat arcuate and generally more acutely pointed posteriorly than females. Females often more swollen posteriorly. Posterior ridge high and sharp, often doubled. Posterior slope covered with low, parallel corrugations or ridges. Low corrugations also may be on posterior portion of shell disk. Umbos slightly inflated and very slightly elevated above hinge line. Periostracum may be shiny or dull, yellowish or tan to green, covered with faint, fine green rays that are broken or in form of zigzags. Pseudocardinal teeth short and triangular; lateral teeth slightly curved. No interdentum present and umbo cavity shallow. Shell nacre variable, ranging from white to purple, but may be tan or pale red. (Modified from Simpson 1914, Parmalee and Bogan 1998)

DISTRIBUTION. Distributed throughout Mobile Basin in Alabama, Georgia, Mississippi, and Tennessee (Parmalee and Bogan 1998). Specimens from Gulf Coast drainages, west of Apalachicola Basin, tentatively identified as M. *acutissimus* (Williams *et al.*, in review). However, comparative anatomical and genetic studies may prove them to represent an undescribed species. Several populations of M. *acutissimus* in Alabama appear healthy, including Sipsey Fork in Bankhead National Forest and Sipsey River. However, now gone from much of former distribution and existing populations isolated.

HABITAT. Lotic areas in a wide variety of stream types from small, upland streams to large Coastal Plain rivers. Most frequently encountered in swift, gravel-bottomed shoals or riffles where often tethered to stones by a byssal thread (Haag and Warren 2003).

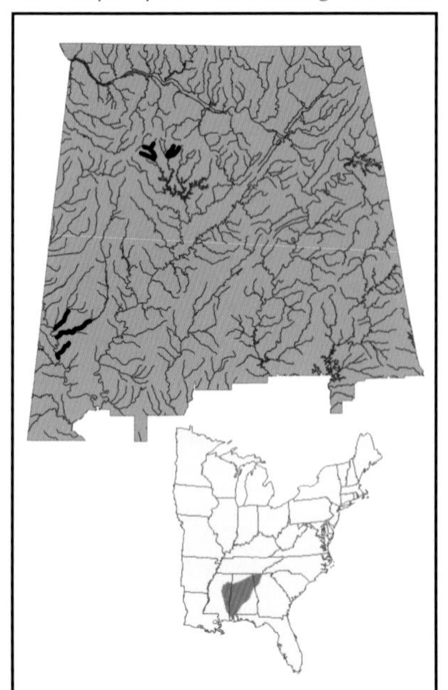

LIFE HISTORY AND ECOLOGY. A long-term brooder; females release glochidia from approximately February to May. Females attract host fish by rapidly flickering mantle lure (matte inky black, with a small, white patch [about two square millimeters, or 0.08 square inches]) located along posterior portion of mantle margin. Primary hosts of glochidia include naked sand, southern sand, redspotted, Tuskaloosa, Johnny, speckled, Gulf, blackbanded, and saddleback darters. Rock darter and blackspotted topminnow are marginal hosts. (Summarized from Haag and Warren 1997, Haag and Warren 2003)

BASIS FOR STATUS CLASSIFICATION. Remain-ing populations usually small, widely scattered, and isolated, making it vulnerable to extinction. Classified as threatened throughout its distribution (Williams *et al.* 1993) and in Alabama (Lydeard *et al.* 1999). **Listed as *threatened* by the U.S. Fish and Wildlife Service in 1993**.

***Prepared by*: Wendell R. Haag**

ALABAMA HICKORYNUT

Obovaria unicolor (Lea)

OTHER NAMES. None.

Photo—Mike Gangloff

DESCRIPTION. Has solid and only slightly inflated shell (max. length = 50 mm [2 in.]), ovate to short elliptical in outline, with rounded anterior and posterior margins, although posterior margin much more broadly rounded. Posterior ridge low and rounded. Shell disk and posterior slope without sculpture. Umbos full and slightly elevated above hinge line. Umbo sculpture in form of weak, imperfectly looped ridges. Periostracum shiny, often greenish in young specimens, becoming yellowish brown to brown with age, and often weakly rayed. Pseudocardinal teeth small and radially arranged; lateral teeth almost straight. Interdentum short, but well developed, and umbo cavity shallow and compressed. Shell nacre usually pinkish, but may be white or bluish. (Modified from Lea 1838, Simpson 1914)

DISTRIBUTION. Taxonomic status and distribution unclear. May be endemic to western Mobile Basin; occurs in Cahaba, lower Black Warrior, and Tombigbee River systems (van der Schalie 1981, Williams *et al.* 1992, McCullagh *et al.* 2002) in Alabama and Mississippi. Although two historical reports from Coosa River system, previous occurrence there doubtful (Hurd 1974). Similar to, and may be conspecific with, *O. jacksoniana*, which has an almost identical distribution in western Mobile Basin. Both species also reported from Pascagoula, Pearl, and Amite River systems in Mississippi and Louisiana, but their relationships to Mobile Basin populations unclear.

HABITAT. Medium-sized to large streams below the Fall Line (van der Schalie 1981, Williams *et al.* 1992). Characteristically found in sandy substrata in areas of low flow, but individuals can be found in practically any habitat type in appropriate streams including: swift, gravel bottomed shoals, deep gravel and sand bottomed runs, silty stream margins, pools, backwater sloughs, and high water side channels (W. R. Haag and M. L. Warren, USDA Forest Service, unpubl. data).

LIFE HISTORY AND ECOLOGY. A long-term brooder; females release glochidia from April to June. Method of host infestation unknown. Glochidial hosts are naked sand, southern sand, and red spotted darters. Marginal hosts include Johnny, Gulf, blackbanded, and dusky darters. (Summarized from Haag and Warren 2003)

BASIS FOR STATUS CLASSIFICATION. Limited distribution and declining population trend make it vulnerable to extinction. In Mobile Basin, only large population found in Sipsey River. Extant in Buttahatchee River but has declined precipitously there in recent years (Jones 1991), and rare in lower Cahaba and Black Warrior Rivers (Pierson 1991, Williams *et al.* 1992, McGregor *et al.* 2000). Although considered endangered in Alabama earlier (Stansbery 1976*a*), more recently classified as a species of special concern throughout its distribution (Williams *et al.* 1993) and in Alabama (Lydeard *et al.* 1999).

Prepared by: **Wendell R. Haag**

SOUTHERN CLUBSHELL

Pleurobema decisum (Lea)

Photo—Jim Williams

OTHER NAMES. None.

DESCRIPTION. Has solid and inflated shell (max. length = 70 mm [2 3/4 in.]), elongate and subtriangular in outline, with an anterior margin usually rounded in younger specimens, but often obliquely truncate in older specimens, and bluntly pointed posterior margin. Dorsal and ventral margins very slightly convex to straight. Posterior slope not greatly elevated, but steep dorsally, becoming rounder and flatter with a ventral progression. A wide, radial swelling located just anterior of posterior ridge. Shell disk and slope without sculpture. Umbos full and elevated well above hinge line, positioned near anterior end, often exceeding anterior margin in old specimens. Periostracum tawny to greenish brown or brown, sometimes weakly rayed. Pseudocardinal teeth irregular and divergent; lateral teeth moderately long, heavy, and very slightly curved. A well-developed interdentum separates them. Umbo cavity shallow. Shell nacre white. (Modified from Lea 1831, Simpson 1914, USFWS 2000)

DISTRIBUTION. Endemic to Mobile Basin in Alabama, Georgia, and Mississippi, including Alabama, Black Warrior, Cahaba, Coosa, Tallapoosa, and Tombigbee River systems (Stansbery 1976*a*). Rarely encountered in large rivers today. Large populations remain only in widely scattered localities in Tombigbee River system (McCullagh *et al.* 2002). Small, isolated populations exist in Alabama, Coosa, and Tallapoosa River systems (McGregor *et al.* 1999, 2000). May be extirpated from Black Warrior and Cahaba River systems.

HABITAT. Lotic areas in medium sized to large streams both above and below the Fall Line. Highest densities occur in deep, gravel and sand-bottomed runs with slow but steady current, but individuals also found in variety of other habitats from swift, shallow shoals to pools (W. R. Haag and M. L. Warren, USDA Forest Service, unpubl. data).

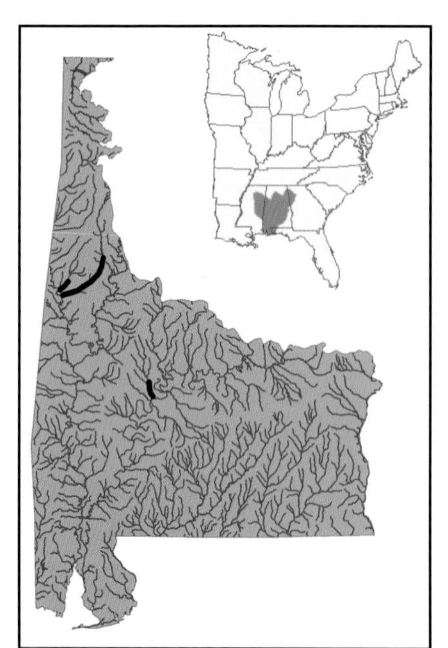

LIFE HISTORY AND ECOLOGY. A short-term brooder; females release glochidia in June and July. Transmission of glochidia to host fishes facilitated by females releasing glochidia in conglutinates that resemble food items of small fishes. Presence of conglutinates shown to elicit feeding responses from fishes both in laboratory and in wild. Conglutinates flat and oval shaped, approximately three by five millimeters (1/16 by1/4 inches), and orange or white. Primary glochidial host is blacktail shiner. Striped shiner is a marginal host. (Summarized from Haag and Warren 2003)

BASIS FOR STATUS CLASSIFICATION. Has experienced dramatic reduction in distribution and fragmentation of its populations. Limited distribution and declining population trend make it vulnerable to extinction. Classified as endangered throughout its distribution (Williams *et al.* 1993) and in Alabama (Stansbery 1976*a*, Lydeard *et al.* 1999). **Listed as *endangered* by the U.S. Fish and Wildlife Service in 1993.**

***Prepared by:* Wendell R. Haag**

FUZZY PIGTOE

Pleurobema strodeanum (Wright)

OTHER NAMES. None.

DESCRIPTION. Shell rather thin and compressed (max. length = 60 mm [2 3/8 in.]), ovate or subtriangular to subelliptical in outline, with a rounded anterior margin and bluntly pointed posterior margin. Posterior ridge poorly defined and posterior slope slightly concave. Shell disk without sculpture, but may be one, or a few, thread-like ridges on the posterior slope, parallel to the posterior ridge. Umbos somewhat full, but barely extend above hinge line. Periostracum shiny on the shell disk, but becomes roughened near the margins, and varies from dark olivaceous brown to almost black, sometimes weakly rayed. Pseudocardinal teeth triangular, two divergent teeth in left valve and one in right. Lateral teeth short and almost straight. Shell nacre bluish white. (Modified from Simpson 1914, Williams and Butler 1994)

Photo—Arthur Bogan

DISTRIBUTION. Occurs in Choctawhatchee, Escambia, and Yellow River drainages in Alabama and Florida (Clench and Turner 1956; Blalock-Herod *et al.*, in press).

HABITAT. Predominately sand substrata in small to large streams with scattered gravel, woody debris, and moderate flow (Clench and Turner 1956).

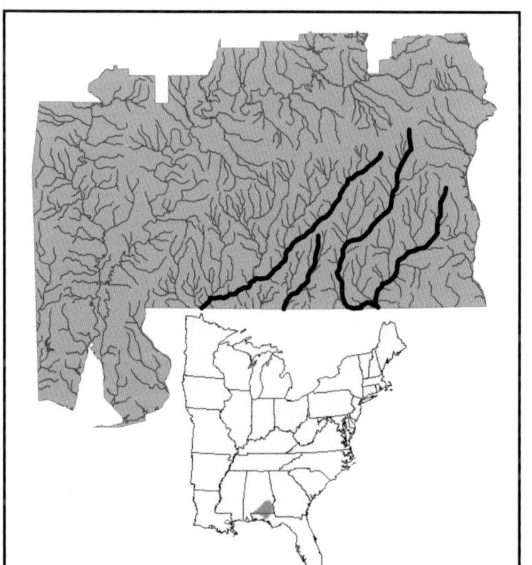

LIFE HISTORY AND ECOLOGY. Nothing known. Presumably a short-term brooder like its congeners.

BASIS FOR STATUS CLASSIFICATION. Limited distribution and dwindling habitat quality make *P. strodeanum* vulnerable to extinction. Was classified as a species of special concern (Williams *et al.* 1993) and a species in need of protection in Alabama (Lydeard *et al.* 1999). Blalock-Herod *et al.* (in press) considered it to be a species of special concern within the Choctawhatchee River system.

***Prepared by:* Stuart W. McGregor**

ALABAMA HEELSPLITTER

Potamilus inflatus (Lea)

Photo—Arthur Bogan

OTHER NAMES. Inflated Heelsplitter.

DESCRIPTION. Has thin, somewhat inflated shell (max. length = 140 mm [5 1/2 in.]). Sexual dimorphism expressed as size difference, with males growing much larger than females. Shell trapezoidal in outline, with anterior margin bluntly pointed, posterior margin obliquely truncate, and ventral margin slightly convex. A large dorsal wing positioned just posterior of umbo is rounded when intact, but often broken. A much smaller dorsal wing, located just anterior of umbo, usually more or less bluntly pointed. Posterior ridge high and widely rounded. Umbos low and compressed, elevated little, if any, above hinge line. Periostracum somewhat dull and olive to dark brown, darkening with age. Young shells often weakly rayed, but rays become obscure as shell darkens. Pseudocardinal teeth weak and elongate; lateral teeth short and curved, distant from pseudocardinal teeth. No interdentum present and umbo cavity shallow. Shell nacre purple. (Modified from Lea 1831, Simpson 1914, USFWS 2000)

DISTRIBUTION. Known from Alabama, Black Warrior, Coosa, and Tombigbee Rivers of Mobile Basin (Roe *et al.* 1997). Known to be extant in Alabama, Black Warrior, and Tombigbee Rivers, but Alabama River population may not be viable. A similar form in Amite, Pearl, and Tangipahoa Rivers of Mississippi and Louisiana has recently been reported to differ genetically and may represent a distinct species. Genetic affinity of Pearl River populations unknown (Roe and Lydeard 1998).

HABITAT. Large rivers in silt and sand substrata in slow to moderate current, and at depths exceeding six meters (19 3/4 feet) (Brown and Curole 1997, USFWS 2000). Can survive in impounded rivers to limited extent (Williams *et al.* 1992).

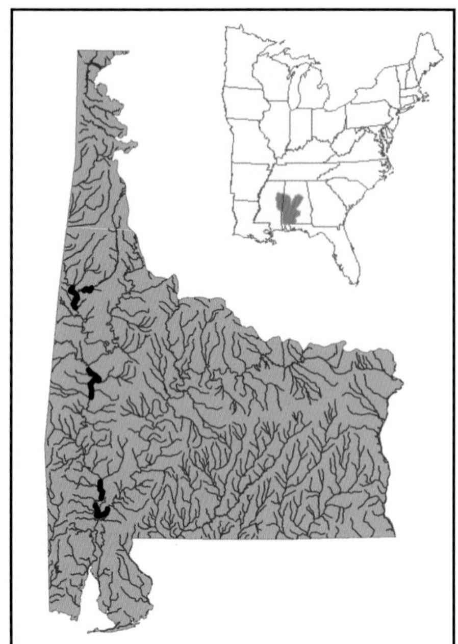

LIFE HISTORY AND ECOLOGY. A long-term brooder; females release glochidia in June and July. A probable host for glochidia is the freshwater drum (Roe *et al.* 1997). Method of host infestation unknown.

BASIS FOR STATUS CLASSIFICATION. Limited distribution and susceptibility to habitat degradation from river dredging operations make *P. inflatus* vulnerable to extinction. Was considered endangered in Alabama (Stansbery 1976*a*), but more recently listed as threatened throughout its distribution (Williams *et al.* 1993) and in Alabama (Lydeard *et al.* 1999). **Listed as *threatened* by the U.S. Fish and Wildlife Service in 1990.**

***Prepared by:* Wendell R. Haag**

CHOCTAW BEAN

Villosa choctawensis (Athearn)

Photo—Arthur Bogan

OTHER NAMES. None.

DESCRIPTION. Has somewhat inflated shell, with slightly thickened margin (may exceed 40 mm [1 5/8 in.] long), ovate in outline, with rounded anterior and posterior margins. Females may be somewhat more truncate or more broadly rounded posteriorly. Posterior ridge low and rounded. Shell disk and posterior slope unsculptured. Umbos broad and full, extending little, if any, above hinge line and positioned well anterior of center. Umbo sculpture consists of thin, undulating ridges. Periostracum shiny and smooth, but may be roughened ventrally and posteriorly. Color is chestnut to dark brown or black, with variable fine, green rays that may be obscure. Two well-developed pseudocardinal teeth occur in left valve and one well developed and two rudimentary pseudocardinal teeth in right valve; lateral teeth short and almost straight. Interdentum moderately wide and umbo cavity moderately deep. Shell nacre white, but may be blotched and brown. (Modified from Athearn 1964, Williams and Butler 1994)

DISTRIBUTION. Includes Choctawhatchee, Escambia, and Yellow River systems in Alabama and Florida (Butler 1989, Williams and Butler 1994).

HABITAT. Occurs in small to medium-sized rivers with sand or silty sand substrata in areas with moderate to swift current (Athearn 1964).

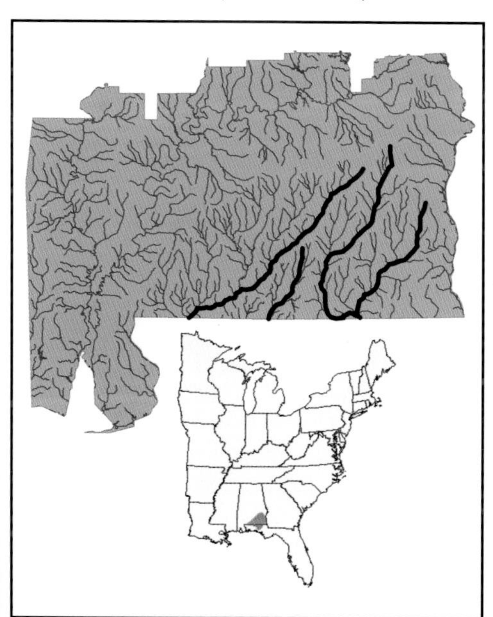

LIFE HISTORY AND ECOLOGY. Little known, but presumably a long-term brooder. Williams *et al.* (in review) reported gravid females in August. Hosts of glochidia unknown.

BASIS FOR STATUS CLASSIFICATION. Limited distribution and habitat degradation within its distribution make *V. choctawensis* susceptible to extinction. Classified as threatened throughout its distribution (Williams *et al.* 1993) and imperiled in Alabama (Lydeard *et al.* 1999). Within drainages, considered a species of special concern in Choctawhatchee River system (Blalock-Herod *et al.*, in press) and endangered in Escambia and Yellow River systems (Williams *et al.*, in review).

Prepared by: **Stuart W. McGregor**

COOSA CREEKSHELL

Villosa umbrans (Lea)

Photo—Jim Williams

OTHER NAMES. None.

DESCRIPTION. Has solid, somewhat inflated, shell (max. length approx. 70 mm [2 3/4 in.]), generally elliptical or somewhat obovate in outline. Anterior margin broadly rounded in both sexes. Males have a rather sharply pointed posterior margin, with point occurring medially. Females have a prominent posterior swelling along posterior ventral margin, with shell truncate beyond this swelling to a point about two thirds of way up posterior margin. Older females may have a strong constriction posterior to marsupial swelling. Ventral margin broadly rounded in males to nearly straight in females. Posterior ridge only slightly developed in both sexes. Umbos low and not elevated above hinge line, located anterior of center in males and near center in females. Periostracum smooth and varying from tan or olive to dark brown, often darkening to nearly black with age. When present, rays generally indistinct in young shells and become even more obscure with age. Pseudocardinal teeth short, compressed and triangular, with two in left valve and one in right valve; lateral teeth thin and slightly curved, with two in left valve and one in right valve. Interdentum usually absent and umbo cavity very shallow. Shell nacre varies from a light lavender, copper or pinkish purple, to very dark purple. (Modified from Parmalee and Bogan 1998)

DISTRIBUTION. Endemic to upper Coosa River system. Although once fairly widespread, now thought to persist only in a few tributaries in uppermost reaches of system, primarily in Georgia.

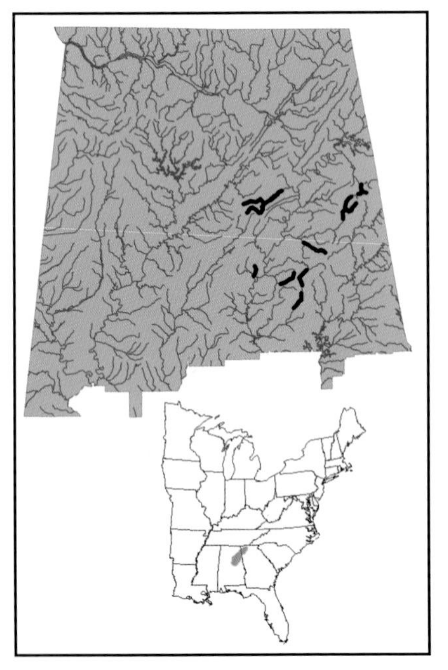

HABITAT. Creeks and small rivers, generally in gravel and sand substrata in shoal and riffle habitats. Sometimes associated with water willow (*Justicia americana*) beds (Parmalee and Bogan 1998).

LIFE HISTORY AND ECOLOGY. A long-term brooder. Reported hosts of its glochidia include bluegill and banded sculpin, with much higher transformation rates occurring on former (P. D. Johnson, Tennessee Aquarium Research Institute, pers. comm., 2002).

BASIS FOR STATUS CLASSIFICATION. May be extant in a few tributaries of Coosa River in Alabama. Appears to be in decline distribution wide. This, along with its limited distribution and specific habitat requirements, makes it vulnerable to extinction. Considered to be of special concern throughout its distribution (Williams *et al.* 1993) and in Alabama (Lydeard *et al.* 1999).

Prepared by: **Robert S. Butler**

TAPERED PIGTOE

Quincuncina burkei (Walker)

OTHER NAMES. None.

Photo—Arthur Bogan

DESCRIPTION. Has small, inflated shell, subelliptical in outline, and reaching a length of 60 millimeters [2 3/8 inches]. Anterior margin broadly rounded and posterior margin narrowly pointed, with point located posterior-ventrally. Dorsal and ventral margins straight to slightly convex. Posterior ridge well defined and posterior slope slightly concave. Sculpturing in form of chevron-like ridges over much of disk and radial ridges on posterior slope. Shell sculpture may be indistinct in some specimens. Umbo inflated little, barely elevated above hinge line. Periostracum dark brown to black, but can be brown or greenish yellow in young specimens. Pseudocardinal teeth well developed and divergent, double in both valves. Two lateral teeth in left valve and usually one in right valve. Interdentum very narrow. Shell nacre varies from light purple to bluish white (Modified from Ortmann and Walker 1922, Clench and Turner 1956, Williams and Butler 1994)

DISTRIBUTION. Endemic to Choctawhatchee River system of southern Alabama and western Florida. Eliminated from much of its historical distribution. In Alabama, known to be extant at nine locations scattered in tributaries of Choctawhatchee drainage, including headwaters of Pea River (Blalock-Herod *et al.*, in review).

HABITAT. Medium sized creeks to large rivers in stable sand or sand and gravel substrata, occasionally occurring in silty sand in slow to moderate current (Williams and Butler 1994).

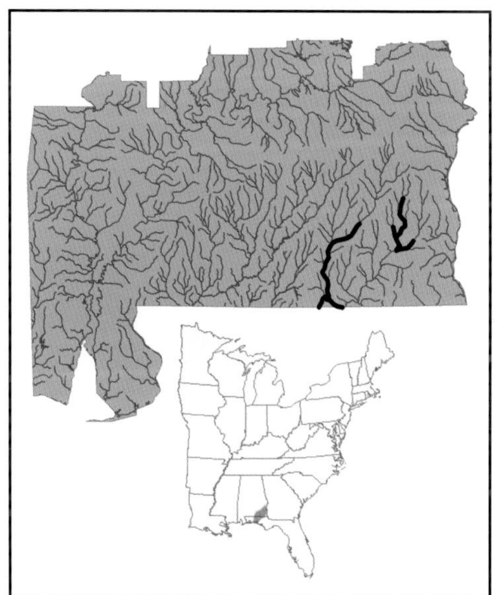

LIFE HISTORY AND ECOLOGY. Little known, but presumably a short-term brooder. Ortmann and Walker (1922) reported a female gravid with eggs in May, with all four gills used as marsupia, and subcylindrical conglutinates. Glochidial hosts unknown.

BASIS FOR STATUS CLASSIFICATION. Limited distribution, rarity, and reduction of quality habitat within distribution make it susceptible to extinction. Considered threatened throughout distribution (Williams *et al.* 1993). Lydeard *et al.* (1999) considered it imperiled in Alabama. Blalock-Herod *et al.* (in review) considered it endangered.

Prepared by: **Holly Blalock-Herod**

ALABAMA CREEKMUSSEL

Strophitus connasaugaensis (Lea)

OTHER NAMES. None.

Photo—From Parmalee and Bogan 1998

DESCRIPTION. Shell thin and slightly inflated (max. length = 120 mm [4 3/4 in.]), elongate and somewhat rhomboid in outline, often narrower anteriorly. Dorsal and ventral margins straight to slightly convex, anterior margin narrowly rounded and posterior margin obliquely truncate. Posterior ridge full, but widely rounded. Shell disk and posterior ridge without sculpture. Umbos full and moderately high, elevated slightly above hinge line. Umbo sculpture consists of a few strong ridges oriented parallel to growth lines. Periostracum yellowish green in juveniles, becoming dull brown with age. Occasionally patterned with weak dark green rays, mostly on posterior half of shell. Pseudocardinal teeth irregular and compressed, only slightly elevated; lateral teeth rudimentary, represented by slightly raised, rounded ridges in each valve. No interdentum present and umbo cavity open and shallow. Shell nacre bluish gray, often with salmon in umbo cavity. (Description modified from Simpson 1914, Parmalee and Bogan 1998)

DISTRIBUTION. A Mobile Basin endemic, where it appears to be restricted to eastern reaches. Extant in widely scattered, isolated localities.

HABITAT. Small to medium-sized rivers and shallow embayments of larger rivers, usually in fine gravel, sand, or silt in shallow water (Parmalee and Bogan 1998).

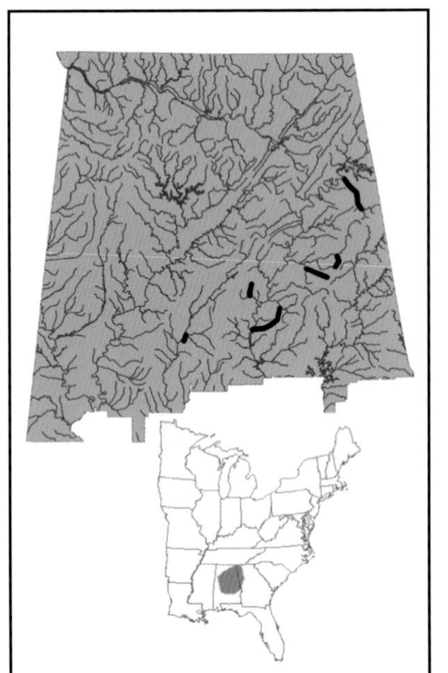

LIFE HISTORY AND ECOLOGY. Nothing known, but its congeners are long-term brooders. Suggested that *Strophitus undulatus*, of Mississippi River Basin, capable of completing its life history without benefit of a host (Lefevre and Curtis 1910). However, unknown if this characteristic is shared with any other members of genus. Hosts of glochidia unknown.

BASIS FOR STATUS CLASSIFICATION. Although widespread in Mobile Basin, populations isolated and often small. Limited distribution and declining population trend make *S. connasaugaensis* vulnerable to extinction. Listed as a species of special concern throughout its distribution (Williams *et al.* 1993) and in Alabama (Lydeard *et al.* 1999).

Prepared by: **Jeffrey T. Garner**

DOWNY RAINBOW

Villosa villosa (Wright)

OTHER NAMES. None.

Photo—Arthur Bogan

DESCRIPTION. Has thin shell (max. length = 70 mm [2 3/4 in.]), elliptical in outline, with straight, parallel dorsal and ventral margins, rounded anterior margin, and rounded to bluntly pointed posterior margin. Females more inflated and slightly more rounded posteriorly than males. Posterior ridge well defined and angular near the umbo, but becomes flatter and less distinct ventrally. Shell disk and posterior ridge unsculptured. Umbos elevated slightly to moderately above hinge line. Umbo sculpture consists of weak double-looped ridges. Periostracum has a cloth-like texture and is brown or black with green, blue, or yellow rays. Pseudocardinal teeth strong and triangular, two in left valve and one in right; lateral teeth long, thin, and straight. No interdentum, and umbo cavity shallow. Shell nacre variable, most commonly white or bluish white (Modified from Brim Box and Williams 2000, Clench and Turner 1956)

DISTRIBUTION. Known from eastern Gulf Coast drainages, from Escambia River east throughout upper peninsular Florida, north to St. Mary's River drainage in Florida and Georgia. In Alabama, extant in Uchee Creek system of Chattahoochee River drainage and possibly Eight-Mile Creek of Choctawhatchee River system (Williams *et al.*, in review; Blalock-Herod *et al.*, in review; Brim Box and Williams 2000; J.T. Garner, Ala. Div. Wildl. Freshwater Fish., unpubl. data).

HABITAT. A variety of habitats, varying from spring-fed creeks to backwaters, with silt, mud, sand, or gravel substrata. May be found in tannic or clear water (Brim Box and Williams 2000, Clench and Turner 1956.)

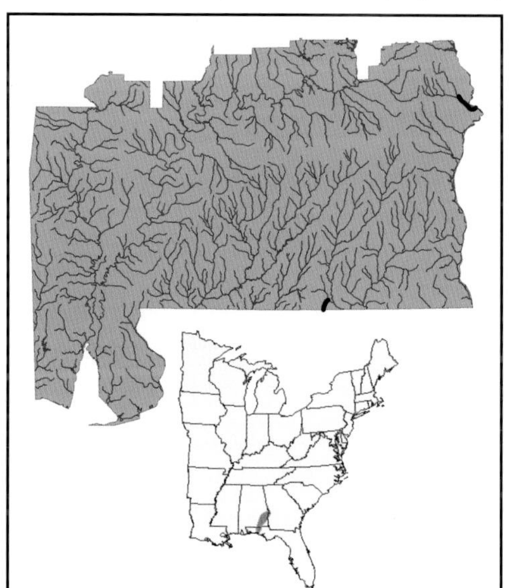

LIFE HISTORY AND ECOLOGY. A long-term brooder, reportedly gravid from April to September. Bluegill and largemouth bass are reported glochidial hosts (Keller and Ruessler 1997).

BASIS FOR STATUS CLASSIFICATION. Limited distribution and rarity make *V. villosa* vulnerable to extirpation from Alabama. Listed as a species of special concern throughout its distribution (Williams *et al.* 1993), but was not included on the list of Alabama species (Lydeard *et al.* 1999.)

Prepared by: **Jeffrey J. Herod**

COMMON AND SCIENTIFIC NAMES OF FISH PARASITIZED BY GLOCHIDIA OF IMPERILED MUSSELS OF ALABAMA

Family Acipenseridae
Shovelnose Sturgeon *Scaphirhynchus platorynchus*

Family Amiidae
Bowfin *Amia calva*

Family Anguillidae
American Eel *Anguilla rostrata*

Family Catostomidae
Northern Hog Sucker *Hypentelium nigricans*
River Redhorse *Moxostoma carinatum*
Shorthead Redhorse *Moxostoma macrolepidotum*
White Sucker *Catostomus commersoni*

Family Centrarchidae
Black Crappie *Pomoxis nigromaculatus*
Bluegill *Lepomis macrochirus*
Green Sunfish *Lepomis cyanellus*
Largemouth Bass *Micropterus salmoides*
Longear Sunfish *Lepomis megalotis*
Orangespotted Sunfish *Lepomis humilus*
Redeye Bass *Micropterus coosae*
Rock Bass *Ambloplites rupestris*
Smallmouth Bass *Micropterus dolomieu*
Spotted Bass *Micropterus punctatus*
Warmouth *Lepomis gulosus*
White Crappie *Pomoxis annularis*

Family Clupeidae
Gizzard Shad *Dorosoma cepedianum*

Family Cottidae
Banded Sculpin *Cottus carolinae*
Black Sculpin *Cottus baileyi*
Mottled Sculpin *Cottus bairdi*

Family Cyprinidae
Alabama Shiner *Cyprinella callistia*
Bigeye Chub *Notropis amblops*
Blacktail Shiner *Cyprinella venusta*
Blotched Chub *Hybopsis insignis*
Bluntnose Minnow *Pimephales notatus*
Centrall Stoneroller *Campostoma anomalum*
Common Carp *Cyprinus carpio*
Common Shiner *Luxilus cornutus*
Creek Chub *Semotilus atromaculatus*
Fathead Minnow *Pimephales promelas*
Largescale Stoneroller *Campostoma oligolepis*
Northern Redbelly Dace *Phoxinus eos*
Popeye Shiner *Notropis ariommus*
River Chub *Nocomis micropogon*
Rosyface Shiner *Notropis rubellus*
Saffron Shiner *Notropis rubricroceus*
Sailfin Shiner *Pteronotropis hypselopterus*
Silver Shiner *Notropis photogenis*
Spotfin Shiner *Cyprinella spiloptera*
Streamline Chub *Erimystax dissimilis*
Striped Shiner *Luxilus chrysocephalus*
Telescope Shiner *Notropis telescopus*
Tennessee Shiner *Notropis leuciodus*
Warpaint Shiner *Luxilus coccogenis*
White Shiner *Luxilus albeolus*
Whitetail Shiner *Cyprinella galactura*

Family Esocidae
Northern Pike *Esox lucius*

Family Fundulidae
Banded Killifish *Fundulus diaphanus*
Blackspotted Topminnow *Fundulus olivaceus*
Northern Studfish *Fundulus catenatus*
Plains Killifish *Fundulus zebrinus*

Family Ictaluridae
Brown Bullhead *Ameiurus nebulosus*
Flathead Catfish *Pylodictis olivaris*
Tadpole Madtom *Noturus gyrinus*
Yellow Bullhead *Ameiurus natalis*

Family Moronidae
White Bass *Morone chrysops*

Family Percidae
Arrow Darter *Etheostoma sagitta*
Banded Darter *Etheostoma zonale*
Barcheek Darter *Etheostoma obeyense*
Blackbanded Darter *Percina nigrofasciata*
Blotchside Logperch *Percina burtoni*
Brown Darter *Etheostoma edwini*
Channel Darter *Percina copelandi*
Dirty Darter *Etheostoma olivaceum*
Dusky Darter *Percina sciera*
Emerald Darter *Etheostoma baileyi*
Fantail Darter *Etheostoma flabellare*
Gilt Darter *Percina evides*
Greenside Darter *Etheostoma blennioides*
Gulf Darter *Etheostoma swaini*
Johnny Darter *Etheostoma nigrum*
Logperch *Percina caprodes*
Mobile Logperch *Percina kathae*
Naked Sand Darter *Ammocrypta beani*
Rainbow Darter *Etheostoma caeruleum*
Redline Darter *Etheostoma rufilineatum*
Redspotted Darter *Etheostoma artesiae*
Roanoke Darter *Percina roanoka*
Rock Darter *Etheostoma rupestre*
Saddleback Darter *Percina vigil*
Sauger *Stizostedion canadense*
Snubnose Darter *Etheostoma simoterum*
Southern Sand Darter *Ammocrypta meridiana*
Speckled Darter *Etheostoma stigmaeum*
Spotted Darter *Etheostoma maculatum*
Striped Darter *Etheostoma virgatum*
Stripetail Darter *Etheostoma kennicotti*
Tangerine Darter *Percina aurantiaca*
Tuskaloosa Darter *Etheostoma douglasi*
Yellow Perch *Perca flavescens*
Walleye *Stizostedion vitreum*
Warrior Darter *Etheostoma bellator*
Wounded Darter *Etheostoma vulneratum*

Family Poeciliidae
Eastern Mosquitofish *Gambusia holbrooki*

Family Sciaenidae
Freshwater Drum *Aplodiotus grunniens*

ALABAMA FRESHWATER MUSSEL WATCH LIST

MODERATE CONSERVATION CONCERN

Taxa with conservation problems because of insufficient data, OR because of two of four of the following: small population; limited, disjunct, or peripheral distribution; decreasing population trend/population viability problem; specialized habitat need/habitat vulnerability due to natural/human caused factors. Research and/or conservation action recommended.

Order Unionoida
Family Unionidae

COMMON NAME	*SCIENTIFIC NAME*	*PROBLEM(s)*
COOSA FIVERIDGE	*Amblema elliottii*	Limited distribution/Decreasing population trend
APALACHICOLA FLOATER	*Anodonta heardi*	Insufficient data on distribution
ROCK POCKETBOOK	*Arcidens confragosus*	Decreasing population trend/ Limited distribution
ROUND PEARLSHELL	*Glebula rotundata*	Insufficient data on distribution
WAVYRAYED LAMPMUSSEL	*Lampsilis fasciola*	Decreasing population trend/ Habitat vulnerability
ROUGH FATMUCKET	*Lampsilis straminea straminea*	Limited distribution/Habitat vulnerability
WHITE HEELSPLITTER	*Lasmigona complanata complanata*	Limited distribution/Decreasing population trend
ALABAMA HEELSPLITTER	*Lasmigona complanata alabamensis*	Limited distribution/Decreasing population trend
PONDMUSSEL	*Ligumia subrostrata*	Insufficient data on distribution
SOUTHERN HICKORYNUT	*Obovaria jacksoniana*	Insufficient data on distribution
OHIO PIGTOE	*Pleurobema cordatum*	Decreasing population trend/ Habitat vulnerability
FUZZY PIGTOE	*Pleurobema strodeanum*	Decreasing population trend/ Habitat vulnerability
EASTERN FLOATER	*Pyganodon cataracta*	Insufficient data on distribution
MONKEYFACE	*Quadrula metanevra*	Decreasing population trend/ Habitat vulnerability
PURPLE PIMPLEBACK	*Quadrula refulgens*	Insufficient data on distribution
SOUTHERN CREEKMUSSEL	*Strophitus subvexus*	Decreasing population trend/ Habitat vulnerability
LILLIPUT	*Toxolasma parvus*	Insufficient data on distribution
IRIDESCENT LILLIPUT	*Toxolasma paulus*	Insufficient data on distribution
FAWNSFOOT	*Truncilla donaciformis*	Decreasing population trend/ Habitat vulnerability
FLORIDA PONDHORN	*Uniomerus carolinianus*	Insufficient data on distribution
FLORIDA FLOATER	*Utterbackia peggyae*	Insufficient data on distribution
RAINBOW	*Villosa iris*	Decreasing population trend/ Limited distribution
ALABAMA RAINBOW	*Villosa nebulosa*	Decreasing population trend/ Habitat vulnerability
PAINTED CREEKSHELL	*Villosa taeniata*	Decreasing population trend/ Habitat vulnerability

Freshwater Snails

EXTINCT

Taxa that historically occurred in Alabama, but are no longer alive anywhere within their former distribution.

Order Neotaenioglossa
Family Pleuroceridae

SHORT-SPIRE ELIMIA *Elimia brevis*
SPINDLE ELIMIA *Elimia capillaris*
CLOSED ELIMIA *Elimia clausa*
FUSIFORM ELIMIA *Elimia fusiformis*
SHOULDERED ELIMIA *Elimia gibbera*
HIGH-SPIRED ELIMIA *Elimia hartmaniana*
CONSTRICTED ELIMIA *Elimia impressa*
HEARTY ELIMIA *Elimia jonesi*
TEARDROP ELIMIA *Elimia lachryma*
RIBBED ELIMIA *Elimia laeta*
WRINKLED ELIMIA *Elimia macglameriana*
ROUGH-LINED ELIMIA *Elimia pilsbryi*
PUPA ELIMIA *Elimia pupaeformis*
BOT ELIMIA *Elimia pupoidea*
PYGMY ELIMIA *Elimia pygmaea*
COBBLE ELIMIA *Elimia vanuxemiana*
EXCISED SLITSHELL *Gyrotoma excisa*
STRIATE SLITSHELL *Gyrotoma lewisii*
PAGODA SLITSHELL *Gyrotoma pagoda*
RIBBED SLITSHELL *Gyrotoma pumila*
PYRAMID SLITSHELL *Gyrotoma pyramidata*
ROUND SLITSHELL *Gyrotoma walkeri*
AGATE ROCKSNAIL *Leptoxis clipeata*
OBLONG ROCKSNAIL *Leptoxis compacta*
INTERRUPTED ROCKSNAIL *Leptoxis foremani*
MAIDEN ROCKSNAIL *Leptoxis formosa*
ROTUND ROCKSNAIL *Leptoxis ligata*
LIRATE ROCKSNAIL *Leptoxis lirata*
KNOB ROCKSNAIL *Leptoxis minor*
BIGMOUTH ROCKSNAIL *Leptoxis occultata*
COOSA ROCKSNAIL *Leptoxis showalteri*
SQUAT ROCKSNAIL *Leptoxis torrefacta*
STRIPED ROCKSNAIL *Leptoxis vittata*

Family Hydrobiidae

CAHABA PEBBLESNAIL *Clappia cahabensis*
UMBILICATE PEBBLESNAIL *Clappia umbilicata*
OLIVE MARSTONIA *Pyrgulopsis olivacea*

Family Pomatiopsidae

DIXIE SEEP SNAIL *Pomatiopsis hinkleyi*

Order Basommatophora
Family Planorbidae

SHOAL SPRITE *Amphigyra alabamensis*
CARINATE FLAT-TOP SNAIL *Neoplanorbis carinatus*
ANGLED FLAT-TOP SNAIL *Neoplanorbis smithi*
LITTLE FLAT-TOP SNAIL *Neoplanorbis tantillus*
UMBILICATE FLAT-TOP SNAIL *Neoplanorbis umbilicatus*

FRESHWATER SNAILS

EXTIRPATED

Taxa that historically occurred in Alabama, but are now absent; may be rediscovered in the state, or be reintroduced from populations existing outside the state.

Order Neotaenioglossa
Family Pleuroceridae

SMOOTH MUDALIA *Leptoxis virgata*
KNOBBY ROCKSNAIL *Lithasia curta*

SMOOTH MUDALIA

Leptoxis virgata (I. Lea)

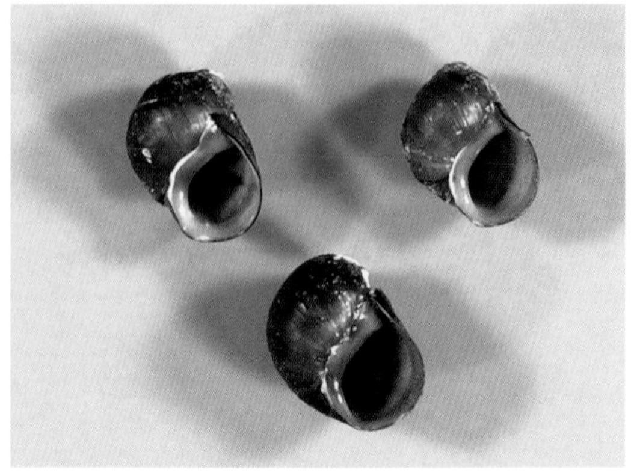

Photo—Paul Johnson

OTHER NAME. Virgate Mudalia.

DESCRIPTION. Has moderately thick shell (approaching 20 mm [3/4 in.] long), subglobose in outline, with a short, unsculptured spire. Body whorl strongly convex and unshouldered. Sutures are little impressed. Aperture oval with dorsal margin compressed against body whorl. Shell nacre white to deep purple, with bands visible inside aperture, when present. Columella curved, with basal lip broad and gently curved to cover umbilicus. Periostracum yellow to reddish brown, with some individuals having faint purple to brown bands. Operculum dark red to brown. Best distinguished from *L. praerosa* by tissue color, with foot of *L. virgata* bright orange, and foot of *L. praerosa* light gray.

DISTRIBUTION. Endemic to Tennessee River drainage, where historically occurred from headwaters downstream to Jackson County, Alabama. Eliminated from Tennessee River, Alabama, following construction of Nickajack and Guntersville Dams. Extant in Clinch, Powell, French Broad, Holston, Nolichucky, and Hiawassee River systems in Tennessee, Virginia, and North Carolina.

HABITAT. Shoal habitats, with moderate to heavy current. Cannot survive in lentic waters. Generally found grazing on biofilm of rock surfaces, but occasionally found in areas with substrata composed of high percentage of sand. Does not occur in areas with heavy silt deposition. Found in hydro-dam tailwaters (Hiawassee River).

LIFE HISTORY AND ECOLOGY. Lays eggs from March through May in flowing water (P. D. Johnson, unpubl. data). Females attach clutches of eggs directly to firm substrata. Individuals held in captivity for two years (P. D. Johnson, unpubl. data), but species probably lives at least three years in wild.

BASIS FOR CLASSIFICATION. Secure in Tennessee River headwaters, but has been extirpated from waters of Alabama. Occurs in approximately 40 percent of its original distribution, and remaining populations are reproducing.

Prepared by: **Paul D. Johnson**

KNOBBY ROCKSNAIL

Lithasia curta (I. Lea)

Photo—Paul Johnson

OTHER NAME. None

DESCRIPTION. Has moderately thick shell (max. length = 20 mm [3/4 in.]), ovately conic in outline, with three rows of irregularly shaped tubercles on dorsum of body whorl. Spire short and usually eroded, with convex whorls and an irregular, but well-formed suture. Body whorl strongly convex and shouldered. Aperture obliquely trapezoidal with anterior margin very thin and elongate, attached well above midline of body whorl. Periostracum dark yellow to green or brown, with up to four faint bands. Nacre white to a light pink or tan, with bands usually visible when present. Operculum dark red to brown.

DISTRIBUTION. Endemic to Tennessee River system, where historically known only from Tennessee River at Muscle Shoals and adjacent Shoal Creek, Lauderdale County. Has not been collected in Alabama since construction of Wilson Dam and likely extirpated from state. However, appears to be extant in tailwaters of Kentucky Dam, suggesting that its historic distribution extended at least to mouth of Tennessee River.

HABITAT. Appears to be restricted to tailwaters of Kentucky Dam, where current is moderate to strong.

LIFE HISTORY AND ECOLOGY. Unknown.

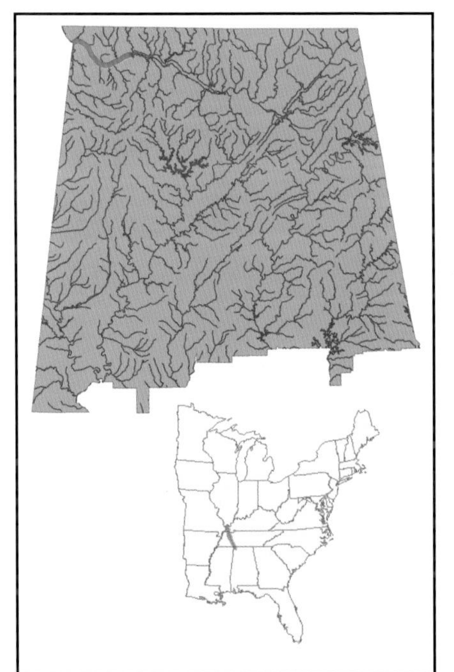

BASIS FOR CLASSIFICATION. Not reported from Alabama since impoundment of the Tennessee River.

Prepared by: **Paul D. Johnson**

Freshwater Snails

EXTIRPATED
CONSERVATION ACTION UNDERWAY

Taxa that historically occurred in Alabama, were absent for a period of time, and currently are being reintroduced, or have a plan for being reintroduced, into the state from populations outside the state.

Order Neotaenioglossa
Family Pleuroceridae

SPINY RIVERSNAIL *Io fluvialis*
INTERRUPTED ROCKSNAIL *Leptoxis downiei*

SPINY RIVERSNAIL

Io fluvialis (Say)

Photo—Arthur Bogan

OTHER NAMES. None.

DESCRIPTION. Highly variable with regard to shell morphology. Spiny form typically found in large rivers and occurred in Alabama historically. Shell thick (max. length = approx. 50 mm [2 in.]), fusiform in outline, with a tapered, elevated spire. Whorls somewhat convex to inflated and sutures not impressed. Body whorl sculptured by wrinkled growth lines and marked by series of spines in middle of whorl. Spines vary from raised knobs to large spines that are of equal height and width. Aperture elliptical and about one-half length of shell, with a well-developed basal channel. Inside of aperture whitish and often marked with dull reddish lines. Periostracum olive green to brown. Columella concave and imperforate. (Modified from Tryon 1873, Bogan and Parmalee 1983)

DISTRIBUTION. Endemic to Tennessee River system. Downstream limit reportedly Muscle Shoals (Adams 1915), but not reported from Alabama since Tennessee River impounded. Recently reintroduced into tailwaters of Nickajack Dam, Marion County, Tennessee (P. D. Johnson, Southeastern Aquatic Research Institute, pers. comm., 2002). If reintroduction successful, should result in recolonization of upper reaches of Guntersville Reservoir.

HABITAT. Large rocks and bedrock outcrops in well-oxygenated, shallow water with fast current (Bogan and Parmalee 1983).

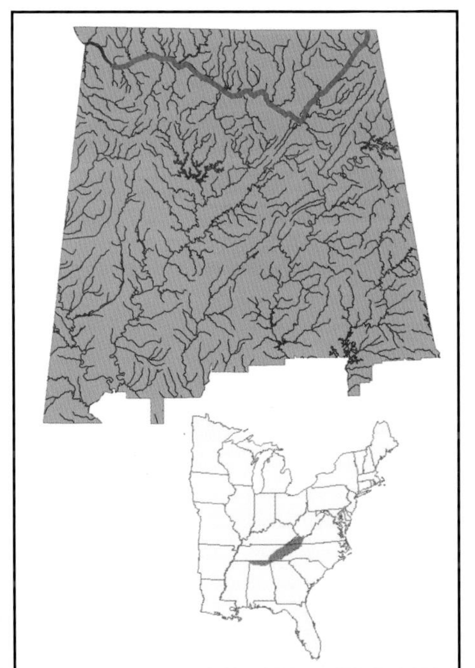

LIFE HISTORY AND ECOLOGY. Reported to lay two to six clutches of eggs during late April and May, with clutches containing 16 to 100 eggs each. Laid without a sand-grain covering in diagonal a pattern with rows of one to five eggs that begin to hatch after about 15 days (Bogan and Parmalee 1983).

BASIS FOR STATUS CLASSIFICATION. Extirpated from Alabama with impoundment of Tennessee River. However, if planned reintroductions into Nickajack Dam tailwaters are successful, should expand distribution downstream into Alabama. Listed as endangered in Alabama (Stein 1976).

***Prepared by:* Arthur E. Bogan**

INTERRUPTED ROCKSNAIL

Leptoxis formani (Lea)

Photo—Paul Johnson

OTHER NAMES. Downie's Round Riversnail.

DESCRIPTION. Has moderately thick shell (approaching 20 mm [3/4 in.] long), elongately globose in outline, generally with no more than three whorls present. Small folds (incised striae) cover body whorl from suture to base, but are weaker basally. Dorsal edge of each suture accented by several heavy plicae that are irregularly formed and varied in size. Plicae do not extend far onto body whorl, and may be nearly absent in some individuals. Folds surrounding body whorl are nearly straight basally, but are more undulating as they approach suture and cross plicae. Periostracum light brown to burnt orange, but folds may be darker, giving shell a mottled appearance. Juveniles tend to be more lightly colored (nearly orange). Body weakly convex and shell spire elevated considerably with increasing size, an unusual character in genus. Sutures deeply impressed, but weak shoulders diminish overall appearance. Columella dark purple to nearly white, darkening basally, and folded to cover umbilicus completely. Juveniles have tightly coiled whorls, display strong folds and plicae, distinguishing *L. formani* juveniles from other *Leptoxis* species. Operculum red to maroon, with coarse growth lines widely spaced across surface. (Modified from Lea 1868, Goodrich 1922)

DISTRIBUTION. Endemic to Coosa River system and historically occurred from its headwaters in northwestern Georgia: Conasauga, Coosawattee, and Etowah River systems, downstream to Elmore County, Alabama. Historically present in lower reaches of Terrapin Creek, Cherokee County. Species extant in a 12 kilometers (7.4 miles) reach of Oostanaula River, Gordon and Floyd Counties, Georgia.

HABITAT. In Oostanaula River, occurs in shoal and shallow run habitats. Shows preference for currents between 20 and 40 centimeters per second (8-15 inches per second) and most abundant in areas less than 40 centimeters (15 inches) deep (Johnson and Evans 2000). Preferred substratum a mixture of smooth gravel, cobble, and boulders, free of fine sediments (Johnson and Evans 2000). Does not occur on bedrock or aquatic vegetation.

LIFE HISTORY AND ECOLOGY. Females lay multiple clutches of 11 to 24 eggs from March until mid-May and eggs hatch in seven to 10 days at 20°C (68°F) (P. D. Johnson, unpubl. data). Egg clutches laid in a concentric circular pattern, near air-water interface or in well-oxygenated riffle habitats. Do not reproduce until its second year and may live up to five years (P. D. Johnson, unpubl. data). Grazes on silt-free gravel, cobble, and boulders.

BASIS FOR STATUS CLASSIFICATION. Vulnerable to extinction due to highly restricted distribution, small population size, and specific habitat requirements. May be most imperiled *Leptoxis* in Mobile Basin. Considered extinct until rediscovery in 1997. Listed as endangered in Alabama (Stein 1976). Eliminated from more than 95 percent of historical distribution. **Now a *candidate* for federal protection by the U.S. Fish and Wildlife Service.**

***Prepared by:* Paul D. Johnson**

Freshwater Snails

PRIORITY I
HIGHEST CONSERVATION CONCERN

Taxa critically imperiled and at risk of extinction/extirpation because of extreme rarity, restricted distribution, decreasing population trend/population viability problems, and specialized habitat needs/habitat vulnerability due to natural/human-caused factors. Immediate research and/or conservation action required.

Order Architaenioglossa

Family Viviparidae

SLENDER CAMPELOMA *Campeloma decampi*
CYLINDRICAL LIOPLAX *Lioplax cyclostomaformis*
TULOTOMA *Tulotoma magnifica*

Family Hydrobiidae

FLAT PEBBLESNAIL *Lepyrium showalteri*
ARMORED MARSTONIA *Pyrgulopsis pachyta*
MOSS PYRG *Pyrgulopsis scalariformis*

Family Pleuroceridae

ANTHONY'S RIVERSNAIL *Athearnia anthonyi*
PRINCESS ELIMIA *Elimia bellacrenata*
COCKLE ELIMIA *Elimia cochliaris*
LACY ELIMIA *Elimia crenatella*
ROUND-RIB ELIMIA *Elimia nassula*
ENGRAVED ELIMIA *Elimia perstriata*
PLICATE ROCKSNAIL *Leptoxis plicata*
CORPULENT HORNSNAIL *Pleurocera corpulenta*
ROUGH HORNSNAIL *Pleurocera foremani*

SLENDER CAMPELOMA

Campeloma decampi (Binney)

Photo—Arthur Bogan

OTHER NAMES. None.

DESCRIPTION. Shell moderately thick (may approach 40 mm [1 5/8 in.] long), ovately conic in outline, with a tapered and pointed spire. Whorls slightly convex and sutures slightly impressed. Body whorl broadly rounded and without shoulders. Fine microscopic striations adorn shell and very early whorls may be carinate. Juveniles generally carinate throughout and resemble some members of *Lioplax* in form. Aperture broadly ovate, with a rounded basal lip. Periostracum yellowish brown, darkening slightly with age.

DISTRIBUTION. Known only from northern tributaries of Tennessee River in small area of north-central Alabama, which includes parts of Limestone, Madison, and Jackson Counties (Burch 1989). Known to be extant in Limestone, Piney, and Round Island Creeks, Limestone County.

HABITAT. Primarily in slow to moderate current, often along stream margins. May be found in gravel, mud deposits in water willow (*Justicia virginiana*) beds, or on marginal clay ledges. Apparently tolerant of moderately silty conditions. Individuals usually spend most of time burrowed into substratum.

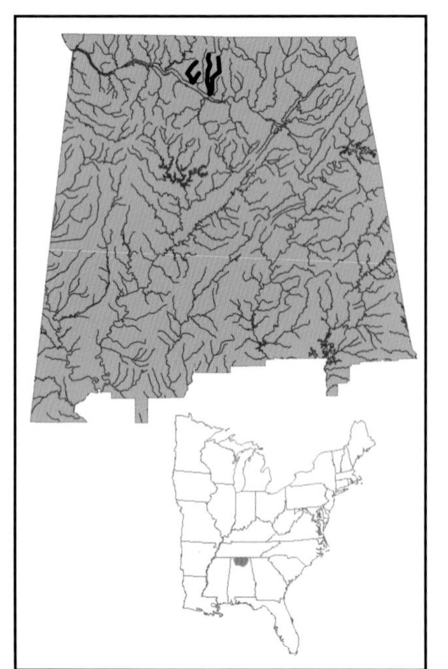

LIFE HISTORY AND ECOLOGY. Ovoviviparous, as all viviparids. Little else known.

BASIS FOR STATUS CLASSIFICATION. Limited distribution, declining population trend, and vulnerability to human perturbations to habitat make C. *decampi* susceptible to extinction. Distribution reduced to approximately one-third of its historic size. All three streams where extant lie in areas of intense agriculture, making them at risk of pesticide and/or fertilizer pollution, as well as degradation to habitat from excessive irrigation and sedimentation. **Listed as *endangered* by the U.S. Fish and Wildlife Service in 2000**.

***Prepared by:* Jeffrey T. Garner**

CYLINDRICAL LIOPLAX

Lioplax cyclostomaformis (Lea)

Photo—Paul Johnson

OTHER NAMES. None.

DESCRIPTION. Shell solid (approaches 30 mm [1 1/8 in.] long) ovately conic in outline, with an extended spire. Whorls somewhat flattened and sutures moderately impressed. Body whorl convex, but not shouldered. Early whorls carinate and remainder of shell may be adorned with microscopic beaded striations. Aperture subovate to subquadrate, with a sigmoidal outer margin and well-rounded basal lip. Operculum has subcentral nucleus. Periostracum light to dark olive. Inside aperture shell bluish. (Modified from Clench and Turner 1955)

DISTRIBUTION. Endemic and historically widespread in Mobile Basin. Known from Alabama River, Dallas County; Black Warrior River, Jefferson County; Cahaba and Little Cahaba Rivers, Jefferson and Bibb Counties; Coosa River, Shelby and Elmore Counties; Prairie Creek, Marengo County; Valley Creek, Jefferson County; Little Wills Creek, Etowah County; Choccolocco Creek, Talladega County; and Yellowleaf Creek, Shelby County (Clench and Turner 1955). In Coosa River headwaters of Georgia, known from Othcalooga Creek, Bartow County; Coahulla Creek, Whitfield County, and Armuchee Creek, Floyd County (Clench and Turner 1955). A single collection reported from Tensas River, Madison Parish, Louisiana (Clench 1962) may be erroneous, since there are no previous or subsequent records outside of Alabama-Coosa system, and searches of Tensas River in Louisiana have produced no evidence of species or its typical habitat (Paul Hartfield, U.S. Fish and Wildlife Service, unpubl. data; Vidrine 1996). Appears extant only in approximately 24 kilometers (14.9 miles) of Cahaba River above the Fall Line in Shelby and Bibb Counties, Alabama (Bogan and Pierson 1993*a*).

HABITAT. Unusual for a viviparid. Occurs in shoals of streams and rivers, living in mud under large rocks in rapid current. Other *Lioplax* species usually found in exposed situations, or in mud or muddy sand along river margins.

LIFE HISTORY AND ECOLOGY. Little known. Believed to be ovoviviparous and a filter feeder, as other viviparids. Life spans reported to range from three to 11 years in various species of Viviparidae (Heller 1990).

BASIS FOR STATUS CLASSIFICATION. Vulnerable to extinction due to extremely limited distribution, specialized habitat requirements, and declining population trend. Impoundment of Black Warrior, Alabama, and Coosa Rivers eliminated species from much of its distribution and left remaining populations fragmented. Degraded water quality from point and nonpoint sources believed to have eliminated most of population fragments not directly eliminated by impoundments. Listed as endangered in Alabama (Stein 1976). **Listed as *endangered* by the U.S. Fish and Wildlife Service in 1996.**

***Prepared by:* Paul D. Hartfield**

TULOTOMA

Tulotoma magnifica (Conrad)

Photo—Paul Johnson

OTHER NAME. Alabama Live-bearing Snail.

DESCRIPTION. One of largest and most distinct freshwater snails in North America. Shell globose (length may exceed 30 mm [1 1/8 in.]), whorls strongly concave with prominent shoulders, sutures impressed. Body whorl slightly to strongly convex. Shell sculpture consists of a row of large, evenly spaced nodules encircling shell whorls at midline. In some specimens, an additional row of reduced nodules present on ventral side of main body whorl below midline. Aperture ovate and obliquely angled, with ventral margin strongly recessed. Occasional specimens umbilicate. For a snail of its size, columellar lip narrow and completely white. Periostracum dark yellow to green or brown, with most individuals having dull brown bands. Shell often coated with dark layer of mineral deposits, making snail appear black in color. Nacre inside aperture white, with bands visible when present.

DISTRIBUTION. Endemic to Mobile Basin. Historically found throughout much of Coosa and Alabama River systems. Known to be extant in: Coosa River below Jordan Dam, Elmore County; Kelly Creek, St. Clair County; Ohatchee and Cane Creeks, Calhoun County; and Hatchet and Weogulfka Creeks, Coosa County (Hershler *et al.* 1990).

HABITAT. Flowing waters of large creeks and rivers. Remaining tributary populations generally restricted to lotic habitats in lower reaches, adjacent to Coosa River. Species does not occur in reservoir backwaters. Generally found beneath large flat rocks not in direct contact with river bottom, and having a large interstitial space beneath. Primary predator appears to be freshwater drum.

LIFE HISTORY AND ECOLOGY. Ovoviviparous with female's mantle cavity functioning as a marsupium for newly hatched juveniles. Once juveniles reach two to four millimeters (1/16 to 1/8 inch) in length, they crawl from mantle. Juveniles released from early spring to early fall. As with other viviparid snails, assumed that detritus is primary food source. However, likely also feeds on algae and bacteria attached to rocks. Often occurs in large clusters of individuals under large flat rocks, sometimes numbering in thousands under a single large boulder (one to two meters [3 1/4 to 6 1/2 feet) long). Likely that individuals live three to five years.

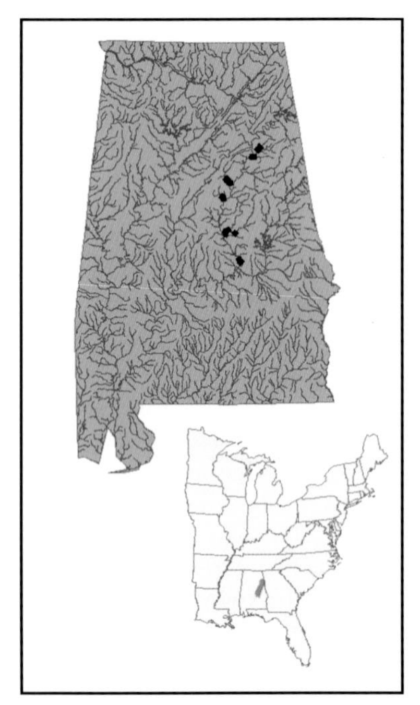

BASIS FOR STATUS CLASSIFICATION. Lim-ited distribution, declining population trend (outside of Jordan Dam tailwaters), and specialized habitat requirements make *T. magnifica* vulnerable to extinction. Eliminated from a majority of its habitat from construction and operation of hydroelectric dams on Coosa River. Poor land use practices, resulting in increased sedimentation were likely detrimental to tributary populations. Listed as endangered in Alabama (Stein 1976), but populations in Coosa River at Wetumpka have recovered due to restoration of minimum flow downstream of Jordan Dam. **Listed as *endangered* by the U.S. Fish and Wildlife Service in 1991.**

***Prepared by:* Paul D. Johnson**

FLAT PEBBLESNAIL

Lepyrium showalteri (Lea)

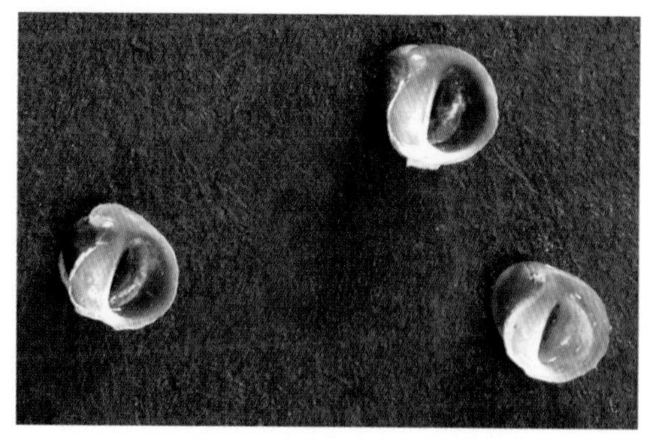

Photo—Arthur Bogan

OTHER NAMES. Coosa Scale Shell, Cahaba Scale Shell.

DESCRIPTION. A small snail, but with large, distinct shell relative to other hydrobiid species (max. length = 4.5 mm [3/16 in.]; width = 5 mm [1/4 in.]). Shell ovate and flattened, with a depressed spire and expanded body whorl giving it a neritid-like appearance, with spire barely extending above body whorl. Shell unsculptured and umbilical area imperforate. Aperture greatly expanded and oriented ventrally. Operculum paucispiral, with nucleus located near lower left margin. Male verge, a distinguishing character of most hydrobiids, unpigmented, long, flattened, and bladelike. However, species easily identified by distinctive shell. (Modified from Thompson 1984)

DISTRIBUTION. Endemic to Mobile Basin. Historically known from Coosa River, Shelby and Talladega Counties; Cahaba River, Bibb and Dallas Counties; and Little Cahaba River, Bibb County, Alabama (Thompson 1984). Recent surveys failed to locate any surviving populations outside of Cahaba River drainage. Currently known from one site on Little Cahaba River, Bibb County, and from a single shoal series on Cahaba River, above the Fall Line, Shelby County.

HABITAT. Clean, smooth stones in rapid current of small to large rivers.

LIFE HISTORY AND ECOLOGY. Eggs laid singly, in capsules on hard surfaces (Thompson 1984). Little else known.

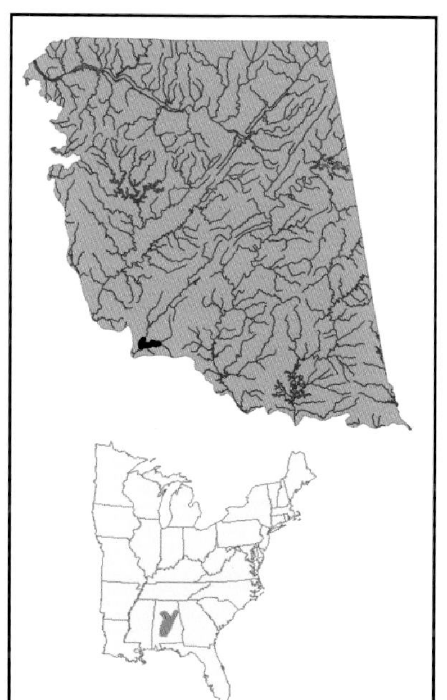

BASIS FOR STATUS CLASSIFICATION. Extremely limited distribution, declining population trend, and specialized habitat requirements make *L. showalteri* vulnerable to extinction. Eliminated from Coosa River system with its impoundment. Sedimentation and other forms of water pollution from point and nonpoint sources believed to have eliminated species from much of its historic distribution in Cahaba River drainage. Listed as endangered in Alabama (Stein 1976). **Listed as *endangered* by the U.S. Fish and Wildlife Service in 1996.**

***Prepared by:* Paul D. Hartfield**

ARMORED MARSTONIA

Pyrgulopsis pachyta (Thompson)

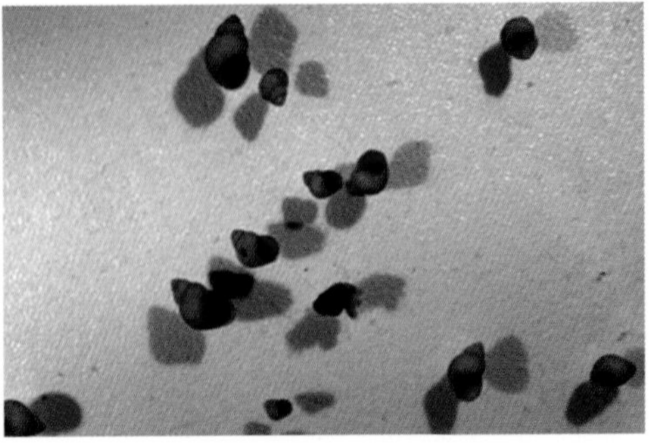
Photo—Jeffrey T. Garner

OTHER NAMES. Thick-shelled Marstonia.

DESCRIPTION. Shell small (max. length = 4 mm [1/8 in.]), but thick, ovately conic in outline, with a raised spire. Whorls moderately convex and sutures impressed. Body whorl broadly rounded and without shoulders. Aperture ovate, complete across parietal wall and with a well-rounded basal lip. Umbilicus imperforate or narrowly rimate. Periostracum tawny, with dull luster. Male verge stocky, with a stout penis projecting from right distal margin. An apical lobe present and bears two small apical glands; not as enlarged as in most other species of *Pyrgulopsis*. (Modified from Thompson 1977)

DISTRIBUTION. Endemic to Limestone Creek system, Limestone County, Alabama, where occurs in lower 13 unimpounded miles (21 kilometers) of Limestone Creek. Also occurs in lower eight unimpounded miles (13 kilometers) of Piney Creek, largest Limestone Creek tributary.

HABITAT. Densest concentrations occur in submerged masses of tree rootlets and bryophytes along stream margins, but also may occur on rocks and woody debris, as well as aquatic macrophytes.

LIFE HISTORY AND ECOLOGY. Unknown.

BASIS FOR STATUS CLASSIFICATION. Limited distribution, specialized habitat requirements, and susceptibility to habitat degradation make *P. pachyta* vulnerable to extinction. Limestone Creek lies in an area of intense agriculture, making species susceptible to pesticide and fertilizer pollution, excessive irrigation, and sedimentation. Listed as endangered in Alabama (Stein 1976). **Listed as *endangered* by the U.S. Fish and Wildlife Service in 2000.**

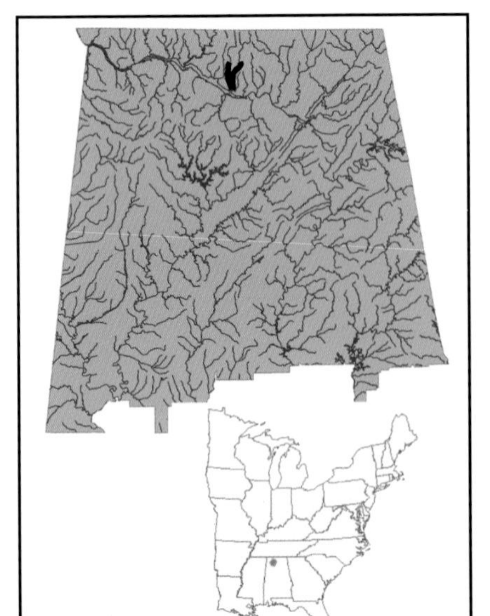

***Prepared by:* Jeffrey T. Garner**

MOSS PYRG

Pyrgulopsis scalariformis (Wolf)

Photo—Arthur Bogan

OTHER NAMES. None.

DESCRIPTION. Shell moderately thick (approaches 5 mm [1/4 in.] long), pupiform, and narrowly conic in outline, with a high spire. Whorls flat to very slightly convex. A well-developed carina usually adorns lower margin of each whorl, but may be weak or absent in some specimens. Carina gives body whorl an angular outline, but specimens lacking a carina have broadly rounded body whorl. Aperture ovate and complete across parietal wall, with a thickened, well-rounded basal lip. Umbilicus narrowly rimate when present. Periostracum light brown. Male verge somewhat narrow, with a large penis projecting from right side of distal margin. Apical lobe short and broad, slightly constricted basally giving a somewhat reniform appearance. Terminal gland positioned on distal margin of lobe. (Modified from Hershler 1994)

DISTRIBUTION. Appears to have been widely distributed in the Mississippi, Ohio, and Tennessee River systems, at least prehistorically (Hershler 1994), and was described from Pleistocene deposits in Illinois. However, apparently has been collected alive only from three localities: Meramec River, Missouri; Shoal Creek, Lauderdale County, Alabama (Hershler 1994); and Flint River, Madison County, Alabama. Shoal Creek population appears to be extirpated.

HABITAT. Primarily in submerged masses of tree rootlets and bryophytes along steep banks adjacent to flowing water.

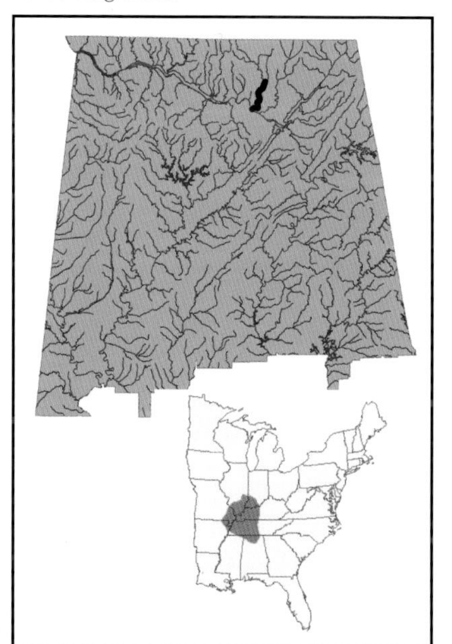

LIFE HISTORY AND ECOLOGY. Unknown.

BASIS FOR STATUS CLASSIFICATION. Extremely limited distribution, declining population trend, and vulnerability to habitat degradation make *P. scalariformis* vulnerable to extinction. Flint River population may be only one remaining. Terrain surrounding Flint River primarily used for agriculture, but currently under threat of increasing residential use, as suburbs of Huntsville expand eastward. Additionally, a golf complex was recently constructed adjacent to reach of creek where species occurs. All of these make snail susceptible to introduction of pesticides and fertilizers, as well as excessive irrigation and sedimentation.

***Prepared by:* Jeffrey T. Garner**

ANTHONY'S RIVERSNAIL

Athearnia anthonyi (Redfield)

Photo—Arthur Bogan

OTHER NAMES. None.

DESCRIPTION. Shell thick and ponderous (max. length = approx. 25 mm [1 in.]), ovate in outline, with a short spire, although whorls above body whorl often badly eroded. Body whorl flat to slightly convex and strongly shouldered, usually with series of large, irregular, obtuse tubercles that often appear as little more than undulations of shoulder. Shell aperture ovate, with a thin basal lip, often showing some purple coloration within. Basal lip reflected to partially or completely cover umbilical area. Periostracum of shell yellowish green to dark brown, usually darkening with age. Purplish or brownish bands often encircle body whorl, becoming indistinct as shell darkens. Juveniles distinct, being as wide as long, with a pointed spire and basal lip and strong carina around body whorl producing a saucer-shaped outline. Carina gradually disappears with age. (Modified from Tryon 1873, USFWS 1996)

DISTRIBUTION. Historically occurred in Tennessee River from Knoxville, Tennessee, downstream to Muscle Shoals, extending into some large and medium tributaries. Only two populations known to be extant. One population in a short reach of Tennessee River proper, from just upstream of mouth of Sequatchie River at Tennessee River Mile (TRM) 423, Marion County, Tennessee, to just downstream of Bridgeport Island (TRM 411), Jackson County, Alabama. This population extends up Sequatchie River for at least several miles (USFWS 1996). Other population located in lower Limestone Creek, Limestone County, Alabama, extending from just above impounded mouth at Limestone Creek Mile (LCM) 4 upstream to LCM 13.

HABITAT. Typically in lotic areas, but occasionally found in pools adjacent to shoals. Main stem Tennessee River population located in water three to four meters (13 feet) deep in riverine habitat downstream of Nickajack Dam. Preferred substrata vary from gravel to boulders. Often found on submerged woody debris and aquatic vegetation in and around lotic habitats.

LIFE HISTORY AND ECOLOGY. New recruits into Limestone Creek population appear in May. Little else known.

BASIS FOR STATUS CLASSIFICATION. Vulnerable to extinction due to specialized habitat requirements and declining population trend. Eliminated from majority of distribution when Tennessee River impounded, destroying most of habitat in main stem and isolated fragments of population in tributaries. Other poor land use practices probably played role in decline of species in tributaries. Listed as endangered in Alabama (Stein 1976). **Listed as *endangered* by the U.S. Fish and Wildlife Service in 1994**. In 2001, included in list of species approved for a Nonessential Experimental Population in riverine reach of Tennessee River, downstream of Wilson Dam. A federal permit is in place to transplant 4,000 individuals from Limestone Creek population in spring 2003.

***Prepared by:* Jeffrey T. Garner**

PRINCESS ELIMIA

Elimia bellacrenata (Haldeman)

Photo—Paul Johnson

OTHER NAMES. None.

DESCRIPTION. Shell of moderate thickness (approx. 20 mm [3/4 in.] long), narrowly conic in outline, with a delicate, elongate spire that tapers to a fine point. Whorls flattened and sutures little impressed. Upper whorls carinate, often with a line of crenulations below carina. Body whorl convex and unshouldered. Aperture elongate oval, with a narrowly rounded basal lip. Periostracum reddish brown to black and shell reddish inside aperture.

DISTRIBUTION. Endemic to Cahaba River drainage. Historically reported in springs and small spring-fed creeks in Shelby, Bibb, and Tuscaloosa Counties. In latest survey of Cahaba River system, was collected at only one site on Shoal Creek, Shelby County (Bogan and Pierson 1993*a*).

HABITAT. Where extant in Shoal Creek, about 10 meters (33 feet) wide, with slow to moderate current. Substrata are limestone slabs, concrete rubble, gravel, and sand. Collection site had heavy algal growths suggesting nutrient pollution.

LIFE HISTORY AND ECOLOGY. Unknown.

BASIS FOR STATUS CLASSIFICATION. Vulnerable to extinction due to limited distribution, specialized habitat requirements, and declining population trend.

Prepared by: **Paul D. Hartfield**

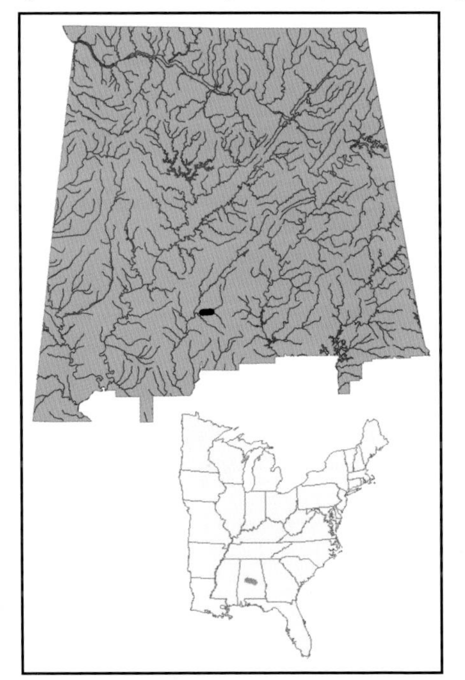

COCKLE ELIMIA

Elimia cochilaris (Lea)

Photo—Malcolm Pierson

OTHER NAMES. None.

DESCRIPTION. Shell thin and small, elongately conic in outline, with a high spire. Sutures greatly impressed and whorls ornamented with a carina medially and striations above and below. Upper striation may be crenulate. Aperture small and ovate. Periostracum dark brown. (Modified from Lea 1868)

DISTRIBUTION. Endemic to Cahaba River system. Historically known from springs in Jefferson, Tuscaloosa, and Bibb Counties, Alabama, but known to be extant only in a small spring tributary to Little Cahaba River in Bibb County.

HABITAT. Springs and small spring-fed streams.

LIFE HISTORY AND ECOLOGY. Unknown.

BASIS FOR STATUS CLASSIFICATION. Vulnerable to extinction due to extremely limited distribution, specialized habitat requirements, and declining population trend.

Prepared by: **Paul D. Hartfield**

LACY ELIMIA

Elimia crenatella (Lea)

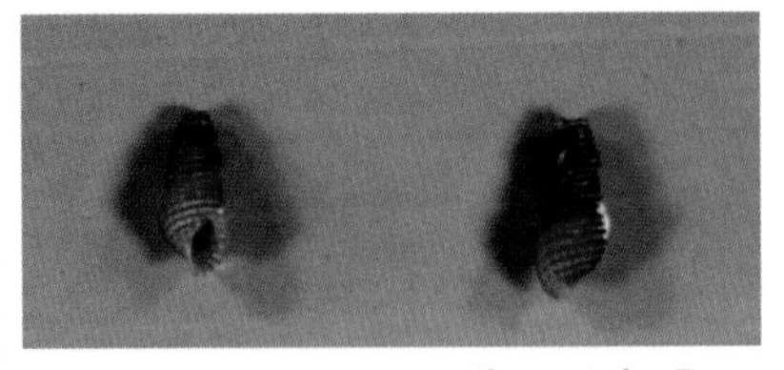

Photo—Arthur Bogan

OTHER NAMES. Crenate Riversnail.

DESCRIPTION. Perhaps smallest species in genus (max. length = 10 mm [3/8 in.]). Shell elongately conic in shape and completely covered with heavy striae. Shell often weakly plicate, with individual plicae more numerous on upper whorls and nearly absent on body whorl. Because of snail's small size, striae are often folded in upper whorls. Whorl count can reach 10 but usually only four or five of most mature whorls survive erosion. Body whorl slightly convex to nearly flat and unshouldered. Sutures usually distinctly impressed. Aperture ovate and a small columellar lip barely visible on medial margin. Outer margin of aperture regularly crenulate, following pattern of individual striae. No umbilicus present. Shell nacre inside aperture appears nearly clear, but often highlighted by some purple coloring of basal lip and at base of individual striae. Operculum almost as wide as high and its spiral lines are paleomelanian. Periostracum brown to nearly black and upper whorls appear darker than body whorl. Color gradient caused by smaller spacing between individual striae on smaller whorls. (Modified from Tryon 1873, Goodrich 1936, USFWS 1998)

DISTRIBUTION. Endemic to Coosa River system in Alabama. Historically occurred in Coosa River between Etowah and Chilton Counties, and in several tributaries, including Big Wills Creek, Etowah and Dekalb Counties; Kelly Creek, St. Clair and Shelby Counties; and Choccolocco and Tallasseehatchee Creeks, Talladega County (Goodrich 1936). Extirpated from all known historic sites, but recently located in Cheaha, Weewoka, and Emauhee Creeks, Talladega County (Bogan and Pierson 1993*b*). Best remaining population in lower section of Cheaha Creek.

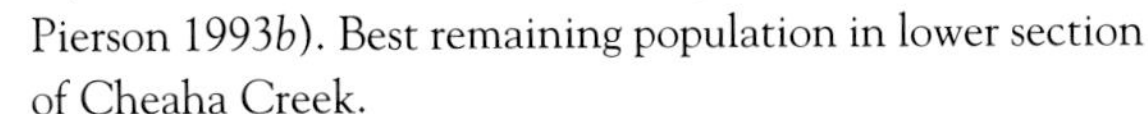

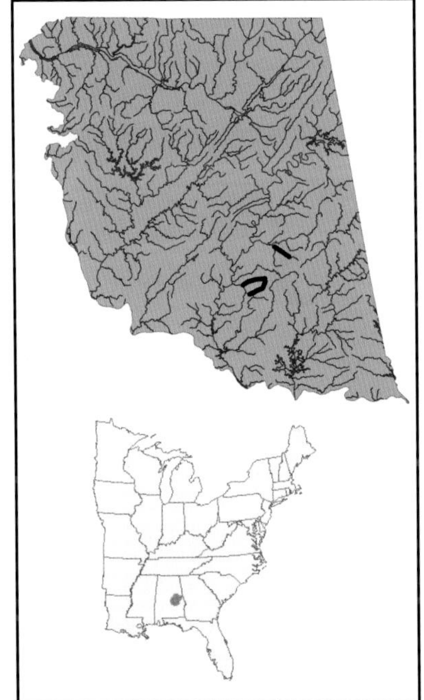

HABITAT. Requires shoal habitat, with minimal siltation. Often displays a contiguous distribution with clusters of individuals situated on larger rocks in shoal habitats (P. Hartfield, U.S. Fish and Wildlife Service, pers. comm., 2002). In Coosa River, reportedly found in slower moving, shallow water on downriver side of large rocks.

LIFE HISTORY AND ECOLOGY. Dioecious. Females believed to begin laying eggs in February and continue until May. Although no specific life history information exists, species may be similar to congeners. Most *Elimia* species lay eggs in short strings of one to three, in shallow water on bottom half of rocks exposed to direct current (Mancini 1978, P. D. Johnson, unpubl. data). Feed on algae and bacteria attached to rock surfaces.

BASIS FOR STATUS CLASSIFICATION. Vulnerable to extinction due to restricted distribution, specialized habitat requirements, and declining population trend. Has been eliminated from over 95 percent of its historic distribution and was considered extinct prior to 1993 (Bogan and Pierson 1993*b*). Majority of historic habitat eliminated with impoundment of Coosa River. Loss of species from some tributaries likely caused by habitat and water quality degradation. Listed as endangered in Alabama (Stein 1976). **Listed as *threatened* by the U.S. Fish and Wildlife Service in 1998.**

***Prepared by:* Paul D. Johnson**

ROUND-RIB ELIMIA
Elimia nassula (Conrad)

OTHER NAMES. None.

Photo—Arthur Bogan

DESCRIPTION. Shell moderately thin (may approach 20 mm [3/4 in.] long), elongately conic in outline, with a high spire. Whorls convex or somewhat angulate and sutures impressed. Whorls plicate and crossed by numerous spiral, elevated lines. Sculpture persists from spire through body whorl. Umbilical area imperforate. Aperture open and elongately ovate. Periostracum dark brown to black and may have darker color bands. (Modified from Tryon 1873)

DISTRIBUTION. Endemic to north-central and northwestern Alabama. Found in springs and spring branches in Colbert and Madison Counties (Burch 1989). Recent field work produced *E. nassula* from five springs: Buzzard Roost and Tuscumbia Springs, Colbert County; Wheeler Spring, Lawrence County; Big Spring, Madison County; and Cave Spring, Morgan County (J. T. Garner, Ala. Div. Wildl. Freshwater Fish., unpubl. data). However, recent "improvements" to Tuscumbia Spring, which included diverting water from the channel while substrata were removed with heavy equipment, may have destroyed that population.

HABITAT. Only springs and spring branches on gravel, cobble, and submerged macrophytes. Appear to have little tolerance for silty conditions.

LIFE HISTORY AND ECOLOGY. Unknown.

BASIS FOR STATUS CLASSIFICATION. Limited distribution, highly specific habitat requirements, and declining population trend make *E. nassula* vulnerable to extinction. Big Spring lies in an urban setting, so may be susceptible to similar "improvements" as recently befell Tuscumbia Spring.

***Prepared by:* Arthur E. Bogan**

ENGRAVED ELIMIA

Elimia perstriata (Lea)

Photo—Arthur Bogan

OTHER NAMES. None.

DESCRIPTION. Shell moderately thin (max. length = 10 mm [3/8 in.]), elongately conic in outline, with a tall spire. Whorls convex, with impressed sutures. Apical whorls carinate and granulate. Striations adorn shell surface. Whorls below apex less carinate and body whorl may be smooth. Umbilical area imperforate. Aperture small, elliptical, and angular at base. Periostracum reddish brown. Inside aperture, shell reddish. (Modified from Tryon 1873)

DISTRIBUTION. Tryon (1873) erroneously listed habitat as Coosa River, Alabama, and Huntsville, Tennessee [*sic*]. Occurs in springs and small streams of northern Alabama (Burch 1989). Recent field work has produced species from Big Spring Branch and Indian Creek, Madison County, and Fox and Sandy Creeks, Lawrence County (J. T. Garner, Ala. Div. Wildl. and Freshwater Fish., unpubl. data).

HABITAT. May occur on sand, gravel, or cobble substrata in springs and small streams. Appears to have little tolerance for silty conditions.

LIFE HISTORY AND ECOLOGY. Unknown. Some specimens of highly variable *E. acuta* approach form of *E. perstriata*. Although the two occur within same general area, typical *E. perstriata* always found in pure populations, and two forms do not appear to intergrade (J. T. Garner, Ala. Div. Wildl. and Freshwater Fish., unpubl. data).

BASIS FOR STATUS CLASSIFICATION. Limited distribution, specific habitat requirements, and declining population trend make *E. perstriata* vulnerable to extinction. Springs and small streams in which it occurs are easily disturbed and destroyed. Big Spring Branch and Indian Creek are susceptible to negative impacts of urbanization in and around Huntsville. Populations in Fox and Sandy Creeks are in agricultural areas, so could be affected by pesticide and fertilizer pollution, as well as excessive sedimentation and irrigation.

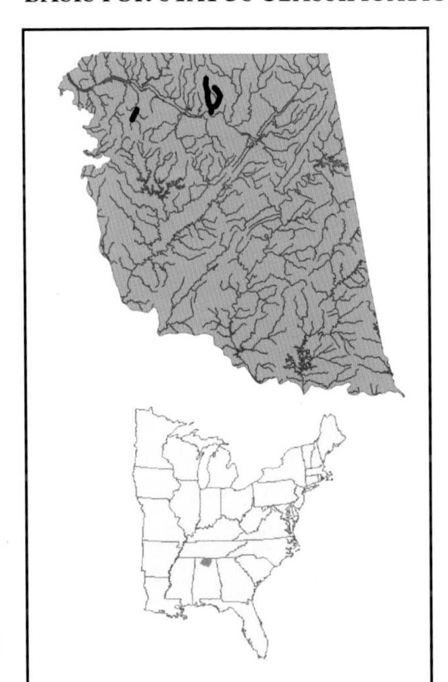

***Prepared by:* Arthur E. Bogan**

PLICATE ROCKSNAIL

Leptoxis plicata (Conrad)

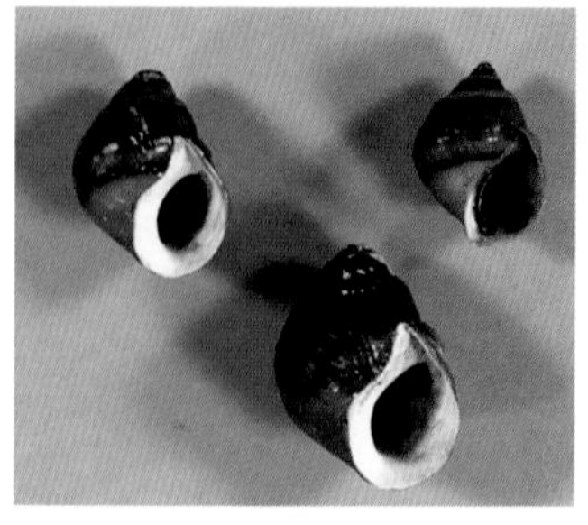

Photo—Paul Johnson

OTHER NAMES. Pleated Round Riversnail, Smith's Round Riversnail.

DESCRIPTION. Shell moderately thick (max. length = 20 mm [3/4 in.]) and subglobose in outline. Whorls strongly convex and shouldered, with a raised ridge on shoulder. Small, regularly spaced plicae set on ridge of most specimens. Periostracum generally dark yellowish green, with up to three dark bands, but generally becomes darker with age and bands become obscure. Columella mostly white, but may have light brown area on lower edge. Columellar lip reflected to completely cover umbilicus. Aperture ovate in shape and most commonly bluish white to pink, with no internal banding. Operculum dark red and moderately thick, with curved left margin and straight right margin. Juveniles same general shape, have tightly coiled whorls, and often display a weak carina on mid-whorl. Banding pattern dominates juvenile coloring, making them appear nearly all black. (Modified from Goodrich 1922)

DISTRIBUTION. Endemic to Black Warrior River system, historically occurring from headwaters to near confluence with Tombigbee River, Greene County. Now restricted to 15 shoals in a 30-kilometer (18.6-mile) reach of Locust Fork, Jefferson County, between Kimberly and Sayre (Johnson 2002).

HABITAT. Occurs in shoals, on silt-free bedrock, cobble, and boulders, in less than one meter (3 1/4 feet) of water. Often found on undersides of large, flat boulders (Johnson 2002). Grazes on exposed rock surfaces. More commonly found marginally than at mid-channel.

LIFE HISTORY AND ECOLOGY. Females deposit eggs late February through late April. Eggs laid on river bottom in shallow water (less than 50 centimeters [20 inches] deep) with exposure to direct current (P.D. Johnson, unpubl. data). Eggs laid simply and not in well-formed clutches. Juveniles grow rapidly and can reach 10 millimeters (3/8 inch) their first year. Have survived four years in captivity (P. D. Johnson, unpubl. data).

BASIS FOR STATUS CLASSIFICATION. Vulnerable to extinction due to limited distribution, decreasing population trend, and specific habitat requirements. Tombigbee-Black Warrior Waterway Project destroyed over half of available riverine habitat within its distribution. Poor water quality apparently caused extirpation from Mulberry Fork. Has disappeared from upper Locust Fork since late 1980s, but causes of decline are uncertain, possibly including sedimentation, organic pollution, and hydrologic disruption. Eliminated from over 90 percent of distribution, and populations at over half of 15 shoals where extant appear to be declining (Johnson 2002). Listed as endangered in Alabama (Stein 1976).

However, it responds well to captive propagation, and recovery efforts under Mobile Basin Recovery Plan (USFWS 2000) are currently ongoing. **Listed as *endangered* by the U.S. Fish and Wildlife Service in 1998.**

***Prepared by:* Paul D. Johnson**

CORPULENT HORNSNAIL

Pleurocera corpulenta (Anthony)

OTHER NAMES. Stout Riversnail.

Photo—Paul Johnson

DESCRIPTION. Has moderately thin shell (may approach 20 mm [3/4 in.] long), ovately conic in outline, with a short spire. Shell whorls (six or seven) convex, with impressed sutures. Body whorl inflated, with a convex outer margin. Shell unsculptured, but often marked by longitudinal growth lines. Aperture narrowly ovate, narrow at top and broader basally. Aperture about one-half shell length. Columella very curved below middle, white, thickened at base, and marked by a broad sinus in basal area. Periostracum brown; may have bands that can be seen inside aperture. (Modified from Tryon 1873)

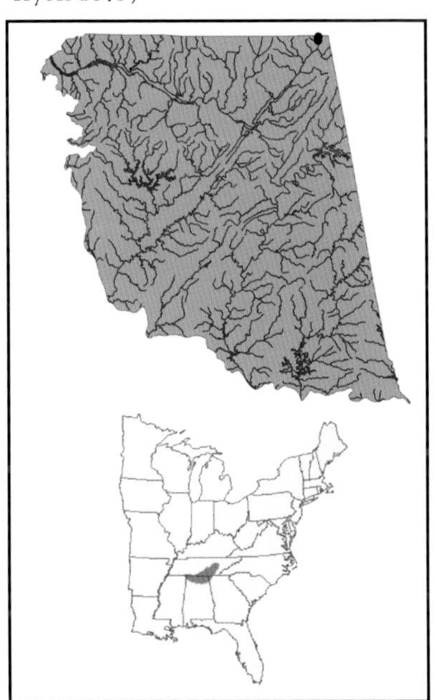

DISTRIBUTION. Known only from Tennessee River, between Bridgeport, Jackson County, and Muscle Shoals, Alabama, and Battle Creek, Marion County, Tennessee, which is not far from Bridgeport (Burch 1989). Only known extant population in tailwaters of Nickajack Dam, in vicinity of Bridgeport.

HABITAT. Gravel and cobble substrata in moderate to swift current. In Nickajack Dam tailwaters, occurs at depths of two to four meters (6.5 to 13 feet).

LIFE HISTORY AND ECOLOGY. Unknown.

BASIS FOR STATUS CLASSIFICATION. Extremely limited distribution, specialized habitat requirements, and declining population trend make *P. corpulenta* vulnerable to extinction. Distribution was severely restricted by impoundment of Tennessee River. Apparently was common at Muscle Shoals prior to impoundment of river, but listed as endangered in Alabama (Stein 1976).

***Prepared by:* Arthur E. Bogan**

ROUGH HORNSNAIL

Pleurocera foremani (Lea)

OTHER NAMES. Foreman's High-spired Riversnail.

DESCRIPTION. Shell thick (>30 mm [1 1/8 in.] long), elongately conic in outline with a moderately high spire. Whorls rather flattened and sutures irregularly impressed. Body whorl may be widely convex to broadly angled. Lower whorls circled by row of closely set tubercles. Aperture elongately ovate, angular above, and channeled basally. Periostracum yellowish brown. Inside of aperture white. (Modified from Tryon 1873)

Photo—Paul Johnson

DISTRIBUTION. Endemic to Mobile Basin, where it was historically known from Coosa and Cahaba Rivers (Goodrich 1944). Extant in lower Yellowleaf Creek, Shelby County, and Coosa River downstream of Jordan Dam.

HABITAT. Primarily on clean gravel and cobble in moderate currents at depths of approximately one meter (3 1/4 feet). However, has been collected from silty bedrock at three-meter (9 3/4-foot) depth just downstream of shoals on Coosa River at Wetumpka.

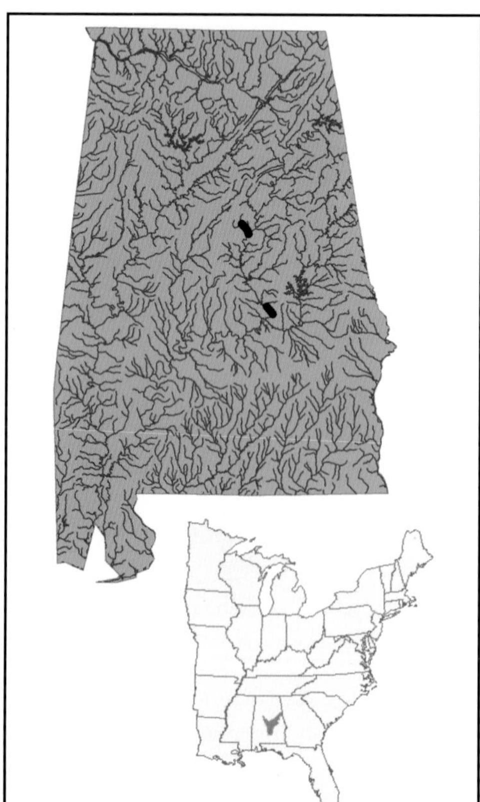

LIFE HISTORY AND ECOLOGY. Unknown.

BASIS FOR STATUS CLASSIFICATION. Vulnerable to extinction due to limited distribution, declining population trend, and specialized habitat requirements. Distribution greatly reduced by impoundment of Coosa River and degraded water quality in Cahaba River. Listed as endangered in Alabama (Stein 1976).

***Prepared by:* Paul D. Hartfield**

Freshwater Snails

PRIORITY 2
HIGH CONSERVATION CONCERN

Taxa imperiled because of three of four of the following: rarity; very limited, disjunct, or peripheral distribution; decreasing population trend/population viability problems; specialized habitat needs/habitat vulnerability due to natural/human-caused factors. Timely research and/or conservation action needed.

Order Neotaenioglossa
Family Pleuroceridae

AMPLE ELIMIA *Elimia ampla*
LILYSHOALS ELIMIA *Elimia annettae*
BLACK MUDALIA *Elimia melanoides*
PUZZLE ELIMIA *Elimia varians*
SQUAT ELIMIA *Elimia variata*
ROUND ROCKSNAIL *Leptoxis ampla*
SPOTTED ROCKSNAIL *Leptoxis picta*
PAINTED ROCKSNAIL *Leptoxis taeniata*
ARMORED ROCKSNAIL *Lithasia armigera*
WARTY ROCKSNAIL *Lithasia lima*
MUDDY ROCKSNAIL *Lithasia salebrosa*
RUGGED HORNSNAIL *Pleurocera alveare*
SKIRTED HORNSNAIL *Pleurocera pyrenella*

AMPLE ELIMIA

Elimia ampla (Anthony)

Photo—Paul Johnson

OTHER NAMES. Large Cahaba Riversnail.

DESCRIPTION. Shell relatively thin, but large (max. length >30 mm [1 1/8 in.]), elongately conic in outline, with a high spire. Whorls flattened and sutures irregularly and deeply impressed. Body whorl large and unshouldered. Shell unsculptured, but may have a roughened texture. Aperture large and elongately ovate. No umbilicus present. Periostracum tawny to olive, often with multiple dark bands. Inside aperture shell may be rose colored. (Modified from Tryon 1873)

DISTRIBUTION. Endemic to Cahaba River drainage. Historically reported from a short reach of river between Centreville and Lily Shoals, Bibb County. Today, known to survive between Centreville and Booth Ford, Shelby County, where uncommon.

HABITAT. Moderate to fast current over bedrock, boulder, and cobble substrata. Also found beneath rock ledges of undercut areas (Bogan and Pierson 1993*a*).

LIFE HISTORY AND ECOLOGY. Unknown.

BASIS FOR STATUS CLASSIFICATION. Vulnerable to extinction due to limited distribution, rarity, and declining population trend. Listed as endangered in Alabama (Stein 1976).

***Prepared by:* Paul D. Hartfield**

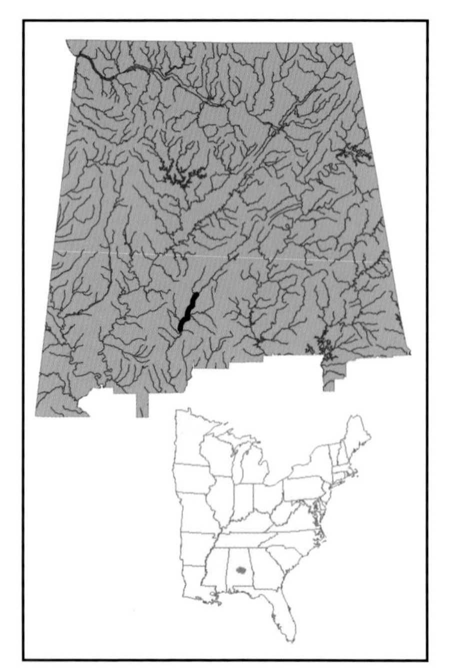

LILYSHOALS ELIMIA

Elimia annettae (Goodrich)

Photo—Paul Johnson

OTHER NAMES. Annette's Riversnail.

DESCRIPTION. Shell moderately thick (max. length >20 mm [3/4 in.]), elongately conic in outline, with a moderately high spire. Whorls flattened and sutures moderately impressed. Body whorl convex and unshouldered. No sculpture present. Aperture ovate, with basal lip that is slightly reflected. Periostracum shiny, yellowish brown with darker bands. (Modified from Goodrich 1941)

DISTRIBUTION. Endemic to Cahaba River system, occurring in main stem from Centreville to Bibb/Shelby County line, and in lower Little Cahaba to, and including mouth of, Six Mile Creek (Bogan and Pierson 1993*a*).

HABITAT. All types found at Lily Shoal on main stem Cahaba, but appears to prefer areas of moderate current (Goodrich 1941).

LIFE HISTORY AND ECOLOGY. Unknown.

BASIS FOR STATUS CLASSIFICATION. Vulnerable to extinction due to limited distribution and declining population trend. Listed as endangered in Alabama (Stein 1976).

***Prepared by:* Paul D. Hartfield**

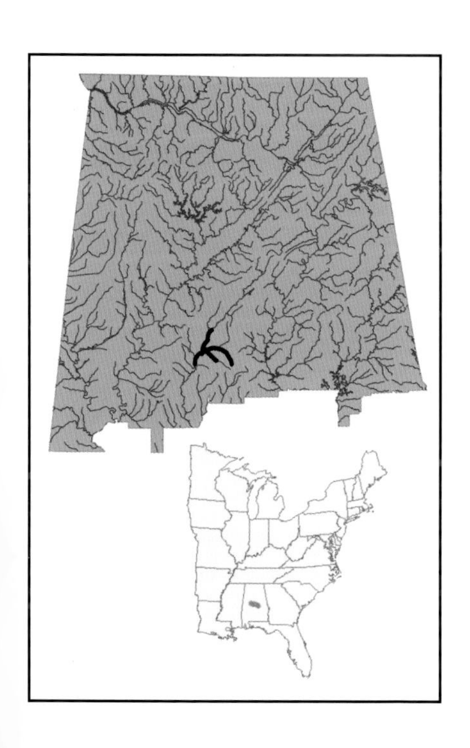

BLACK MUDALIA

Elimia melanoides (Conrad)

OTHER NAME. Black Warrior Riversnail.

DESCRIPTION. Shell small, but moderately thick (approaching 10 mm [5/8 in.] long), ovately conic in outline, with whorls flat and very weakly shouldered, and sutures little impressed. Periostracum usually black, but may be yellow to green or brown, sometimes highlighted with two or three bands. Columella wide and angled at center, white to pink, and folded basally to completely cover umbilical area. Aperture ovate and operculum very thin and light brown. (Modified from Tryon 1873, Goodrich 1922) Recently considered to belong to *Leptoxis*, but moved to *Elimia* following phylogenetic analysis (Minton *et al.*, in prep).

Photo—Arthur Bogan

DISTRIBUTION. Endemic to Black Warrior River system, where historically occurred throughout upper reaches of drainage. Extant in middle and upper reaches of Locust Fork in Blount and Marshall Counties.

HABITAT. Margins of channels on rocks or vegetation. Appears tolerant of minor sedimentation, but may be sensitive to hydrologic disruption.

LIFE HISTORY AND ECOLOGY. Unknown

BASIS FOR STATUS CLASSIFICATION. Limited distribution, specific habitat requirements, and declining population trend make *E. melanoides* vulnerable to extinction. Has been eliminated from more than 80 percent of its historic distribution and was listed as extinct (Turgeon *et al.* 1998) before being rediscovered during the 1990s. Listed as endangered in Alabama (Stein 1976).

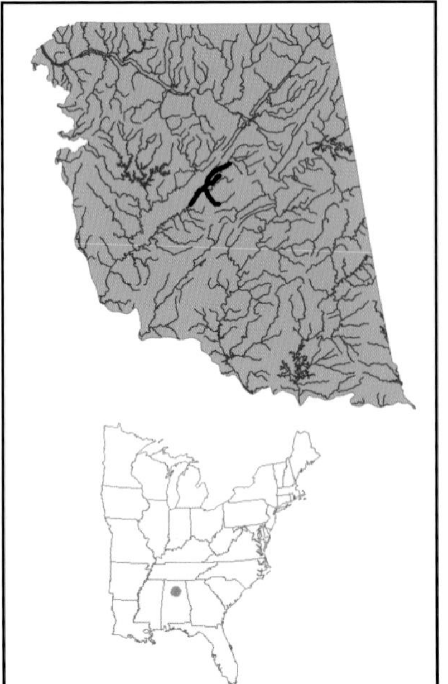

***Prepared by:* Paul D. Johnson**

PUZZLE ELIMIA

Elimia varians (Lea)

Photo—Arthur Bogan

OTHER NAMES. Variable Cahaba Riversnail.

DESCRIPTION. Shell (>30 mm [1 1/8 in.] long) elongately conic, with convex whorls that may be slightly shouldered, and impressed sutures. Body whorl somewhat elongate and convex, but may be slightly flattened. Shell sculpture highly variable, with some individuals being plicate, some striate, and some smooth. Aperture somewhat small, relative to shell size, and elliptical, with a bluntly pointed basal lip. Periostracum yellowish to light brown, often banded. Inside aperture, shell white and banded. Shells of young pyramidal in shape. (Modified from Tryon 1873)

DISTRIBUTION. Endemic to Cahaba River, where historically occurred from Pratt's Ferry to about 12 kilometers (7.4 miles) downstream of Centreville, Bibb County. Recently, only reported from Cahaba River at a few sites in Bibb County between Marvel and Centreville.

HABITAT. Cobble and bedrock substrata in shoals and runs.

LIFE HISTORY AND ECOLOGY. Unknown.

BASIS FOR STATUS CLASSIFICATION. Vulnerable to extinction due to limited distribution and declining population trend. Listed as endangered in Alabama (Stein 1976).

Prepared by: **Paul D. Hartfield**

SQUAT ELIMIA

Elimia variata (Lea)

OTHER NAMES. Variegated Cahaba Riversnail.

Photo—Arthur Bogan

DESCRIPTION. Shell moderately thick (max. length approx. 20 mm [3/4 in.]) and varies in shape from pyramidal or ovately conic in smaller streams to inflated and nearly cylindrical in larger streams. Spire short and unsculptured, with slightly convex whorls and irregularly impressed sutures. Body whorl enlarged, with a convex outer margin. Aperture large relative to shell size and broadly ovate. Periostracum yellowish brown and may have dark bands. (Modified from Tryon 1873)

DISTRIBUTION. Endemic to Cahaba River system. Historically reported from Peavine, Buck, and Town Creeks, Shelby County; Little Cahaba River, Jefferson County; and Cahaba River between mouth of Buck Creek and Lily Shoals, Shelby and Bibb Counties. Recently reported from Buck and Shoal Creeks, Shelby County; Little Cahaba and Six Mile Creeks, Bibb County; and Cahaba River at Booths Ford, Shelby County.

HABITAT. Cobble, gravel, and bedrock substrata in shoals with moderate current in small to medium streams, and the Cahaba River.

LIFE HISTORY AND ECOLOGY. Unknown.

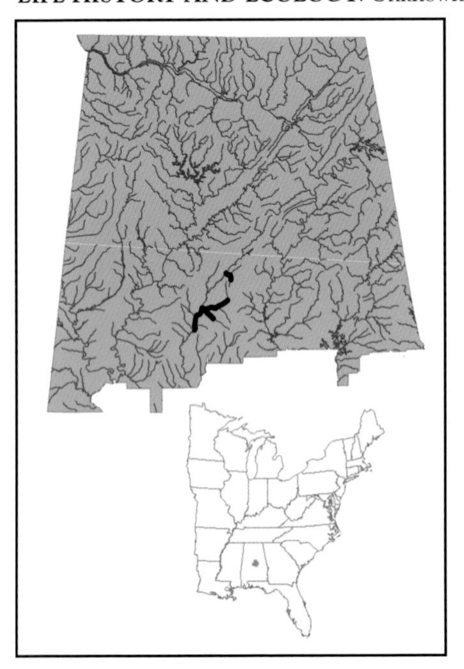

BASIS FOR STATUS CLASSIFICATION. Vulnerable to extinction due to limited distribution and declining population trend. Listed as endangered in Alabama (Stein 1976).

***Prepared by:* Paul D. Hartfield**

ROUND ROCKSNAIL

Leptoxis ampla (Anthony)

Photo—Malcolm Pierson

OTHER NAMES. Large Round Riversnail, Mimic Riversnail.

DESCRIPTION. Shell globose (max. length = 10 mm [5/8 in.]). Whorl number usually less than three and spire generally severely, sometimes completely, eroded. Body whorl strongly convex and sometimes weakly shouldered. Aperture generally ovate in large specimens, but may be elliptical, in small shells. A broad and reflected columellar lip covers umbilicus. Columella usually purple, but may be white or red, and broadened and flattened on lower edge. Up to four dark color bands usually accent periostracum that is dark yellow to green or brown. Operculum dark red to brown with rounded apex. (Modified from Goodrich 1922)

DISTRIBUTION. Endemic to Mobile Basin, where historically occurred throughout Cahaba and Coosa River systems (USFWS 1998, Goodrich 1922). Tributaries of Coosa River from which it has been reported include: Kelly and Big Canoe Creeks, St. Clair County; Ohatchee Creek, Calhoun County; Yellowleaf Creek, Shelby County; and Waxahatchee Creek, Shelby and Chilton Counties (Goodrich 1922). Appears to have been extirpated from Coosa River Basin. Current distribution limited to a series of Cahaba River shoals, Bibb and Shelby Counties, and from lower reaches of Shade and Six Mile Creeks and Little Cahaba River, Bibb County.

HABITAT. Shoals of main channel and larger tributaries of Cahaba River system. Preferred substrata are gravel, cobble, and boulders at depths of less than one meter (3 1/4 feet). Highly sensitive to sedimentation.

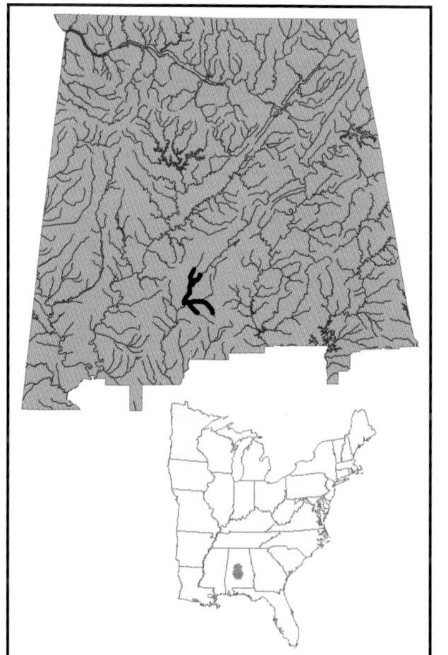

LIFE HISTORY AND ECOLOGY. Little known. However, with other *Leptoxis* (*L. plicata* and *L. foremani*) from Mobile River Basin, females lay eggs from March until mid-May (P. D. Johnson, unpubl. data). Individuals live approximately two or three years and graze on rock surfaces.

BASIS OF STATUS CLASSIFICATION. Limited distribution, specialized habitat requirements, and a declining population trend make *L. ampla* vulnerable to extinction. Has been extirpated from more than 80 percent of historic distribution. Impoundment of Coosa River apparently resulted in its elimination from that system. Sedimentation, sediment toxicity, and poor water quality are possible causes of its decline in Cahaba River system and tributaries. Listed as endangered in Alabama (Stein 1976). **Listed as *threatened* by the U.S. Fish and Wildlife Service in 1998.**

***Prepared by:* Paul D. Johnson**

SPOTTED ROCKSNAIL

Leptoxis picta (Conrad)

OTHER NAMES. Painted Riversnail.

Photo—Arthur Bogan

DESCRIPTION. Shell moderately thick, growing to approximately 20 millimeters (3/4 inch) long, ovately conic in shape, with a short spire that usually is eroded. Whorls flattened and sutures hardly impressed. Shell unsculptured. Body whorl enlarged, broadly converse, and unshouldered, with a large, semicircular aperture that has a well-rounded basal lip. Periostracum tawny to dark brown and occasionally green, usually with thin, continuous or interrupted color bands. (Modified from Tryon 1873)

DISTRIBUTION. Endemic to Mobile Basin, where historically occurred in lower Coosa River and Alabama River downstream to Claiborne, Monroe County, and lower Cahaba River below the Fall Line. Recently found only in Alabama River downstream of Jones Bluff, Millers Ferry, and Claiborne Locks and Dams in Autauga, Dallas, Wilcox, Monroe, and Clarke Counties.

HABITAT. Requires flowing water over gravel, cobble, or bedrock substrata. Often found associated with limestone outcrops.

LIFE HISTORY AND ECOLOGY. Unknown. Appears somewhat sedentary, moving little within appropriate habitats.

BASIS FOR STATUS CLASSIFICATION. Limited distribution and rarity make species vulnerable to extinction. Apparently extirpated from extensive portions of Alabama and Coosa Rivers as a result of impoundment. Changes in flow and water quality associated with past operation of Jordan Dam may have contributed to extirpation from extreme lower Coosa River. Causes of extirpation from lower Cahaba River and flowing portions of Alabama River are unknown, but deterioration of water quality suspected to be a contributing factor. In early 1990s, was locally abundant downstream of Claiborne Lock and Dam. However, more recent surveys were able to locate only a few specimens there. Listed as endangered in Alabama (Stein 1976).

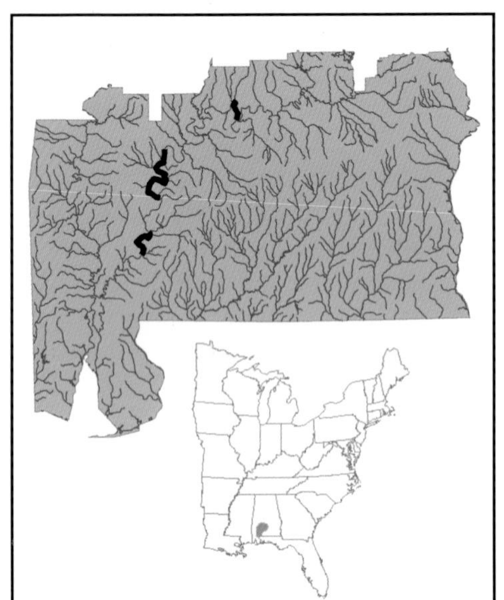

Prepared by: **Paul D. Hartfield**

PAINTED ROCKSNAIL

Leptoxis taeniata (Conrad)

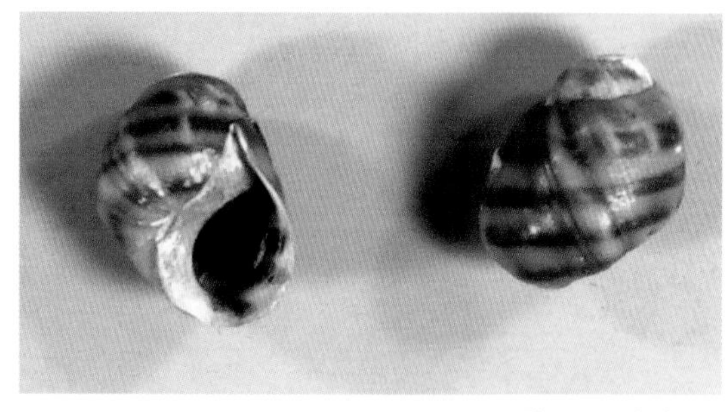
Photo—Paul Johnson

OTHER NAMES. Banded Round Riversnail.

DESCRIPTION. Shell heavy (max. length = 20 mm [3/4 in.]), subglobose in outline, with whorls broadly convex and heavily shouldered. Number of visible whorls seldom more than three, with shell erosion usually severe. Body whorl covered with very fine growth lines often crossed by very fine striae. Often has heavy folds on dorsal margin of each whorl, but some specimens completely smooth. Periostracum dark yellow to olive brown, with most individuals having four, contiguous or irregularly broken, concentric lines encircling body whorl. Aperture ovate, with white nacre accented with brown to purple banding inside. Columella wide and rounded, covering umbilicus basally, usually white, but may vary from pink to red or purple. Operculum elongate and reddish brown. Juveniles smooth surfaced with prominent bands, but whorls loosely coiled. (Modified from Goodrich 1922)

DISTRIBUTION. Endemic to Mobile Basin, historically occurring in Coosa River from St. Clair County, Alabama, to its mouth, in Alabama River headwaters downstream to Monroe County and in Cahaba River downstream of Fall Line. Believed extant only in lower Choccolocco, Buxahatchee, and Ohatchee Creeks in Talladega, Shelby, and Calhoun Counties, respectively (Bogan and Pierson 1993*a*,*b*; USFWS 2000).

HABITAT. Appears more tolerant of siltation than other *Leptoxis* in Mobile Basin. However, usually found attached to bedrock ledges and boulders in areas of flowing water where they graze on biofilm.

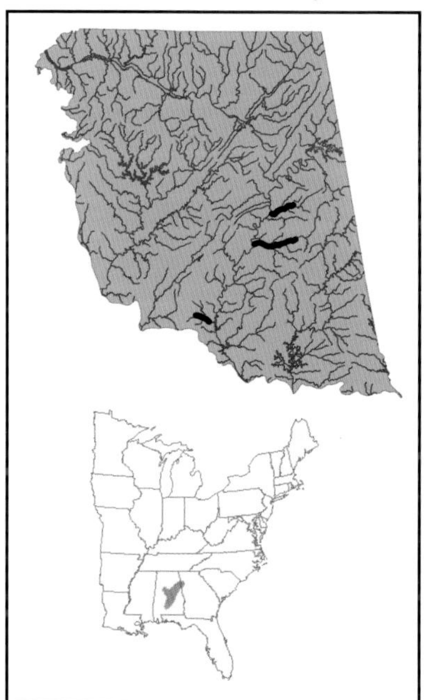

LIFE HISTORY AND ECOLOGY. Specific information nonexistent. However, females likely attach egg clutches to firm substrata in late winter to early spring, with new recruits appearing by early summer.

BASIS FOR STATUS CLASSIFICATION. Limited distribution, rarity, and specialized habitat requirements make *L. taeniata* vulnerable to extinction. Once had largest distribution of any *Leptoxis* in Mobile Basin, but has been eliminated from more than 85 percent of its former distribution.

Prepared by: **Paul D. Johnson**

ARMORED ROCKSNAIL

Lithasia armigera (Say)

OTHER NAMES. Knobby Ohio Riversnail.

DESCRIPTION. Shell thick (up to 25 mm [1 in.] long), broadly conic in outline, with a tapering spire. Shell with about six whorls, slightly wrinkled, slightly convex, with sutures slightly impressed. A revolving series of five or six prominent tubercles ornament body whorl, but diminish on spire, being covered by succeeding whorls. Body whorl may be marked by a second row of smaller and more obtuse subsutural tubercles. Aperture almost trapezoidal in outline, bluish white inside. A parietal callus is located near suture of body whorl and penultimate whorl, and there is a channel at base of aperture. Periostracum brownish yellow, becoming darker with age, possibly with obsolete revolving reddish brown lines on body whorl. (Modified from Tryon 1873, Bogan and Parmalee 1983)

Photo—Arthur Bogan

DISTRIBUTION. Endemic to Ohio River system, where widespread (Burch 1989). Historically occurred in Tennessee River and Shoal Creek in vicinity of Florence, Lauderdale County, Alabama (Goodrich 1940, Bogan and Parmalee 1983). Reportedly extirpated from Alabama with impoundment of Tennessee River (Stein 1976), but subsequently discovered in tailwaters of Wilson Dam.

HABITAT. Large rivers on gravel, cobble, and boulder substrata in strong current. Often found on undersides of large, flat rocks.

LIFE HISTORY AND ECOLOGY. Unknown. However, some species of *Lithasia* (*e.g.*, *L. salebrosa* and *L. geniculata*) lay pink eggs in spiral rows on exposed surfaces of cobble and boulders in moderate to swift current.

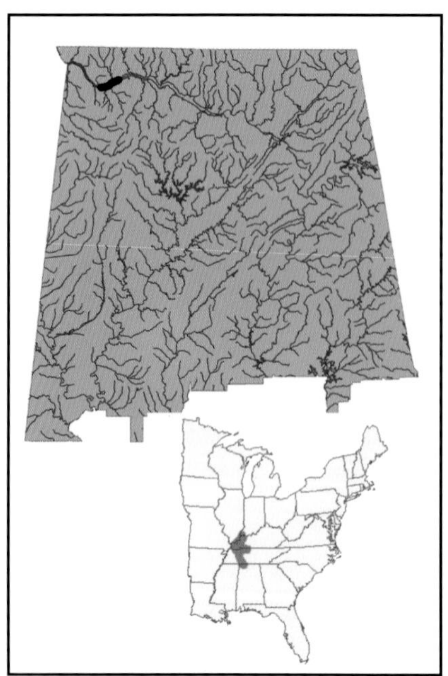

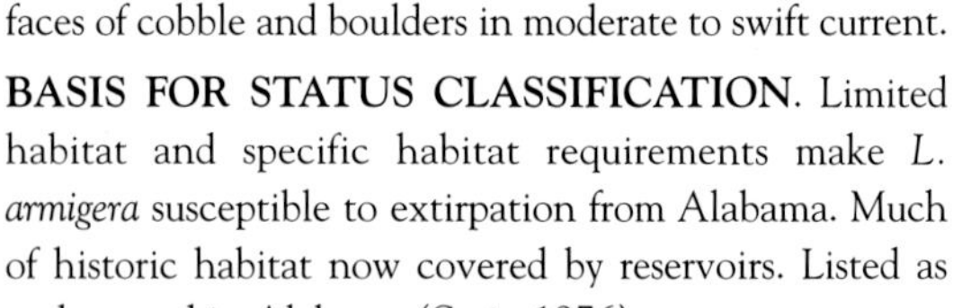

BASIS FOR STATUS CLASSIFICATION. Limited habitat and specific habitat requirements make *L. armigera* susceptible to extirpation from Alabama. Much of historic habitat now covered by reservoirs. Listed as endangered in Alabama (Stein 1976).

***Prepared by:* Arthur E. Bogan**

WARTY ROCKSNAIL

Lithasia lima (Conrad)

OTHER NAMES. Elk River File Snail.

DESCRIPTION. Shell relatively thick (≥ 20 mm [3/4 in.] long), ovately conic in outline, with a tapered, often eroded, spire. Shell whorls (about five) slightly convex, with sutures impressed little. Body whorl varies from rounded to angular and marked by one to six rows of subequal, equally spaced granules or tubercles that are not higher than wide. Tubercles often evident on second and third whorls. Aperture nearly half length of shell, contracted, and acutely angular above and obtusely pointed basally. Periostracum yellowish green to olive, darkening with age. Inside aperture may be marked with purple bands. (Modified from Tryon 1873, Bogan and Parmalee 1983)

Photo—Arthur Bogan

DISTRIBUTION. Known from Elk River and several other tributaries of middle reaches of Tennessee River (Stein 1976, Parmalee and Bogan 1983, Burch 1989). Recently collected from Elk River, Alabama and Tennessee, as well as Bear, Cypress, and Sugar Creeks in Colbert, Lauderdale, and Limestone Counties, Alabama, respectively. Recent systematic work on *Lithasia* group suggested taxonomic problems at species level (Minton 2002). Distribution of what is perceived as *L. lima* could change as these problems are addressed.

HABITAT. Gravel and cobble substrata in moderate to swift current. Appears intolerant of silty conditions.

LIFE HISTORY AND ECOLOGY. Unknown. However, some species of *Lithasia* (*e.g.*, *L. salebrosa* and *L. geniculata*) lay pink eggs in spiral patterns on exposed surfaces of cobble and boulders in moderate to swift current.

BASIS FOR STATUS CLASSIFICATION. Vulnerable to extinction due to limited distribution and specific habitat requirements. Listed as endangered due to impoundment and sedimentation from gravel washing in Elk River Basin (Stansbery 1971, Stein 1976).

***Prepared by:* Arthur E. Bogan**

MUDDY ROCKSNAIL

Lithasia salebrosa (Conrad)

OTHER NAMES. Rugged Riversnail.

Photo—Arthur Bogan

DESCRIPTION. Has thick shell (may approach 25 mm [1 in.] long), globose in outline with a very short, often eroded, spire. Body whorl has a series of elevated nodes and generally has a double series of smaller nodes below. Sutures impressed. Aperture about half length of shell and compressed oval in outline, with a callus located at dorsal margin and another near lower margin. Periostracum a yellowish green or brown and may be banded. When present, bands are visible inside aperture. (Modified from Tryon 1873, Bogan and Parmalee 1983)

DISTRIBUTION. Endemic to Tennessee and Cumberland River systems (Burch 1989). In Alabama, known from Tennessee River at Muscle Shoals and Elk River, as well as Cypress Creek, Lauderdale County. Known to be extant in state only in Wilson Dam tailwaters of Tennessee River (Isom *et al.* 1979, Bogan and Parmalee 1983).

HABITAT. Large rivers on gravel, cobble, and boulder substrata in moderate to swift current. Appears intolerant of silty conditions.

LIFE HISTORY AND ECOLOGY. Little known. Observed laying pink eggs in spiral rows on exposed surfaces of cobble and boulders in moderate to swift current in tailwaters of Wilson and Pickwick Dams.

BASIS FOR STATUS CLASSIFICATION. Vulnerable to extinction due to limited distribution and specific habitat requirements. Listed as endangered in Alabama (Stansbery 1971, Stein 1976) due to loss of habitat by impoundment of Tennessee River.

***Prepared by:* Arthur E. Bogan**

RUGGED HORNSNAIL

Pleurocera alveare (Conrad)

OTHER NAMES. Hollow Riversnail.

Photo—Arthur Bogan

DESCRIPTION. Shell thick and short (approaching 20 mm [3/4 in.] long), conic in outline with an elevated spire often ribbed at apex. Whorls flattened, with a line of broad, compressed tubercles at base of penultimate whorl. Body whorl sharply angled, with angle covered by prominent tubercles. Prominent lines, usually five in number, are located on slightly convex base of body whorl. Tubercles may become elongate, approaching form of plications. Aperture obliquely elliptical and about one-half length of shell. Periostracum yellowish brown, darkening with age. Some specimens may have darker spots between tubercles on body whorl. (Modified from Tryon 1873)

DISTRIBUTION. Widespread in Ohio River system and occurs in some rivers of Ozark region. In Alabama, reported from Tennessee River at Muscle Shoals, and Elk River and Cypress Creek, Lauderdale County (Goodrich 1940). Apparently extant in Alabama only in Tennessee River just downstream of Wilson and Wheeler Dams (Isom *et al.* 1979; J. T. Garner, Ala. Div. Wildl. Freshwater Fish., unpubl. data).

HABITAT. Gravel, cobble, and boulder substrata, as well as submerged logs, in moderate to swift current of small to large rivers.

LIFE HISTORY AND ECOLOGY. Unknown.

BASIS FOR STATUS CLASSIFICATION. Limited distribution and specific habitat requirements make *P. alveare* vulnerable to extirpation from Alabama. Although suggested as being extirpated from the Tennessee River, was actually listed as endangered in Alabama (Stein 1976). Has subsequently been found living in the tailwaters of both Wilson and Wheeler Dams.

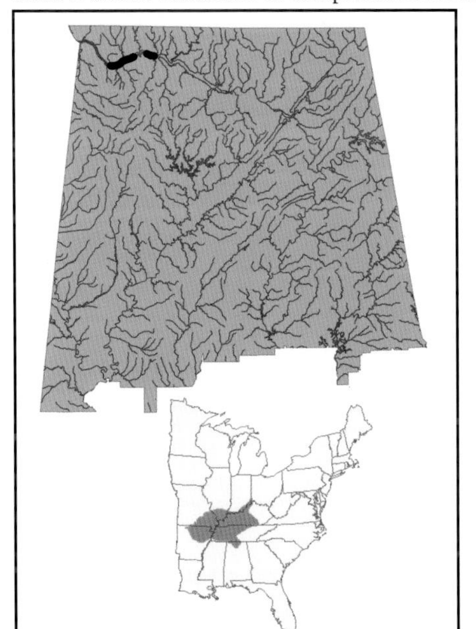

Prepared by: **Arthur E. Bogan**

SKIRTED HORNSNAIL

Pleurocera pyrenella (Conrad)

OTHER NAMES. None.

DESCRIPTION. Shell thick (may approach 40 mm [1 5/8 in.] long), elongately conic in outline, with a high, gradually tapering spire. Whorls flattened and sutures hardly impressed. Body whorl ponderous, with outer margin sharply angled to narrowly rounded. Shell sculpture highly variable, with some shells both carinate and strongly striate, and some carinate only on early whorls and otherwise smooth. Aperture rectangular and patulous, with a moderately to strongly reflected basal lip. Periostracum dark brown to black. Juveniles usually carinate and/or striate. (Modified from Tryon 1873)

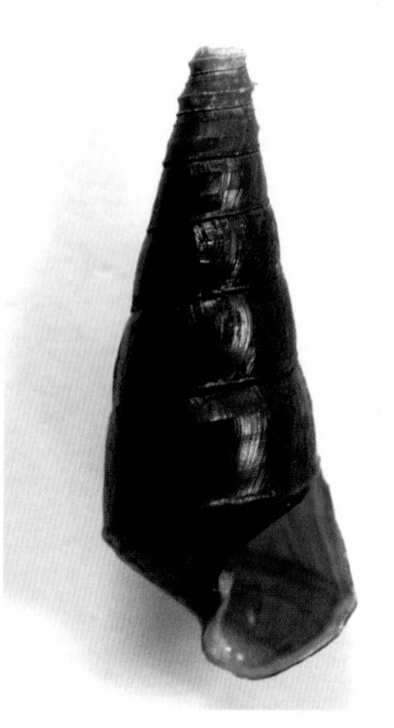

Photo—Arthur Bogan

DISTRIBUTION. Endemic to tributaries of Tennessee River in north-central Alabama. Was historically reported from tributaries in parts of Limestone, Madison, and Morgan Counties (Burch 1989). However, none found south of Tennessee River, Morgan County, during a survey in mid-1990s (Aquatic Resources Center 1997). Although separable from *Pleurocera brumbyi/currierianum* complex based on aperture shape, additional taxonomic work needed to better understand group.

HABITAT. Variety; generally in areas without swift current. Found in pools, runs, and occasionally riffles, as well as swampy streams and impounded springs. Appears tolerant of silty conditions.

LIFE HISTORY AND ECOLOGY. Unknown.

BASIS FOR STATUS CLASSIFICATION. Limited distribution and vulnerability to habitat degradation make *P. pyrenella* vulnerable to extinction. Area in which it occurs almost exclusively used for intense agriculture, making its habitat susceptible to degradation from input of pesticides and fertilizers, as well as excessive irrigation and sedimentation. Also, urban sprawl of Huntsville, Alabama, encroaching on distribution and may present future problems with point and nonpoint source pollution.

Prepared by: **Jeffrey T. Garner**

ALABAMA FRESHWATER SNAIL WATCH LIST

MODERATE CONSERVATION CONCERN

Taxa with conservation problems because of insufficient data, OR because of two of four of the following: small population; limited, disjunct, or peripheral distribution; decreasing population trend/population viability problem; specialized habitat need/habitat vulnerability due to natural/human-caused factors. Research and/or conservation action recommended.

COMMON NAME	SCIENTIFIC NAME	PROBLEM(s)
	Order Architaenioglossa	
	Family Viviparidae	
OVATE CAMPELOMA	*Campeloma geniculum*	Insufficient data on distribution
FURROWED LIOPLAX	*Lioplax sulculosa*	nsufficient data on distribution
	Family Ampullariidae	
FLORIDA APPLESNAIL	*Pomacea paludosa*	Insufficient data on distribution
	Order Neotaenioglossa	
	Family Pleuroceridae	
ACUTE ELIMIA	*Elimia acuta*	Limited distribution/Habitat vulnerability
MUD ELIMIA	*Elimia alabamensis*	Insufficient data on distribution
WALNUT ELIMIA	*Elimia bellula*	Insufficient data on distribution
RUSTY ELIMIA	*Elimia bentonensis*	Insufficient data on distribution
RIPPLED ELIMIA	*Elimia caelatura*	Limited distribution/Habitat vulnerability
PRUNE ELIMIA	*Elimia chiltonensis*	Insufficient data on distribution
SLACKWATER ELIMIA	*Elimia clenchi*	Limited distribution/Habitat vulnerability
COMMA ELIMIA	*Elimia comma*	Limited distribution/Habitat vulnerability
GRAPHITE ELIMIA	*Elimia curvicostata*	Limited distribution/Habitat vulnerability
CYLINDER ELIMIA	*Elimia cylindracea*	Decreasing populationtrend/ Habitat vulnerability
STATELY ELIMIA	*Elimia dickinsoni*	Limited distribution/Habitat vulnerability
BANDED ELIMIA	*Elimia fascians*	Insufficient data on distribution
GLADIATOR ELIMIA	*Elimia hydei*	Limited distribution/Habitat vulnerability
SLOWWATER ELIMIA	*Elimia interveniens*	Limited distribution/Habitat vulnerability
PANEL ELIMIA	*Elimia laqueata*	Limited distribution/Habitat vulnerability
OAK ELIMIA	*Elimia mutabilis*	Insufficient data on distribution
CAPER ELIMIA	*Elimia olivula*	Decreasing population trend/ Habitat vulnerability
SOOTY ELIMIA	*Elimia paupercula*	Insufficient data on distribution
SPRING ELIMIA	*Elimia pybasi*	Insufficient data on distribution
COMPACT ELIMIA	*Elimia showalteri*	Insufficient data on distribution
DENTED ELIMIA	*Elimia taitiana*	Insufficient data on distribution
SLOUGH ELIMIA	*Elimia viennaensis*	Insufficient data on distribution
SPIRAL HORNSNAIL	*Pleurocera brumbyi*	Insufficient data on distribution
SHORTSPIRE HORNSNAIL	*Pleurocera curta*	Insufficient data on distribution
NOBLE HORNSNAIL	*Pleurocera nobilis*	Insufficient data on distribution
BROKEN HORNSNAIL	*Pleurocera postelli*	Insufficient data on distribution
UPLAND HORNSNAIL	*Pleurocera showalteri*	Insufficient data on distribution
SULCATE HORNSNAIL	*Pleurocera trochiformis*	Insufficient data on distribution
BROOK HORNSNAIL	*Pleurocera vestita*	Insufficient data on distribution
TELESCOPE HORNSNAIL	*Pleurocera walkeri*	Insufficient data on distribution

ALABAMA FRESHWATER SNAIL WATCH LIST, continued

COMMON NAME	SCIENTIFIC NAME	PROBLEM(s)
	Family Hydrobiidae	
MANITOU CAVESNAIL	*Antrorbis breweri*	Insufficient data on distribution
GLOBE SILTSNAIL	*Birgella subglobosa*	Insufficient data on distribution
ALLIGATOR SILTSNAIL	*Notogillia wetherbyi*	Insufficient data on distribution
COOSA PYRG	*Pyrgulopsis hershleri*	Insufficient data on distribution
TEARDROP SNAIL	*Rhapinema dacryon*	Insufficient data on distribution
GOLDEN PEBBLESNAIL	*Somatogyrus aureus*	Insufficient data on distribution
ANGULAR PEBBLESNAIL	*Somatogyrus biangulatus*	Insufficient data on distribution
KNOTTY PEBBLESNAIL	*Somatogyrus constrictus*	Insufficient data on distribution
COOSA PEBBLESNAIL	*Somatogyrus coosaensis*	Insufficient data on distribution
STOCKY PEBBLESNAIL	*Somatogyrus crassus*	Insufficient data on distribution
TENNESSEE PEBBLESNAIL	*Somatogyrus currierianus*	Insufficient data on distribution
HIDDEN PEBBLESNAIL	*Somatogyrus decipens*	Insufficient data on distribution
OVATE PEBBLESNAIL	*Somatogyrus excavatus*	Insufficient data on distribution
CHEROKEE PEBBLESNAIL	*Somatogyrus georgianus*	Insufficient data on distribution
FLUTED PEBBLESNAIL	*Somatogyrus hendersoni*	Insufficient data on distribution
GRANITE PEBBLESNAIL	*Somatogyrus hinkleyi*	Insufficient data on distribution
ATLAS PEBBLESNAIL	*Somatogyrus humerosus*	Insufficient data on distribution
DWARF PEBBLESNAIL	*Somatogyrus nanus*	Insufficient data on distribution
MOON PEBBLESNAIL	*Somatogyrus obtusus*	Insufficient data on distribution
TALLAPOOSA PEBBLESNAIL	*Somatogyrus pilsbryanus*	Insufficient data on distribution
COMPACT PEBBLESNAIL	*Somatogyrus pumilus*	Insufficient data on distribution
PYGMY PEBBLESNAIL	*Somatogyrus pygmaeu*	Insufficient data on distribution
QUADRATE PEBBLESNAIL	*Somatogyrus quadratus*	Insufficient data on distribution
MUD PEBBLESNAIL	*Somatogyrus sargenti*	Insufficient data on distribution
ROLLING PEBBLESNAIL	*Somatogyrus strengi*	Insufficient data on distribution
CHOCTAW PEBBLESNAIL	*Somatogyrus substriatus*	Insufficient data on distribution
OPAQUE PEBBLESNAIL	*Somatogyrus tennesseensis*	Insufficient data on distribution
GULF COAST PEBBLESNAIL	*Somatogyrus walkerianus*	Insufficient data on distribution
SCULPIN SNAIL	*Stiobia nana*	Insufficient data on distribution
	Order Heterostropha	
	Family Valvatidae	
TWO-RIDGE VALVATA	*Valvata bicarinata*	Insufficient data on distribution
	Order Basommatophora	
	Family Planorbidae	
DISC SPRITE	*Micromenetus brogniartianus*	Insufficient data on distribution
THICKLIP RAMS-HORN	*Planorbula armigera*	Insufficient data on distribution
	Family Ancylidae	
HOOD ANCYLID	*Ferrissia mcneili*	Insufficient data on distribution
CREEPING ANCYLID	*Ferrissia rivularis*	Insufficient data on distribution
CYMBAL ANCYLID	*Laevapex diaphanous*	Insufficient data on distribution
DOMED ANCYLID	*Rhodacmea elatior*	Insufficient data on distribution
WICKER ANCYLID	*Rhodacmea filosa*	Insufficient data on distribution
KNOBBY ANCYLID	*Rhodacmea hinkleyi*	Insufficient data on distribution

References Cited

Adams, C. C. 1915. The variations and ecological distribution of the snails of the genus *Io*. Memoirs Acad. Nat. Sci. 12:1-184.

Ahlstedt, S. A. 1995. Status survey for federally listed endangered freshwater mussels in the Paint Rock River system, northeastern Alabama. Walkerana 8:63-80.

_____, and C. Saylor. 1995. Status survey of the little-wing pearly mussel, *Pegias fabula* (Lea, 1838). Walkerana 8:81-105.

Aquatic Resources Center. 1997. Survey for selected species of gastropods in the Tennessee River drainage of northern Alabama. Unpubl. report for U.S. Fish and Wildl. Serv., Asheville, NC.

Athearn, H. D. 1964. Three new unionids from Alabama and Florida and a note on *Lampsilis jonesi*. The Nautilus 77:134-139.

_____. 1970. Discussion of Dr. Heard's paper. *Pp*. 28-31 *in* A. H. Clarke, ed. Proceedings of the American Malacological Union symposium on rare and endangered mollusks. Malacologia 10:28-31.

Baird, M. S. 2000. Life history of the spectaclecase, *Cumberlandia monodonta* Say, 1819 (Bivalvia, Unionoidea, Margaritiferidae). M.S. Thesis, Southwest Missouri State Univ., Springfield, MO. 108 pp.

Baker, F. C. 1928. The fresh water Mollusca of Wisconsin. Part II. Pelecypoda. Bull.Wisconsin Geological and Natural History Survey 70:1-495.

Blalock-Herod, H. N. 2000. Community ecology of three freshwater mussel species (Bivalvia: Unionidae) from the New River, Suwanee drainage, Florida. Unpubl. M.S. Thesis, Univ. of Florida. 72 pp.

_____, J. J. Herod, and J. D. Williams. 2000. Freshwater mussels of the Choctawhatchee River drainage in Alabama and Florida, Gainesville, FL. U. S. Geological Survey Florida Caribbean Science Center Open-File Report. 60 pp.

_____, _____, and _____. 2002. Evaluation of conservation status, distribution and reproductive characteristics of an endemic Gulf Coast freshwater mussel, *Lampsilis australis* (Bivalvia: Unionidae). Biodiversity and Conservation 11: 1877-1887.

_____, _____, _____, B. N. Wilson, and S. W. McGregor. In press. A historical and current perspective of the freshwater mussel fauna (Bivalvia: Unionidae) of the Choctawhatchee River drainage in Alabama and Florida.

Bogan, A. E., and P. W. Parmalee. 1983. Tennessee's rare wildlife, volume II: the mollusks. Tennessee Wildlife Resources Agency, Nashville, TN. 123 pp.

_____, and J. M. Pierson. 1993*a*. Survey of the aquatic gastropods of the Cahaba River Basin, Alabama: 1992. Unpubl. report, Alabama Natural Heritage Program, Montgomery, AL. 82 pp.

_____. 1993*b*. Survey of the aquatic gastropods of the Coosa River Basin, Alabama: 1992. Unpubl. report, Alabama Natural Heritage Program, Montgomery, AL. 109 pp.

Brim Box, J., and J. D. Williams. 2000. Unionid mollusks of the Apalachicola Basin in Alabama, Florida and Georgia. Bull. Ala. Mus. Nat. Hist. 21:1-143.

Britton, J. C., and S. L. H. Fuller. 1979. The freshwater bivalve Mollusca (Unionidae, Sphaeriidae, Corbiculidae) of the Savannah River Plant, South Carolina. Unpubl. report, Savannah River Plant, U.S. Department of Energy, Aiken, SC. 37 pp.

Brown, K. M., and J. P. Curole. 1997. Longitudinal changes in the mussels of the Amite River: endangered species, effects of gravel mining and shell morphology. Pp. 236-246 *in*: K. S. Cummings, A. C. Buchanan, C. A. Mayer and T. J. Naimo, eds. Conservation and management of freshwater mussels II: initiatives for the future. Proc. UMRCC symposium, October 1995, St. Louis, MO. Upper Mississippi River Conservation Committee, Rock Island, IL. 293 pp.

Bruenderman, S. A., and R. J. Neves. 1993. Life history of the endangered fine-rayed pigtoe *Fusconaia cuneolus* (Bivalvia: Unionidae) in the Clinch River, Virginia. Am. Malacological Bull. 10:83-91.

Burch, J. B. 1975. Freshwater unionacean clams (Mollusca: Pelecypoda) of North America. U. S. Environmental Protection Agency, Identification Manual No. 11. 204 pp.

_____. 1989. North American freshwater snails. Malacological Publications, Hamburg, MI. 365 pp.

Butler, R. S. 1989. Distributional records for freshwater mussels (Bivalvia: Unionidae) in Florida and south Alabama, with zoogeographic and taxonomic notes. Walkerana 3:239-261.

Clarke, A. H. 1981. The tribe Alasmidontini (Unionidae: Alasmidontinae), Part 1: *Pegias*, *Alasmidonta* and *Arcidens*. Smithsonian Contributions to Zoology No. 326. 101 pp.

Clench, W. J. 1962. New records for the genus *Lioplax*. Occasional Papers on Mollusks. Museum of Comparative Zoology 2:288.

_____, and R. D. Turner. 1955. The North American genus *Lioplax* in the family Viviparidae. Occasional Papers on Mollusks. Museum of Comparative Zoology 2:1-20.

_____, and R. D. Turner. 1956. Freshwater mollusks of Alabama, Georgia and Florida from the Escambia to the Suwannee River. Bull. Florida State Museum, Biological Sciences 1:97-239.

Coker, R. E., A. F. Shira, H.W. Clark, and A. D. Howard. 1921. Natural history and propagation of fresh-water mussels. Bull. U.S. Bureau of Fisheries 37:77-181.

Cummings, K. S., and C. A. Mayer. 1992. Field guide to freshwater mussels of the Midwest. Illinois Natural History Survey Manual 5. 194 pp.

Fuller, S. L. H. 1974. Clams and mussels (Mollusca: Bivalvia). *Pp.* 215-273 *in*: C. W. Hart and S. L. H. Fuller, eds. Pollution ecology of freshwater invertebrates. Academic Press, New York, NY. 389 pp.

_____, and D. J. Bereza. 1973. Recent additions to the naiad fauna of the eastern Gulf drainage (Bivalvia: Unionida: Unionidae). Assoc. Southeastern Biologists Bull. 20:53. (Abstract)

Garner, J. T., and S. W. McGregor. 2001. Current status of freshwater mussels (Unionidae, Margaritiferidae) in the Muscle Shoals area of the Tennessee River in Alabama (Muscle Shoals revisited again). American Malacological Bull. 16:155-170.

Gooch, C. H., W. J. Pardue, and D. C. Wade. 1979. Recent mollusk investigations on the Tennessee River: 1978. Unpubl. report, Tennessee Valley Authority, Muscle Shoals, AL, and Chattanooga, TN. 126 pp.

Goodrich, C. 1922. The Anculosae of the Alabama River drainage. Univ. of Michigan Museum of Zoology Miscellaneous Publications No. 7. 63 pp.

_____. 1936. *Goniobasis* of the Coosa River, Alabama. Univ. of Michigan Museum of Zoology Miscellaneous Publications No. 31. 62 pp.

_____. 1940. The Pleuroceridae of the Ohio River drainage system. Occasional Papers of the Museum of Zoology, Univ. of Michigan No. 417. 21 pp.

_____. 1941. Distribution of the gastropods of the Cahaba River, Alabama. Occasional Papers of the Museum of Zoology, Univ. of Michigan No. 428. 30 pp.

_____. 1944. Pleuroceridae of the Coosa River Basin. The Nautilus 58:40-48.

Gordon, M. E. 1991. Survey of freshwater mollusca of Shoal Creek, Tennessee and Alabama. Tennessee Wildl. Resour. Agency, Nashville, TN, and Tennessee Cooperative Fishery Unit, Tennessee Technological Univ., Cookeville, TN. Unpubl. Rept. 14 pp.

Haag, W. R., R. S. Butler, and P. D. Hartfield. 1995. An extraordinary reproductive strategy in freshwater bivalves: prey mimicry to facilitate larval dispersal. Freshwater Biology 34:471-476.

_____, and M. L. Warren, Jr. 1997. Host fish and reproductive biology of six freshwater mussel species from the Mobile Basin, USA. J. North Am. Benthological Soc. 16:576-585.

_____. 1999. Mantle lures of freshwater mussels elicit attacks from fishes. Freshwater Biology 42:35-40.

_____. 2003. Host fishes and infection strategies of freshwater mussels in large Mobile Basin streams, USA. J. North Am. Benthological Soc. 22:78-91.

_____, and M. Shillingsford. 1999. Host fishes and host attracting behavior of *Lampsilis altilis* and *Villosa vibex* (Bivalvia: Unionidae). Am. Midl. Natur. 141:149-157.

_____, K. Wright and L. Shaffer. 2002. Occurrence of the rayed creekshell, *Anodontoides radiatus* in the Mississippi River Basin: implications for conservation and biogeography. Southeastern Naturalist 1:169-178.

Hartfield, P. D., and R. L. Jones. 1989. The status of *Epioblasma penita*, *Pleurobema curtum* and *P. taitianum* in the East Fork Tombigbee River 1988. Mississippi Museum of Natural Science Technical Report No. 4. 44 pp.

Heller, J. 1990. Longevity in molluscs. Malacologia 31:259-295.

Hershler, R. 1994. A review of the North American freshwater snail genus *Pyrgulopsis* (Hydrobiidae). Smithsonian Contributions to Zoology No. 554. 115 pp.

_____, J. M. Pierson, and R. S. Krotzer. 1990. Rediscovery of *Tulotoma magnifica* (Conrad) (Gastropoda: Viviparidae). Pro. Biological Society of Washington 103:815-824.

Howells, R. G., R. W. Neck, and H. D. Murray. 1996. Freshwater mussels of Texas. Texas Parks and Wildlife Press, Austin, TX. 218 pp.

Hurd, J. C. 1974. Systematics and zoogeography of the unionacean mollusks of the Coosa River drainage of Alabama, Georgia and Tennessee. Ph.D. Diss., Auburn Univ., Auburn, AL. 240 pp.

Isom, B. G. 1969. The mussel resource of the Tennessee River. Malacologia 7:397-425.

_____, S. D. Dennis, and C. Gooch. 1979. Rediscovery of pleurocerids (Gastropoda) near Muscle Shoals, Tennessee River, Alabama. The Nautilus 93:69-70.

_____, and P. Yokley. 1973. The mussels of the Flint and Paint Rock River systems of the southwest slope of the Cumberland Plateau in north Alabama. Am. Midl. Natur. 89:442-446.

Jenkinson, J. J. 1973. Distribution and zoogeography of the Unionidae (Mollusca: Bivalvia) in four creek systems in east-central Alabama. M.S. Thesis, Auburn Univ., Auburn, AL. 96 pp.

Johnson, P. D. 2002. An inventory of *Leptoxis plicata*, the Plicate Rocksnail in the Locust Fork of the Black Warrior River, Jefferson County, Alabama. Report for U.S. Fish and Wildl. Serv., Jackson, MS. 22 pp.

_____, and R. R. Evans. 2000. A freshwater gastropod survey of selected drainages in the Chattooga River, Oostanaula River and the Big Cedar Creek basins in northwest Georgia. Report for Georgia Natural Heritage Program. 45 pp.

Johnson, R. I. 1967. Additions to the unionid fauna of the Gulf drainage of Alabama, Georgia and Florida (Mollusca: Bivalvia). Breviora No. 270. 21 pp.

_____. 1969. Further additions to the unionid fauna of the Gulf drainage of Alabama, Georgia and Florida. The Nautilus 83:34-35.

_____. 1970. The systematics and zoogeography of the Unionidae (Mollusca: Bivalvia) of the southern Atlantic Slope region. Bull. Museum of Comparative Zoology 140:263-449.

_____. 1977. Monograph of the genus *Medionidus* (Bivalvia: Unionidae) mostly from the Apalachicolan Region, southeastern United States. Occasional Papers on Mollusks, Museum of Comparative Zoology 4:161-187.

_____. 1978. Systematics and zoology of *Plagiola* (= *Dysnomia* = *Epioblasma*), an almost extinct genus of freshwater mussels (Bivalvia: Unionidae) from middle North America. Bull. Museum of Comparative Zoology 148:239-321.

_____. 1983. *Margaritifera marrianae*, a new species of Unionacea (Bivalvia: Margaritiferidae) from the Mobile-Alabama-Coosa and Escambia River systems, Alabama. Occasional Papers on Mollusks, Museum of Comparative Zoology 4: 299-304.

Jones, R. L. 1991. Population status of endangered mussels in the Buttahatchee River, Mississippi and Alabama, Segment 2, 1990. Mississippi Museum of Natural Science Technical Report No. 14. 51 pp.

_____, and R. J. Neves. 2002*a*. Life history and artificial culture of endangered mussels. Unpubl. report, Virginia Cooperative Fish and Wildlife Research Unit, Blacksburg, VA. 90 pp.

_____. 2002*b*. Life history and propagation of the endangered fanshell pearlymussel, *Cyprogenia stegaria* Rafinesque (Bivalvia: Unionidae). J. North Am. Benthological Soc. 21:76-88.

Keller, A. E., and D. S. Ruessler. 1997. Determination or verification of host fish for nine species of unionid mussels. Am. Midl. Natur. 138:402-407.

Kitchel, H. E. 1985. Life history of the endangered shiny pigtoe pearly mussel, *Fusconaia edgariana*, in the North Fork Holston River, Virginia. M.S. Thesis, Virginia Polytechnic Institute and State Univ., Blacksburg, VA. 124 pp.

Layzer, J. B., and L. M. Madison. 1995. Microhabitat use by freshwater mussels and recommendations for determining their instream flow needs. Regulated Rivers: Research and Management 10:329-345.

Lea, I. 1831. Observations on the naiades, and descriptions of new species of that and other families. Trans. Am. Philosophical Soc. 4 (New Series):63-121.

_____. 1838. Descriptions of new fresh water and land shells. Trans. Am. Philosophical Soc. 6 (New Series):1-154.

_____. 1840. Descriptions of new fresh water and land shells. Proc. Am. Philosophical Soc. 1:284-289.

_____. 1868. Observations on the genus Unio; together with descriptions of new species in the family Unionidae, and descriptions of new species of the Melanidae and Paludinae. Proc. Academy of Natural Sciences, Philadelphia, PA., vol. 12.

Lefevre, G., and W. C. Curtis. 1910. Studies on the reproduction and artificial propagation of freshwater mussels. Bull. U.S. Bureau of Fisheries 30:107-201.

Luo, M. 1993. Host fishes of four species of freshwater mussels and development of an immune response. M.S. Thesis, Tennessee Technological Univ., Cookeville, TN. 32 pp.

Lydeard, C., and R. L. Mayden. 1995. A diverse and endangered aquatic ecosystem of the Southeast United States. Conservation Biology 9:800-805.

_____, J. T. Garner, P. D. Hartfield, and J. D. Williams. 1999. Freshwater mussels in the Gulf region: Alabama. Gulf of Mexico Science 2:125-134.

_____, R. L. Minton, and J. D. Williams. 2000. Prodigious polyphyly in imperiled pearly-mussels (Bivalvia: Unionidae): a phylogenetic test of species and generic designations. *Pp.* 145-158 *in* E. M. Harper, J. D. Taylor, and J. A. Crame, eds. The evolutionary biology of the Bivalvia. Geological Society of London Special Publication 177:145-158.

Madison, L. M., and J. B. Layzer. 2000. Propagation of *Lampsilis abrupta* and *Lampsilis ovata*. Unpubl. report, Tennessee Technological Univ., Cookeville, TN. 8 pp.

Mancini, R. E. 1978. The biology of *Goniobasis semicarinata* (Say) (Gastropoda: Pleuroceridae) in the Mosquito Creek drainage system, southern Indiana. Ph.D. Diss., Univ. of Louisville, Louisville, KY. 92 pp.

Mathiak, H. A. 1979. A river survey of the unionid mussels of Wisconsin, 1973-1977. Sand Shell Press, Horicon, WI. 75 pp.

McCullagh, H. W., J. D. Williams, S. W. McGregor, J. M. Pierson, and C. Lydeard. 2001. The Unionid (bivalvia) fauna of the Sipsey River in northwestern Alabama, an aquatic hotspot. Amer. Malacological Bull. 17 (Nos. 1 & 2):1-15.

McGregor, S. W., and W. R. Haag. In press. Freshwater mussels (Bivalvia: Unionidae) and habitat conditions in the upper Tombigbee River system, Alabama and Mississippi, 1993-2001. Bull. 10, Geological Survey of Alabama.

_____, P. E. O'Neil and J. M. Pierson. 2000. Status of the freshwater mussel (Bivalvia: Unionidae) fauna in the Cahaba River system, Alabama. Walkerana 11: 215-237.

_____, and J. T. Garner. In press. Changes in the freshwater mussel (Bivalvia: Unionidae) fauna of the Bear Creek system of northwest Alabama and northeast Mississippi.

_____, and J. M. Pierson. 1999. Recent freshwater mussel (Bivalvia: Unionacea) records from the North River system, Fayette and Tuscaloosa Counties, Alabama. J. Ala. Acad. Sci. 70:153-162.

_____, and D. N. Shelton. 1995. A qualitative assessment of unionid fauna of the headwaters of the Paint Rock and Flint Rivers of north Alabama and adjacent areas of Tennessee, 1995. Geological Survey of Alabama open file rept. to Alabama Dept. Conserv. Nat. Resour. 23 pp.

Mettee, M. F., P. E. O'Neil, and J. M. Pierson. 1996. Fishes of Alabama and the Mobile Basin. Oxmoor House, Birmingham, AL. 820 pp.

Minton, R. 2002. A cladistic analysis of *Lithasia* (Gastropoda: Pleuroceridae) using morphological characters. The Nautilus 116: 39-49.

_____, J. T. Garner, and C. Lydeard. In preparation. Rediscovery, systematic position, and redescription of "*Leptoxis melanoides*" (Conrad, 1834) (Gastropoda: Pleuroceridae) of the Black Warrior River, Alabama, U.S.A.

Morrison, J. P. E. 1942. Preliminary report on mollusks found in the shell mounds of the Pickwick Landing Basin in the Tennessee River Valley. Bureau of Am. Ethnology Bull. 129:339-392.

Mott, S., and P. Hartfield. 1994. Status review summary of the Alabama pearlshell, *Margaritifera marrianae*. U.S. Fish and Wildl. Serv., Jackson, MS. 6 pp.

Neves, R. J. 1991. Mollusks, Pp. 251-319 *in* K. Terwilliger, coordinator. Virginia's endangered species. Proceedings of a symposium. McDonald and Woodward Publishing Co., Blacksburg, Virginia. 672 pp.

O'Brien, C. A. 1997. Reproductive biology of *Amblema neislerii, Eliptoideus sloatianus, Lampsilis subangulata, Medionidus penicillatus* and *Pleurobema pyriforme* (Bivalvia: Unionidae). Unpubl. report, U.S. Fish and Wildl. Serv., Jacksonville, FL. 70 pp.

_____, and J. Brim Box. 1999. Reproductive biology and juvenile recruitment of the shinyrayed pocketbook, *Lampsilis subangulata* (Bivalvia: Unionidae) in the Gulf Coastal Plain. Am. Midl. Natur. 142:129-140.

_____, and J. D. Wlliams. 2002. Reproductive biology of four freshwater mussels (Bivalvia: Unionidae) endemic to eastern Gulf Coastal Plain drainages of Alabama, Florida and Georgia. Amer. Malacological Bull. 17:147-158.

Oesch, R. D. 1995. Missouri naiades: a guide to the mussels of Missouri. Missouri Department of Conservation, Jefferson City, MO. 271 pp.

Ortmann, A. E. 1909. The breeding season of Unionidae in Pennsylvania. The Nautilus 22:91-95.

_____. 1912. Notes upon the families and genera of the najades. Annals of the Carnegie Museum 8:222-365.

_____. 1914. Studies in najades. The Nautilus 28:63-69.

_____. 1915. Studies in najades. The Nautilus 28:106-108, 141-143.

_____. 1916. The anatomy of *Lemiox rimosus* (Raf.). The Nautilus 30:39-41.

_____. 1919. A monograph of the naiades of Pennsylvania. Part III: systematic account of the genera and species. Memoirs of the Carnegie Museum No. 8. 321 pp.

_____. 1921. The anatomy of certain mussels from the upper Tennessee. The Nautilus 34:81-91.

_____. 1924*a*. Notes on the anatomy and taxonomy of certain Lampsilinae from the Gulf drainage. The Nautilus 37: 99-105, 137-144.

_____. 1924*b*. The naiad-fauna of Duck River in Tennessee. Amer. Midl. Natur. 9:18-62.

_____. 1925. The naiad-fauna of the Tennessee River system below Walden Gorge. Am. Midl. Natur. 9:321-372.

_____, and B. Walker. 1922. On the nomenclature of certain North American naiades. Occasional Papers of the Museum of Zoology Univ. of Michigan No. 112. 75 pp.

Parmalee, P. W., and A. E. Bogan. 1998. The freshwater mussels of Tennessee. Univ. of Tennessee Press, Knoxville, TN. 328 pp.

Pierson, J. M. 1991. Status survey of the southern clubshell, *Pleurobema decisum* (Lea, 1831). Mississippi Museum of Natural Science Technical Report No. 13. 91 pp.

Roe, K. J., A. M. Simons, and P. D. Hartfield. 1997. Identification of a fish host of the inflated heel-splitter *Potamilus inflatus* (Bivalvia: Unionidae) with a description of its glochidium. Am. Midl. Natur. 138:48-54.

_____, and C. Lydeard. 1998. Molecular systematics of the freshwater mussel genus *Potamilus* (Bivalvia: Unionidae). Malacalogia 39:195-205.

_____, P. D. Hartfield, and C. Lydeard. 2001. Phylogeographic analysis of the threatened and endangered superconglutinate-producing mussels of the genus *Lampsilis* (Bivalvia: Unionidae). Molecular Ecology 10:2225-2234.

Rogers, S. O., B. T. Watson, and R. J. Neves. 2001. Life history and population biology of the endangered tan riffleshell (*Epioblasma florentina walkeri*) (Bivalvia: Unionidae). J. North Am. Benthological Soc. 20:582-594.

Shelton, D. N. 1997. Observations on the life history of the Alabama pearl shell, *Margaritifera marrianae* R. I. Johnson, 1983. Pp. 26-29 *in* K. S. Cummings, A. C. Buchanan, C. A. Mayer and T. J. Naimo, eds. Proceedings of a UMRCC Symposium, Conservation and Management of Freshwater Mussels II, Initiatives for the Future. 293 pp.

Simpson, C. T. 1900. New and unfigured Unionidae. Proc. Academy of Natural Sciences Philadelphia, PA. 52:74-86.

_____. 1914. A descriptive catalogue of the naiades, or pearly freshwater mussels. Privately published by Bryant Walker, Detroit, MI. 1540 pp.

Stansbery, D. H. 1970. Eastern freshwater mollusks (I). The Mississippi and St. Lawrence River systems. Malacologia 10: 9-22.

_____. 1971. Rare and endangered molluscs in the eastern United States. Pp. 5-18 *in* S. E. Jorgensen and R. W. Sharpe, eds. Proceedings of a symposium on rare and endangered mollusks (naiads) of the United States. U.S. Fish and Wildl. Serv., Twin Cities, MN. 79 pp.

_____. 1976*a*. Naiad mollusks. Pp. 42-52 *in* H. T. Boschung, ed. Endangered and threatened plants and animals of Alabama. Bull. Ala. Mus. Nat. Hist. 2. 92 pp.

_____. 1976*b*. Status of endangered fluviatile mollusks in central North America: *Pegias fabula* (Lea, 1838). Unpubl. report, The Ohio State Museum of Zoology. 8 pp.

_____. 1983*a*. The status of *Epioblasma penita* (Conrad, 1834) (Mollusca: Bivalvia: Unionoida). Unpubl. report, U.S. Fish and Wildl. Serv., Jackson, MS. 15 pp.

_____. 1983*b*. The status of *Lampsilis perovalis* (Conrad, 1834) (Mollusca: Bivalvia: Unionoida). Unpubl. report, U.S. Fish and Wildl. Serv., Jackson, MS. 20 pp.

_____. 1983*c*. The status of *Pleurobema taitianum* (Lea, 1834) (Mollusca: Bivalvia: Unionoida). Unpubl. report, U.S. Fish and Wildl. Serv., Jackson, MS. 10 pp.

Stein, C. B. 1976. Gastropods. Pp. 21-41 *in* H. Boschung, ed. Endangered and threatened plants and animals of Alabama. Bull. Ala. Mus. Nat. Hist. 2. 92 pp.

Strayer, D. L., and K. J. Jirka. 1997. The pearly mussels of New York State. Memoirs of the New York State Museum No. 26. 113 pp.

Surber, T. 1912. Identification of the glochidia of freshwater mussels. U.S. Bureau of Fisheries Document No. 771. 10 pp.

_____. 1913. Notes on the natural hosts of fresh-water mussels. Bull. U.S. Bureau of Fisheries 32:101-116.

Tennessee Valley Authority. 1986. Cumberlandian Mollusk Conservation Program, Activity 3: identification of fish hosts. Tennessee Valley Authority, Knoxville, TN. 57 pp.

Thompson, F. G. 1977. The hydrobiid snail genus *Marstonia*. Bull. Florida State Mus., Biol. Sci. 21:113-158.

_____. 1984. North American freshwater snail genera of the hydrobiid subfamily Lithoglyphinae. Malacologia 25:109-141.

_____, and R. Hershler. 2002. Two genera of North American snails: *Marstonia* Baker, 1926, resurrected to generic status, and *Floridobia*, new genus (Prosobranchia: Hydrobiidae:Nymphophilinae). The Veliger 45:269-271.

Tryon, J. W., Jr. 1873. Land and fresh-water shells of North America, Part IV. Strepomatidae. Smithsonian Miscellaneous Collections No. 253. 435 pp.

Turgeon, D. D., J. F. Quinn, Jr., A. E. Bogan, E. V. Coan, F. G. Hochberg, W. G. Lyons, P. M. Mikkelsen, C. F. E. Roper, G. Rosenberg, B. Roth, A. Scheltema, M. J. Sweeney, F. G. Thompson, M. Vecchione, and J. D. Williams. 1998. Common and scientific names of aquatic invertebrates from the United States and Canada: mollusks. Second edition. Am. Fisheries Soc. Special Publication 26. Bethesda, MD. 526 pp.

U.S. Fish and Wildife. Service.. 1983*a*. Dromedary pearly mussel (*Dromus dromas*) recovery plan. Atlanta, GA. 58 pp.

_____. 1983*b*. Birdwing pearly mussel (*Conradilla caelata*) recovery plan. Atlanta, GA. 56 pp.

_____. 1997. Anthony's riversnail (*Athearnia anthonyi*) recovery plan. Atlanta, GA. 21 pp.

_____. 2000. Mobile River Basin aquatic ecosystem recovery plan. Atlanta, GA. 129 pp.

_____. 2001. Endangered species facts: scaleshell mussel. Unpubl. report, Twin Cities, MN.

van der Schalie, H. 1934. *Lampsilis jonesi*, a new naiad from southeastern Alabama. The Nautilus 47:125-127.

_____. 1939. Additional notes on the naiades (fresh-water mussels) of the lower Tennessee River. Am. Midl. Natur. 22:452-457.

_____. 1940. The naiad fauna of the Chipola River, in northwestern Florida. Lloydia 3:191-206.

_____. 1981. Mollusks of the Alabama River drainage, past and present. Sterkiana 71:24-41.

Vidrine, M. F. 1996. River survey of freshwater mollusks of Tensas River system in northeastern Louisiana. Report to Natural Heritage Program, Louisiana Department of Wildlife and Fisheries, Baton Rouge, LA. 135 pp.

Weaver, L. R., G. B. Pardue, and R. J. Neves. 1991. Reproductive biology and fish hosts of the Tennessee clubshell *Pleurobema oviforme* (Mollusca: Unionidae) in Virginia. Am. Midl. Natur. 126:82-89.

Weiss, J. L., and J. B. Layzer. 1995. Infections of glochidia on fishes in the Barren River, Kentucky. Am. Malacological Bull. 11:153-159.

Williams, J. D. 1982. Distribution and habitat observations of selected Mobile Basin unionid mollusks. *Pp.* 61-85 *in* A. C. Miller, ed. Report of freshwater mollusks workshop, 19-20 May 1981. U.S. Army Corps of Engineers, Waterways Experiment Station, Vicksburg, MS. 166 pp.

_____, H. N. Blalock-Herod, A. J. Benson, and D. N. Shelton. In review. Distribution and conservation assessment of the freshwater mussel fauna (Bivalvia: Margaritiferidae and Unionidae) in the Escambia and Yellow River drainages in southern Alabama and western Florida.

_____, and R. S. Butler. 1994. Class Bivalvia, freshwater bivalves. *Pp.* 53-128, 740-742 *in* R. Ashton, ed. Rare and endangered biota of Florida. Volume 6. Invertebrates. Univ. Press of Florida , Gainesville, FL. 798 pp.

_____, S. L. H. Fuller, and R. Grace. 1992. Effects of impoundment of freshwater mussels (Mollusca: Bivalvia: Unionidae) in the main channel of the Black Warrior and Tombigbee Rivers in western Alabama. Bull. Ala. Mus. Nat. Hist. 13:1-10.

_____, M. L. Warren, Jr., K. S. Cummings, J. L. Harris, and R. J. Neves. 1993. Conservation status of freshwater mussels of the United States and Canada. Fisheries 18:6-22.

Wilson, C. B. 1916. Copepod parasites of fresh-water fishes and their economic relations to mussel glochidia. Bull. U.S. Bureau of Fisheries 34:331-374.

_____, and H. W. Clark. 1914. The mussels of the Cumberland River and its tributaries. U.S. Bureau of Fisheries Document No. 781. 63 pp.

Yeager, B. L., and R. J. Neves. 1986. Reproductive cycle and fish hosts of the rabbit's foot mussel, *Quadrula cylindrica strigillata* (Mollusca: Bivalvia: Unionidae), in the upper Tennessee River drainage. Am. Midl. Natur. 116:329-340.

_____, and C. F. Saylor. 1995. Fish hosts of four species of freshwater mussels (Pelecypoda: Unionidae) in the upper Tennessee River drainage. Am. Midl. Natur. 133:1-6.

Yokley, P. 1972. Freshwater mussel ecology, Kentucky Lake, Tennessee. Tennessee Game and Fish Comm., Nashville, TN. 133 pp.

Young, D. 1911. The implantation of the glochidium on the fish. Univ. of Missouri Bull., Science Series 2:1-16.

Zale, A. V., and R. J. Neves. 1982*a*. Fish hosts of four species of lampsiline mussels (Mollusca: Unionidae) in Big Moccasin Creek, Virginia. Can. J. Zool. 62:2535-2542.

_____, and R. J. Neves. 1982*b*. Reproductive biology of four freshwater mussel species (Mollusca: Unionidae) in Virginia. Freshwater Invertebrate Biology 1:17-28.

FISHES

Photo: Bruce Bauer

Ashy Darter
Etheostoma cinereum

INTRODUCTION

Alabama's rivers and streams are inhabited by one of the richest fish faunas in North America, numbering around 300 freshwater and 50 estuarine species. This exceptional biodiversity is no accident. A unique combination of geographical, climatic, and hydrologic conditions working through geologic time provides a near ideal environment supportive of aquatic communities. Alabama has a mild average annual temperature (16-20°C [60-68°F]), an abundant supply of rainfall (160 centimeters [63 inches] in the south to 127 centimeters [50 inches] in the north), and 124,000 kilometers (77,000 miles) of perennial and intermittent streams that flow over a complex geologic terrain. This state is drained by well-sustained streams that have excellent surface-water quality and support a variety of aquatic habitats. In addition to its native species, Alabama is inhabited by numerous species that have invaded or been introduced from regional centers of freshwater fish speciation and distribution including the Atlantic Coast, Tennessee River, and Mississippi River.

Alabama's fish diversity has remained relatively constant over the past century, but subtle changes have become increasingly obvious in recent years. High-lift navigational and hydroelectric dams have inhibited upstream spawning migrations of several large river species, such as the anadromous Gulf sturgeon, Alabama shad, and striped bass (Gulf Coast population), into the upper Alabama, Cahaba, Coosa, Tombigbee, and Black Warrior Rivers. Remaining populations of these species are limited to the downstream unimpounded sections of the Alabama, Choctawhatchee, Conecuh, Yellow, and Tombigbee Rivers and the Mobile Delta. Dams with gated spillways have separated once continuous populations of many riverine species such as the paddlefish into disjunct units that are only infrequently reunited when these facilities are inundated by record floods. Maintenance dredging of inland waterways has replaced most permanent sand and gravel bars that once provided ideal feeding and spawning habitat for many riverine species with relatively sterile sand bars that erode during spring floods. Species affected by these types of habitat alteration include the frecklebelly madtom and crystal, freckled, and rock darters in the Alabama, Black Warrior, and Tombigbee Rivers. Within-bank dredging operations occasionally block stream mouths. In these instances, smaller, as well as some game, species are denied river access for long periods of time and, in summer months, are subjected to elevated water temperatures and depressed dissolved oxygen levels.

Tailwater habitat, water quality conditions, and biological communities are adversely affected by pulsed releases from hydroelectric dams. Sediments and eutrophication have adversely impacted fish populations in many watersheds throughout the state. Continued industrial and urban growth in larger cities will likely increase the need for additional water supplies and in-stream waste disposal, both of which can adversely affect fish communities. Resolution of interstate "water wars" could alter river discharges, affect current water-quality regimes, and possibly diminish preferred habitat for fish species in some Alabama rivers. Without careful planning and management, these factors will adversely affect Alabama's diverse fish fauna in future years.

In July 2002, the Alabama Department of Conservation and Natural Resources and Auburn University sponsored a two-day conference to discuss the status of vertebrate and aquatic mussel and snail species inhabiting Alabama. Eighteen biologists from five states and 13 state and federal agencies, universities, industry, and conservation groups reviewed the status of 317 freshwater and estuarine fish species known to occur in state waters. Two species within this group are extinct, nine are extirpated, and 10 are exotics, leaving 296 species of freshwater and estuarine/marine fishes for consideration. The status of 115 species was discussed in detail after which committee members ranked each from Highest to Low Conservation Concern by simple majority vote. Twenty-two species received a ranking of Highest Conservation Concern, 25 were High Conservation Concern, 33 were Moderate Conservation Concern, and 35 were Low Conservation Concern. The remaining fishes (181) were ranked as Lowest Conservation Concern. The inclusion of 11 recognized but as-yet undescribed species in the list offers conclusive proof that additional work is needed before the extent of Alabama's fish diversity is completely known.

The continued perilous nature of Alabama's fish fauna is evidenced by the fact many species treated in the following accounts were consistently listed in earlier publications on endangered, threatened, and special-concern animals of Alabama. Twenty-five of 41 extant species listed as endangered, threatened, or special concern by Ramsey (1976) were ranked in the Highest and High Conservation Concern categories by the 2002 fish committee and seven species were ranked in each of the Moderate and Low categories. Fourteen of 16 species listed by Ramsey (1986*a*) as in need of special attention were ranked in the Highest and High categories by the 2002 fish committee; the remaining two species received a Moderate rating. The good news from this comparison is that no new species have been added to Alabama's list of extirpated species. The bad news is the conservation status of 39 of Alabama's most imperiled species has not changed in 27 years and the current list now includes 47 species, the largest in our state's history.

Most species included in the following accounts have very restricted distributions. Twenty-one species, including several cave- and spring-dwelling forms, only occur in the Tennessee River drainage, 16 are limited to the Mobile River Basin, and six inhabit only coastal drainages. The remaining four species occupy limited areas across two drainages. Bear, Cypress, and Shoal Creeks and the Elk and Paint Rock Rivers are home to most imperiled species in the Tennessee River drainage. Several stream systems in Locust and Sipsey Forks, the upper Black Warrior, Cahaba, Coosa, and Tallapoosa Rivers contain many nominated Mobile River Basin species. Several coastal species, particularly those inhabiting the Chattahoochee drainage, appear to be in serious condition because their distributions have dwindled considerably in the past 20 years.

Future recovery of impaired species will require full cooperation of conservation-minded stakeholders, state and federal agencies, industry and businesses, and the general population. Failure to address this responsibility in a conscientious way soon could mean Alabama will forever loose yet another part of its aquatic biological heritage.

There are a few abbreviations/terms used throughout this section that are unique. Where not obvious, or described here, the definitions of these terms can be found in the glossary at the end of the text. They include and involve: fish measurements - TL = total length, SL = standard length, FL = fork length; water flow measurements - mgd = millions of gallons per day; statistical designations - SD = standard deviation, *n* = sample size. Finally, fish have pharyngeal throat teeth that are bony modifications of the fifth gill arch. Pharyngeal arches contain one to three rows of teeth, with the major row containing the largest and greatest number. A dentition formula based on these teeth can be used to help in identifying various taxa. A formula of 2, 4-4, 2 means that the left arch has two (2) teeth in the minor row and four (4) teeth in the major row. The right arch has four (4) teeth in the major row and two (2) in the minor row.

Maurice F. Mettee

AUTHORS

Dr. Maurice F. (Scott) Mettee, Geological Survey of Alabama, P.O. Box 869999, Tuscaloosa, AL 35486-6999, Co-editor/Co-compiler

Dr. Patrick E. O'Neil, Geological Survey of Alabama, P.O. Drawer 869999, Tuscaloosa AL 35486-6999, Co-editor/Co-compiler

Dr. Henry L. Bart, Tulane University, Royal D. Suttkus Museum of Natural History, Belle Chase, LA 70037

Dr. Paul D. Blanchard, Department of Biology, Samford University, Birmingham, AL 35229

Dr. Herbert T. Boschung, Department of Biological Sciences, Box 870345, University of Alabama, Tuscaloosa, AL 35487-0345.

Mr. Daniel J. Drennen, U.S. Fish and Wildlife, 6578 Dogwood View Parkway, Jackson, MS 39213.

Dr. Robert W. Hastings, Alabama Natural Heritage Program, The Nature Conservancy, Huntingdon College, Montgomery, AL 36106

Dr. Carol E. Johnston, Department of Fisheries and Allied Aquacultures, Auburn University, Auburn, AL 36849

Dr. Bernard R. Kuhajda, Department of Biological Sciences, Box 870345, University of Alabama, Tuscaloosa, AL 35487-0345

Dr. Richard Mayden, Department of Biological Sciences, Box 870345, University of Alabama, Tuscaloosa, AL 35487-0345

Mr. Frank M. Parauka, U.S. Fish and Wildlife Service, Panama City, FL 32405

Mr. J. Malcolm Pierson, Alabama Power Company, GSC #8, Birmingham, AL 35291

Mr. Thomas E. Shepard, Geological Survey of Alabama, P.O. Box 869999, Tuscaloosa, AL 35486-6999

Dr. John R. Shute, Conservation Fisheries, Inc., Knoxville, TN 37919

Mrs. Peggy Shute, TVA National Heritage Project, P.O. Box 1589, Norris TN 37828-1589

Dr. Robert A. Stiles, Department of Biology, Samford University, Birmingham, AL 35229

Dr. Melvin L. Warren, Jr., Center for Bottomland Hardwoods Research, Southern Research Station, USDA Forest Service, 1000 Front Street, Oxford, MS 38655

Fishes

EXTINCT

Taxa that historically occurred in Alabama, but are no longer alive anywhere within their former distribution.

Class Actinopterygii
Order Cypriniformes
Family Catostomidae

HARELIP SUCKER *Moxostoma lacerum*

Order Cyprinodontiformes
Family Fundulidae

WHITELINE TOPMINNOW *Fundulus albolineatus*

FISHES

EXTIRPATED

Taxa that historically occurred in Alabama, but are now absent; may be rediscovered in state, or be reintroduced from populations existing outside state.

Class Actinopterygii
Order Acipenseriformes
Family Acipenseridae

LAKE STURGEON *Acipenser fulvescens*
SHOVELNOSE STURGEON *Scaphirhynchus platorynchus*

Order Lepisosteiformes
Family Lepisosteidae

SHORTNOSE GAR *Lepisosteus platostomus*

Order Hiodontiformes
Family Hiodontidae

GOLDEYE *Hiodon alosoides*

Order Cypriniformes
Family Cyprinidae

SPOTFIN CHUB *Erimonax monachus*
POPEYE SHINER *Notropis ariommus*

Order Siluriformes
Family Ictaluridae

ELEGANT MADTOM *Noturus elegans*

Order Perciformes
Family Percidae

ASHY DARTER *Etheostoma cinereum*
TRISPOT DARTER *Etheostoma trisella*

LAKE STURGEON

Acipenser fulvescens Rafinesque

OTHER NAMES. None.

Photo—Rob Criswell

DESCRIPTION. A large (may exceed 2.7 m [9 ft.] TL), yellowish brown to gray fish with five rows of bony plates along sides and back; ventral surface white to cream. Caudal fin heterocercal, possibly with filamentous extension on small individuals. Head large and shovel-shaped. Mouth located beneath short, conical-shaped snout; lower lip with two pronounced lobes. Four smooth barbels occur on underside of snout in front of mouth. Spiracle present on either side of head near anterior end of gill slit.

DISTRIBUTION. Originally widespread throughout the Mississippi River Basin from Louisiana north into Canada and in tributaries to the Great Lakes (Etnier and Starnes 1993, Mettee *et al.* 1996, Robison and Buchanan 1988). Individuals reported from the Tennessee drainage as far south as northern Alabama. Known in Alabama from collections at two sites in the Coosa River in 1949 (Scott 1951) and three sites in the Tennessee River proper from 1938 to 1941 (Mettee *et al.* 1996).

HABITAT. Large lakes and rivers.

LIFE HISTORY AND ECOLOGY. Life history and spawning behavior well documented in upper Mississippi Basin but unknown in Alabama. Adults living in northern rivers typically migrate more than 160 kilometers (100 miles) upstream to spawn (Etnier and Starnes 1993). Males arrive at shallow-water spawning sites in flowing streams first, followed by gravid females. Lake populations may spawn over rocky ledges in lakes, or migrate a short distance into larger, flowing tributaries. Spawning occurs from April through June when water temperatures range from 12-18°C (54-64°F). A receptive female is usually flanked by two males. Individuals do not spawn every year. Older females produce as many as 3,000,000 eggs in a single year; usual number ranges from 50,000 to 700,000 (Etnier and Starnes 1993). Diet includes crayfishes, mollusks, and insect larvae, particularly midges (Robison and Buchanan 1988). Growth very slow; individuals usually do not reach sexual maturity until they are 15 to 20 years old. Under optimum conditions, may live longer than 100 years.

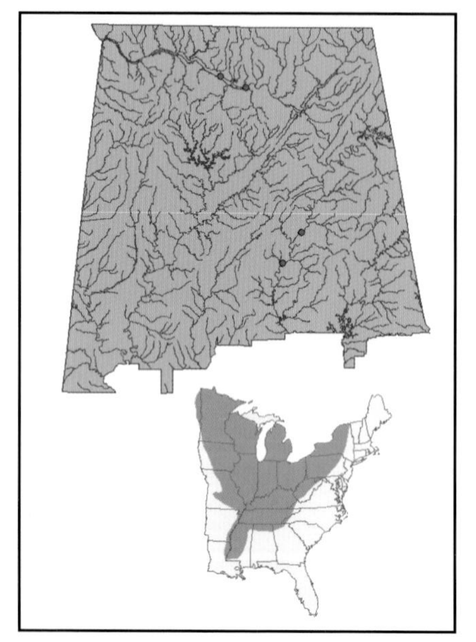

BASIS FOR STATUS CLASSIFICATION. Flesh and eggs have been prized by commercial markets for centuries. Over-fishing produced a "boom or bust" industry for this species in the 1800s and early 1900s. Dams on major rivers have blocked annual spawning migrations. Increased sedimentation from navigational dredging and other nonpoint sources has eliminated many upstream spawning habitats. Hatchery-raised lake sturgeon have been reintroduced into parts of the species' original Mississippi River distribution; the long-term success of these efforts is unknown at this time. May be extirpated from Alabama since specimens have not been collected in more than 60 years.

Prepared by: **Maurice F. Mettee**

SHOVELNOSE STURGEON

Scaphirhynchus platorynchus (Rafinesque)

Photo—*Patrick E. O'Neil*

OTHER NAMES. Hackleback or Sand Sturgeon.

DESCRIPTION. Shovelnose Sturgeons (genus *Scaphirhynchus*) (max. adult length = 0.9 m [3.0 ft.]) can be distinguished from their much larger relatives (genus *Acipenser*) by a broad, flattened, shovel-shaped snout (*vs.* a conical, rounded snout), four lobes on lower lip (*vs.* two lobes), no spiracles on head (*vs.* a spiracle on either side), and four fringed barbels (*vs.* smooth barbels) in front of the inferior mouth. Dorsal area varies from beige or gray to brown; belly is white to cream. Five rows of scutes occur along back and sides of body. Caudal fin heterocercal, possibly with a long filamentous extension on smaller individuals.

DISTRIBUTION. Widespread and occasionally abundant in the Mississippi River Basin. A disjunct population formally inhabited the Rio Grande River. Previously known from the lower Tennessee River in Tennessee (Etnier and Starnes 1993) and northern Alabama. Several hundred pounds were harvested from the Tennessee River in northern Alabama in the late 1800s (Smith 1898). Existing state records include collections near Wilson Dam in 1937 and a 1940 collection near Decatur (Mettee *et al.* 1996).

HABITAT. Large, flowing rivers. Some researchers believe this species may not survive in lakes and impoundments, but it appears to be silt tolerant in Arkansas where individuals seek shelter behind wing dams or other structures during floods and move into river channels in summer (Robison and Buchanan 1988).

LIFE HISTORY AND ECOLOGY. Unknown in Alabama. Gravid adults migrate upstream in the Mississippi River from April to June and spawn in swift current over hard substrates when water temperatures reach 20-21°C (67-70°F) (Becker 1983; Robison and Buchanan 1988). Females may produce from 10,000 to 50,000 eggs. Males reach sexual maturity in four to five years and females in six to seven years. Individuals spawn every two to three years. Primary dietary items in Mississippi River include aquatic insect larvae—particularly midges—terrestrial insects, mollusks, worms, and small fishes (Ross 2001). Estimated life span is 15 years (Etnier and Starnes 1993).

BASIS FOR STATUS CLASSIFICATION. Overharvest for food and caviar decimated populations in the early 1900s, and the fishery eventually collapsed. Construction of high-lift navigational and hydroelectric dams on many U.S. rivers effectively blocked annual migrations to traditional spawning grounds (Coker 1930*a*). Populations never rebounded in the northern part of distribution, although populations appear to be doing well in lower Mississippi River. Collections from two locations in northern Alabama represent the southern extent of distribution in the Tennessee drainage (Mettee *et al.* 1996). Failure to collect this species in more than 60 years could indicate it has been extirpated from state waters.

Prepared by: **Maurice F. Mettee**

SHORTNOSE GAR

Lepisosteus platostomus Rafinesque

Photo—Patrick E. O'Neil

OTHER NAMES. Broadnose Gar, Stubnose Gar, Shortbill Gar, Duckbill Gar, Billy Gar.

DESCRIPTION. An elongated (max. length = 1.0 m [39 in.]), cylindrical-shaped fish with a short, broad snout; upper and lower jaws contain a single row of large, pointed teeth. Body covered with hard ganoid scales; lateral line contains 59 to 64 scales. Dorsal and anal fins have scattered black spots and are located near posterior end of body. Dorsal surface light to dark brown; sides yellow to beige, and ventral surface cream to white. Shortnose and spotted gars are very similar in physical appearance and have been confused in past, but the two can be distinguished by the presence or absence of spots on snout and head. Spotted gar usually has spots on its snout and anterior body, although the spots may be less conspicuous when collected from turbid water; shortnose gar lacks spots on snout and anterior body.

DISTRIBUTION. Widespread and occasionally abundant in the Mississippi River basin, including Tennessee River drainage, and western tributaries to Lake Michigan (Becker 1983, Etnier and Starnes 1993). Has been collected at three locations in northern Alabama, but recent publications on fishes of Alabama (Mettee *et al.* 1996) and Tennessee (Etnier and Starnes 1993) suggest records from upstream sections of Tennessee drainage may be based on poorly marked spotted gar.

HABITAT. Flowing rivers and larger tributaries, but more abundant in reservoirs, oxbow lakes, and backwaters having sand and mud substrates.

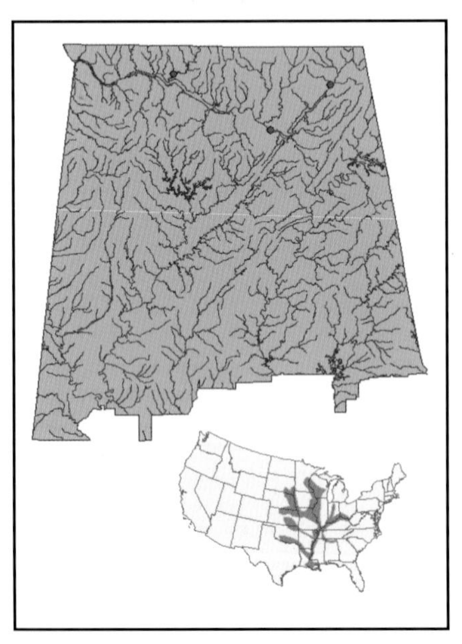

LIFE HISTORY AND ECOLOGY. Unknown in Alabama. This schooling species spawns from May through June in upper part of its distribution when water temperatures reach 19-24°C (66.2-75.2°F) (Becker 1983, Robison and Buchanan 1988). Females usually spawn with two or more males. Greenish yellow eggs contained in a gelatinous mass are deposited on submerged structures (Becker 1983, Pflieger 1975). This opportunistic feeder will consume other fishes, crayfish, and aquatic insects. Life span can be 10 to 12 years in rivers and up to 20 years in captivity (Ross 2001).

BASIS FOR STATUS CLASSIFICATION. Occurrence in northern Alabama is questionable. Existing records could be based on misidentified spotted gar.

Prepared by: **Maurice F. Mettee**

GOLDEYE

Hiodon alosoides (Rafinesque)

Photo—Bill Pflieger

OTHER NAMES. Winnepeg Goldeye, Western Goldeye, Yellow Herring, Toothed Herring, Shad Mooneye, Northern Mooneye.

DESCRIPTION. An elongate (may reach 400 mm [18 in.] long), compressed, deep-bodied fish with a large mouth and sharp teeth. Dorsal area on live individuals bluish green, sides iridescent silver, and venter white. Lateral line contains 57 to 62 scales. Eye distinctly large with a golden iris. Anal fin contains 29 to 34 rays, making it one of longest found on an Alabama fish species. Ventral edge of anal fin convex or lobed on males, and straight to slightly concave on females. Two features distinguish the goldeye from its nearest relative the mooneye. The dorsal fin originates directly over, or slightly behind, anal fin origin on goldeye, and in front of anal fin origin on mooneye. A fleshy abdominal keel extends from the pectoral fin base to the anus on goldeye, and from the pelvic fin base to the anus on mooneye.

DISTRIBUTION. Mississippi River, its major tributaries, and part of Hudson Bay and Mackenzie River drainage in Canada. Small populations previously documented from the upper Tennessee and Cumberland Rivers apparently disappeared after these rivers were impounded (Etnier and Starnes 1993). Extant populations in Tennessee are restricted to the Mississippi River proper. Tennessee Valley Authority biologists collected goldeyes at four locations in northern Alabama from 1937-40 (Mettee *et al*. 1996).

HABITAT. Medium to large silty rivers characterized by a moderate to swift current and a mixture of silt and sand substrates (Etnier and Starnes 1993, Pflieger 1975). Individuals have been reported from reservoirs in northern sections of species' distribution (Becker 1983). Believed to be more silt-tolerant than mooneye. Its apparent disappearance from impounded rivers in Alabama and Tennessee is unexplained (Etnier and Starnes 1993, Pflieger 1975).

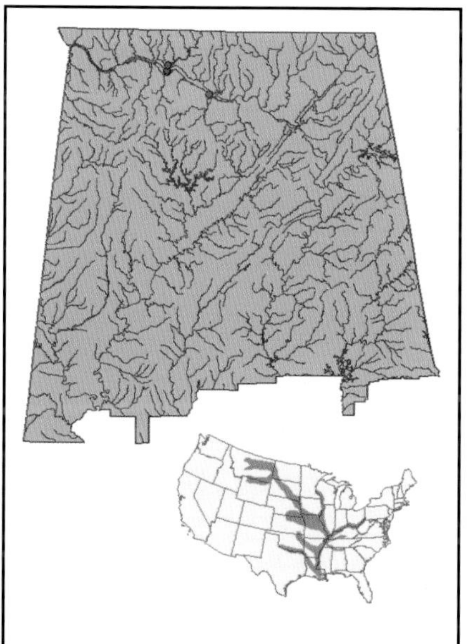

LIFE HISTORY AND ECOLOGY. Spawning occurs in shallow, gravelly shoals and also in pools and backwater areas in early April, possibly at night, in the Mississippi River when water temperatures reach 10-13°C (50-55.4°F) (Robison and Buchanan 1988). Females produce 6,000 to 25,000 eggs that, following fertilization, float until the larvae hatch (Scott and Crossman 1973). Young appear in larval collections taken in May in Mississippi River, but disappear by June (Robison and Buchanan 1988). Larvae feed primarily on zooplankton and transition to aquatic insects and small fishes later in life (Etnier and Starnes 1993). Males become sexually mature in one to two years; females require several additional years to mature. The maximum life span is around 14 years (Scott and Crossman 1973).

BASIS FOR STATUS CLASSIFICATION. Failure to collect this species in more than 60 years suggests it is extirpated from state waters.

Prepared by: **Maurice F. Mettee**

SPOTFIN CHUB

Erimonax monachus (Cope)

Photo—Richard Bryant and Wayne Starnes

OTHER NAMES. Turquoise Shiner, Turquoise Chub.

DESCRIPTION. Very distinctive among minnows with elongate body, small, triangular-shaped head, and moderately elongate and rounded or blunt snout. Adults can reach 90 millimeters (3.5 inches) in length (Jenkins and Burkhead 1993). Eye proportionally small for body size. Mouth inferior and often down curved and upper lip expanded along anterior margin. Distinguished by presence of a barbel, 4-4 pharyngeal teeth, and more than 50 lateral-line scales. Nuptial males with distinct iridescent turquoise to cobalt blue wash extending along upper body. Females are generally tan, gray, or olive green dorsally with silver on remaining body. Lateral line scales vary from 52 to 62, predorsal scale rows vary from 24 to 28. Dorsal fin with eight rays, anal fin with seven to eight rays, and caudal fin with 19 principal rays.

DISTRIBUTION. Thought to have occurred in streams and rivers throughout Tennessee River system, but current distribution limited to upper reaches of Tennessee River in Tennessee, Virginia, and North Carolina. Another population also occurs in the Buffalo River system in the lower reach of the Tennessee River system (Etnier and Starnes 1993). In Alabama, it has only been found in two drainage systems: Little Bear Creek (1937) and Shoal Creek (1884). Hundreds of samples taken over intervening years throughout Tennessee River in Alabama have failed to locate populations within the state.

HABITAT. Medium to large streams of moderate to moderately high gradient and moderate current (Jenkins and Burkhead 1984). Adults are found in glides and runs over flat bedrock with scattered boulders, while young occur over gravel and sand patches associated with eddies around boulders and ledges. Recent collection records (TVA Heritage database) indicate possible fall migration into small tributaries. Streams of low turbidity and high water quality appear necessary for survival.

LIFE HISTORY AND ECOLOGY. Populations in upper Tennessee River have been studied extensively (Etnier and Starnes 1993). Spawning somewhat protracted from mid-May to mid-August. Eggs deposited into crevices or cracks in rocks adjacent to the substrate. Even though periodic, the fractional timing of spawning periods may increase overall fecundity compared with other species of minnows. Since there is no parental care, crevices that remain clean enough for successful hatching may be a limiting factor for natural populations. Observations and morphological evidence suggest it is a sight feeder, selecting minute insect larvae from clean substrates. Midge larvae and blackflies comprise major food items with lesser amounts of mayflies, caddisflies, and beetles.

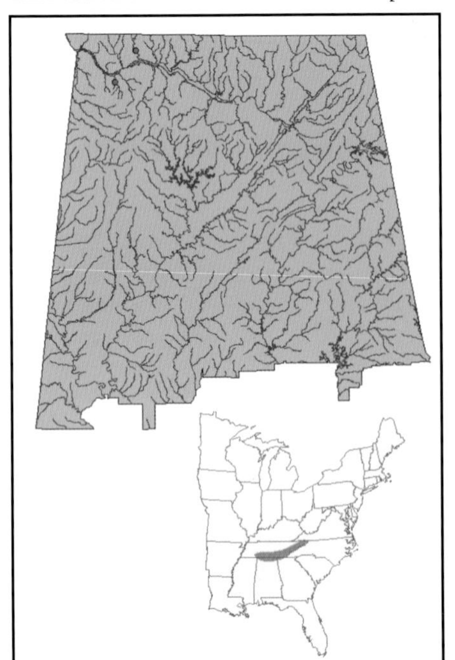

BASIS FOR STATUS CLASSIFICATION. Reasons for decline in some populations are uncertain, but other populations have been adversely affected by dams, runoff from coal mining operations, municipal and industrial wastes, and siltation. Its 65-year absence in Alabama, coupled with no sightings in spite of extensive sampling over the years, leads most to consider this species extirpated in Alabama. **Listed as *threatened* by the U.S. Fish and Wildlife Service in 1977.**

***Prepared by*: Patrick E. O'Neil and Peggy Shute**

POPEYE SHINER

Notropis ariommus (Cope)

OTHER NAMES. None.

Photo—Noel Burkhead

DESCRIPTION. A large (max. size = 100 mm [4 in.] SL) shiner characterized by very large eyes (>1.5 times snout length), a moderately compressed body, and a relatively short snout with a large terminal mouth. Body olive-colored dorsally, and anterior dorsal scales outlined with melanophores. A reflective silver color when alive and viewed from side. A dark lateral band is visible after preservation. Does not develop chromatic breeding colors. Dorsal fin origin located above pelvic fin base. Pharyngeal tooth pattern 2,4-4, 2, usually with nine anal rays and 35-39 lateral line scales. Similar minnows that could be confused with species include telescope, emerald, and striped shiners. Telescope shiner usually has 10 anal rays and a more elongate body. In emerald shiners, dorsal fin origin is behind pelvic fin base. Striped shiners have much smaller eyes and deeper bodies.

DISTRIBUTION. Historically known from major tributaries of Ohio River from Alabama and Georgia north to upper Wabash River drainage in Indiana, and in western Lake Erie drainage in Ohio (Gilbert 1969, 1980). Although widespread, species extirpated in many tributary systems, and present distribution represented by disjunct, localized populations. In Alabama, only record is of five specimens collected in 1884 from Cypress Creek near Florence (Mettee *et al.* 1996). Species known from Elk River in Tennessee (Etnier and Starnes 1993).

HABITAT. Clear, large streams and small rivers over clean gravel substrates. Usually associated with current in runs and flowing pools at depths of 0.5 to 1.5 meters (1.6 to 4.9 feet) (Jenkins and Burkhead 1993). An uncommon species, usually found in low numbers where it occurs.

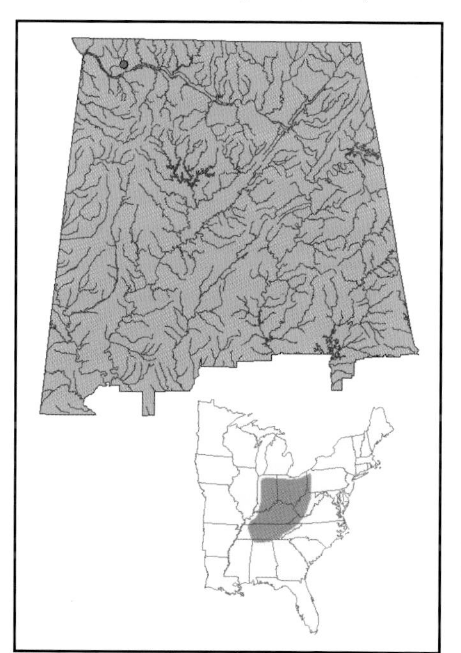

LIFE HISTORY AND ECOLOGY. Little information available. In Virginia, males are reproductively mature at 53 millimeters (2 inches) and females at 49 millimeters (1.9 inches) SL. Spawning occurs in late May to late June based on tubercle size and gonadal development (Jenkins and Burkhead 1993). In Tennessee, diet consists of adult, larval, and pupal insects including coleopterans, dipterans, hymenopterans, trichopterans, and ephemeropterans (Etnier and Starnes 1993). Apparently intolerant of siltation.

BASIS FOR STATUS CLASSIFICATION. Most large tributaries of Tennessee River in Alabama that could potentially harbor species have been well sampled in recent years without success, inferring extirpation from state. Inundation of main channel of Tennessee River as well as lower reaches of major tributaries by Pickwick, Wilson, Wheeler, and Guntersville Dams likely eliminated much habitat. Siltation related to extensive logging throughout Tennessee River Basin in early 1900s also may have contributed to extirpation.

Prepared by: **Thomas E. Shepard**

ELEGANT MADTOM

Noturus elegans Taylor

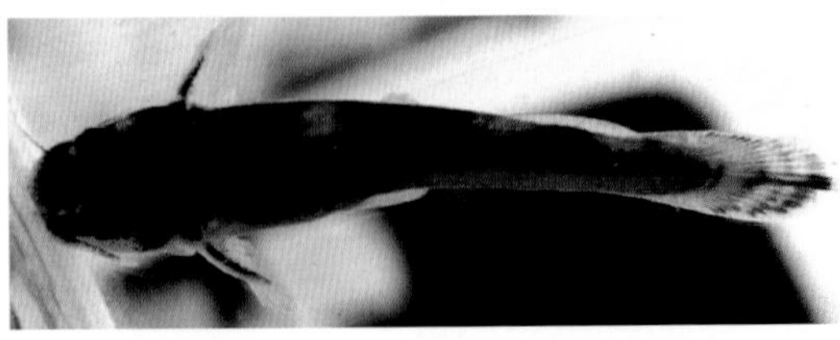
Photo—Richard Mayden

OTHER NAMES. Chucky Madtom.

DESCRIPTION. All madtoms are distinguished from other North American catfish by small adult size (typically <100 mm [4 in. TL]) and by having a long, low, adipose fin joined to, or only slightly separate from, caudal fin. As a member of subgenus *Rabida*, has curved pectoral spines with small teeth protruding from front of spines and large, curved teeth from rear, as well as alternating light and dark dorsal saddles. Can be distinguished from other saddled madtoms by presence of brown blotch at anterior base of dorsal fin that extends along length of dorsal spine, lack of dorsal-fin pigment elsewhere (a few specimens with a weak submarginal band), four pale dorsal saddles (posterior two may be small), and usually 15 or more anal-fin rays. As in other madtoms, males' heads become swollen dorsally during breeding season. Attains total length of 80 millimeters (3.1 inches). Populations in upper Tennessee River Drainage in Alabama and Tennessee represent an undescribed form. Those specimens have a subdued color pattern, two posterior pale saddles are reduced, and body is short and chunky (Taylor 1969, Page and Burr 1991, Etnier and Starnes 1993).

DISTRIBUTION. In Green, Cumberland (one record), and Tennessee River systems in Kentucky, Tennessee, and Alabama. Within Alabama, has been collected from three different localities in Tennessee River Drainage: Piney Creek, Limestone County (1941); West Fork Flint River, Madison County (1969); and Paint Rock River, Jackson County (1981) (Taylor 1969, Etnier and Starnes 1993, Mettee *et al.* 1996). Record in Mettee *et al.* (1996) from Cedar Creek of Bear Creek system based on misidentification of *N. exilis* (B. M. Burr, pers. comm.).

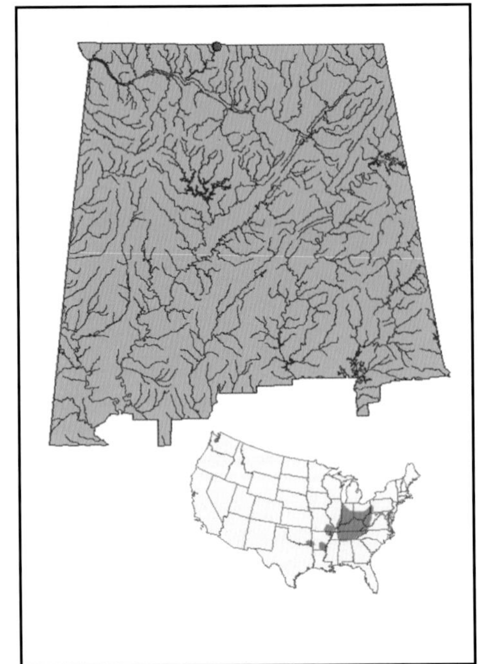

HABITAT. Runs, riffles, and lower end of pools in clear upland creeks and small to moderate-sized rivers with clean sand, gravel, cobble, and bedrock substrates.

LIFE HISTORY AND ECOLOGY. Upper Tennessee River form rare; no more than nine specimens have been collected at one time. No studies have investigated the biology of this undescribed form, but its life history probably similar to elegant madtom in Barren River System in Kentucky (Burr and Dimmick 1981), where nests with eggs and larvae were found under flat rocks above riffles in June at a water temperature of 20°C (68°F). All nests had a guardian parent and up to 30 fry or eggs; two specimens captured were males with empty stomachs. Hypothesized that only male madtoms guard nest and do so without feeding. A clutch of 25 eggs found in a nest hatched in 12 days at 20°C (68°F). Average number of mature oocytes found in eight females (30.8) low compared with other madtoms (more than 50). Diet likely includes benthic aquatic invertebrate larvae and microcrustaceans.

BASIS FOR STATUS CLASSIFICATION. Was collected in Alabama from only three locations represented by 11 specimens; only one or two specimens were collected at two of those sites. Has not been collected in state since 1981 despite attempts by several different organizations specifically searching for it. Most historic sites are in agricultural settings and are impacted by siltation and poor water quality.

Prepared by: **Bernard R. Kuhajda**

ASHY DARTER

Etheostoma cinereum Storer

OTHER NAMES. None

Photo—Bruce Bauer

DESCRIPTION. An uncommon, large (max. size = 100 mm [4 in.] SL), distinctive darter placed in monotypic subgenus *Allohistium.* Has relatively long, pointed snout with thick papillose lips and a broad frenum. Body elongate and slightly compressed with a shallow caudal peduncle. Ground color straw yellow dorsally and cream-colored ventrally. A mid-lateral row of 10 to 13 small black rectangles present, which form an interrupted lateral band on body joining a dark band on head that extends through eye to tip of snout. Rectangles are expanded dorso-ventrally as grayish brown (ashy) diagonal bands on body. Typically are four irregular rows of orange dots on upper sides of body, which form thin horizontal stripes. Seven or eight faint dorsal saddles extend ventrally only as far as uppermost horizontal stripe. Spinous dorsal fin has blood-red marginal band with rust-colored blotches on interradial membranes. In soft dorsal fin, there are several rust-colored blotches at base of interradial membranes, which meld into blood-red bars distally. Soft dorsal fin of mature males greatly elongate, reaching almost to caudal fin base when depressed. Spinous dorsal and anal fins also longer in males than females. Breeding males develop brilliant electric blue on pelvic, anal, and first dorsal fin as well as chin, lower opercle, branchiostegals, breast, belly, and lower caudal peduncle. Populations in Cumberland, Duck, and upper Tennessee River systems somewhat distinctive based on geographic trends in meristic and morphometric characters (Shepard and Burr 1984). Individuals from Cumberland River system unique in having blood-red lips (most intense in breeding males), whereas those from Duck and upper Tennessee River systems do not develop red on lips, even in peak spawning condition. Analysis of mitochondrial DNA data indicates the three populations are genetically divergent and represent unique species (S. L. Powers, Dept. Biological Sciences, Univ. Alabama, Tuscaloosa, pers. comm.).

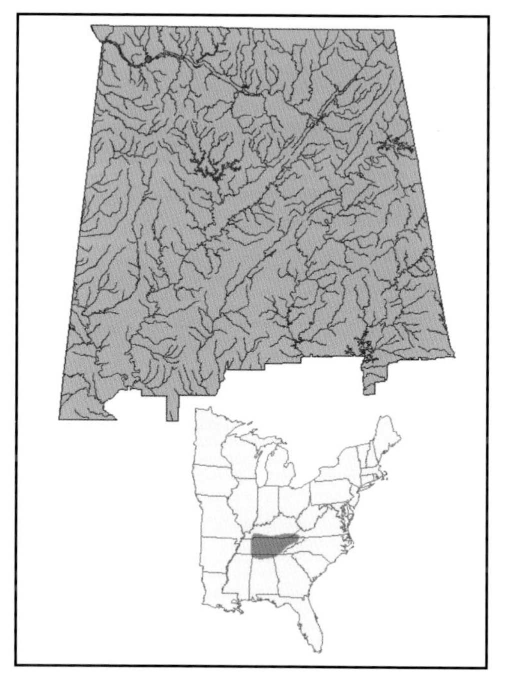

DISTRIBUTION. Known from 15 tributary systems in Cumberland and Tennessee River systems in Kentucky, Tennessee, Alabama, Georgia, and Virginia (Saylor 1980). Has been collected from only nine of these in past 25 years: Rockcastle River, Big South Fork of Cumberland, and Buck Creek, Kentucky; and Little, New (tributary of Big South Fork), Emory, Clinch, Duck, Elk, and Buffalo Rivers, Tennessee (Shepard and Burr 1984, Powers and Mayden 2002). Known from Georgia and Virginia based on single specimens collected from Chickamauga Creek (1955), and the Clinch River (1964), respectively. Clinch River, however, has recently produced species in Tennessee at a locality about 30 river kilometers (19 miles) downstream of Virginia state line and could eventually reappear in Virginia waters. Only Alabama record from original description by D. H. Storer in 1845 based on material collected near Florence, Alabama, by Charles A. Hentz. Exact location where Hentz collected type(s) unknown. Possibly specimen(s) was (were) taken from a large Tennessee River tributary near Florence such as Cypress or Shoal Creek, or from Mussel Shoals area of Tennessee River, all of which could have been inhabited by species. Although species has not been collected in Alabama in more than 150 years, it may yet be rediscovered in state. A single specimen was collected in Elk River by C. F. Saylor and TVA in 1981 near Fayetteville, Tennessee, in a rotenone sample. This section of Elk River, downstream of Tims Ford Dam, is affected by cold water discharges and fluctuating water levels due to power generation. However, since dam was completed in 1970, 11 years before specimen was collected, recruitment must have continued in population after dam began operation. Several miles of the Elk River in Alabama upstream of embayment of Wheeler Lake appear to offer suitable habitat, so this area may present best possibility of finding species in Alabama.

HABITAT. Most frequently collected in medium to large upland streams of moderate gradient, clear, silt-free pools and eddies with sluggish current, at depths of 0.5 to 1.75 meters (1.6 to 5.7 feet). Usually found over substrates of clean sand, gravel, or bedrock in close association with cover in form of boulders, snags, and stands of water willow (Shepard and Burr 1984). Preference for deeper pool habitats associated with cover makes collection with conventional seining techniques difficult and inefficient and may partially explain why species is infrequently collected and usually found in low numbers.

LIFE HISTORY AND ECOLOGY. Maximum life span four years. Males reach reproductive maturity at 50 millimeters (two inches) and females at 55 millimeters (2.1 inches) SL (Shepard and Burr 1984). Most individuals reproductively mature after one year's growth. Spawning period may extend from February to mid-April, with peak spawning in March. During spawning act, both sexes engage in courtship displays leading to male mounting female with spawning on vertical surfaces (Jenkins and Burkhead 1993). Structures such as sides of boulders and water willow stems are probable spawning sites in wild (Etnier and Starnes 1993). Ripe females contain an average of 53 mature ova per gram adjusted body weight. Large females may contain up to 250 mature ova. Major food items are chironomid midge larvae, larvae of burrowing mayfly *Ephemera*, and oligochaetes (Shepard and Burr 1984). Midges represent a greater proportion of diets of small individuals than larger ones. Elongate snout and fleshy, papillose lips may be an adaptation for feeding on burrowing organisms.

BASIS FOR STATUS CLASSIFICATION. Construction of Pickwick, Wilson, Wheeler, and Guntersville Dams on Tennessee River inundated main channel of Tennessee River as well as lower reaches of major tributaries in Alabama and probably eliminated much habitat in Alabama. As a pool-dwelling species, it is especially vulnerable to siltation and may have been impacted by heavy sediment loads associated with logging in early 1900s in Tennessee River Basin. Most major tributaries of Tennessee River in Alabama have been relatively well surveyed in recent years without producing species, and species possibly no longer occurs in state. Species difficult to collect even where good populations exist and easily could be missed by routine sampling.

Prepared by: **Thomas E. Shepard**

TRISPOT DARTER

Etheostoma trisella Bailey and Richards

Photo—Bruce Bauer

OTHER NAMES. None.

DESCRIPTION. A small (max. size = 49 mm [1.9 in.] SL), cylindrical species, which, along with *E. boschungi*, comprises only members of subgenus *Ozarka* within Alabama. Has brown dorsum crossed by three distinct, dark saddles located just anterior to spinous dorsal fin, between spinous and soft dorsal fins, and just posterior to soft dorsal fin. Sides slightly mottled with poorly developed midlateral blotches, and belly white to yellowish. A dark suborbital bar present.

DISTRIBUTION. Endemic to Coosa River drainage of Alabama, Georgia, and Tennessee. Known in Alabama historically from a single specimen collected in 1947 from Cowans Creek, Cherokee County, but habitat in that region was eliminated in 1960 by impoundment of Weiss Reservoir (Bailey and Richards 1963). An additional specimen was taken in 1958 in the Coosa River mainstem, Etowah County, Alabama, but it may have represented a waif surviving in marginal habitat (Ramsey 1986*b*). Subsequent searches in the Coosa River of Alabama did not reveal species (Ramsey 1986*b*, Mettee *et al.* 1996, B. R. Kuhajda, pers. comm.). Also known from uppermost Coosa River in Coosawattee River, Georgia, and Conasauga River, Georgia and Tennessee. Only known reproducing population occupies about 69 river kilometers (43 miles) of upper Conasauga River from above Tennessee-Georgia state line downstream into Whitfield County, Georgia (Etnier and Starnes 1993).

HABITAT. Requires distinct habitats for growth and reproduction. Nonreproductive habitats consist of slackwater areas along moderate size streams, where usually associated with detritus or rooted vegetation. Migration to smaller tributaries begins in late fall and during this migratory period may be associated with riffle areas. Reproductive aggregations form by December adjacent to streams in close association with seepage areas (*e.g.*, flooded marshy areas, pastures, woodlands). During flood season (December to April), darters enter seepage areas to spawn (Ryon 1986, Starnes and Etnier 1993).

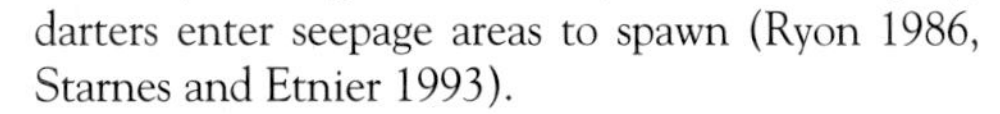

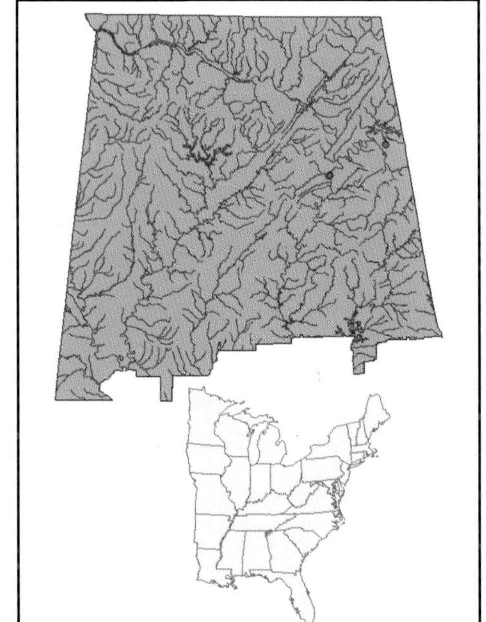

LIFE HISTORY AND ECOLOGY. Within flooded seepage area habitats, spawning behaviors are ritualistic with female broadcasting adhesive eggs that stick to plants or other objects. Male guards spawning site. Eggs hatch in about 30 days at 12 °C (53.6°F). Adults and young presumably emigrate from spawning areas and return downstream to nonreproductive habitats. Longevity less than three years. Diet consists of a variety of invertebrates dominated by midge larvae and mayfly nymphs, but stonefly nymphs, caddisfly larvae, and copepods are important foods during migration (Ryon 1986, Etnier and Starnes 1993).

BASIS FOR STATUS CLASSIFICATION. Southeastern Fishes Council considered trispot darter endangered (Warren *et al.* 2000). No individuals have been taken in Alabama in more than 40 years despite efforts to locate extant populations, thereby inferring extirpation from state.

Prepared by: **Melvin L. Warren, Jr.**

Fishes

PRIORITY 1
HIGHEST CONSERVATION CONCERN

Taxa critically imperiled and at risk of extinction/extirpation because of extreme rarity, restricted distribution, decreasing population trend/population viability problems, and specialized habitat needs/habitat vulnerability due to natural/human-caused factors. Immediate research and/or conservation action required.

Class Actinopterygii
Order Acipenseriformes
Family Acipenseridae

ALABAMA STURGEON *Scaphirhynchus suttkusi*

Order Cypriniformes
Family Cyprinidae

PALEZONE SHINER *Notropis albizonatus*
CAHABA SHINER *Notropis cahabae*
IRONCOLOR SHINER *Notropis chalybaeus*

Order Percopsiformes
Family Amblyopsidae

ALABAMA CAVEFISH *Speoplatyrhinus poulsoni*

Order Scorpaeniformes
Family Cottidae

PYGMY SCULPIN *Cottus paulus*

Order Perciformes
Family Elassomatidae

SPRING PYGMY SUNFISH *Elassoma alabamae*

Family Percidae

SLACKWATER DARTER *Etheostoma boschungi*
HOLIDAY DARTER *Etheostoma brevirostrum*
VERMILION DARTER *Etheostoma chermocki*
BRIGHTEYE DARTER *Etheostoma lynceum*
LOLLIPOP DARTER *Etheostoma neopterum*
WATERCRESS DARTER *Etheostoma nuchale*
RUSH DARTER *Ethesotoma phytophilum*
BOULDER DARTER *Etheostoma wapiti*
GOLDLINE DARTER *Percina aurolineata*
BLOTCHSIDE LOGPERCH *Percina burtoni*
SLENDERHEAD DARTER *Percina phoxocephala*
SNAIL DARTER *Percina tanasi*
WARRIOR BRIDLED DARTER *Percina* sp. cf. *macrocephala*
HALLOWEEN DARTER *Percina* sp.

ALABAMA STURGEON

Scaphirhynchus suttkusi Williams and Clemmer

Photo—Andrew Simmons

OTHER NAMES. Alabama Shovelnose Sturgeon, Buglemouth Trout, Devilfish, Hackleback, Sand Sturgeon, Switchtail.

DESCRIPTION. All sturgeon possess four barbels anterior to a protrusible inferior mouth and have five rows of bony scutes along body. As a member of the genus *Scaphirhynchus*, Alabama sturgeon have a flattened snout, no spiracle, a completely armored caudal peduncle, a long caudal filament extending from upper lobe of caudal fin (may be broken off in adults), and are restricted to freshwater. Differentiated from its closest relative, the allopatric shovelnose sturgeon, by numerous morphological characteristics, including a larger eye, no spines on dorsal tip of snout or just anterior to eye, poorly developed squamation on venter, and a unique brassy, orangish yellow color. No sexual dimorphism. Attains fork length of 78 centimeters (2.6 feet) (Williams and Clemmer 1991, Mayden and Kuhajda 1996).

DISTRIBUTION. Historically restricted to large rivers in Mobile Basin in Alabama and eastern Mississippi, including 1,600 kilometers (1,000 miles) in lower Black Warrior, Cahaba, Coosa, and Tallapoosa Rivers, much of Tombigbee and Tensaw Rivers, and length of Alabama River (Williams and Clemmer 1991, Burke and Ramsey 1995). Over last nine years only 10 specimens captured despite concentrated collecting efforts. Nine specimens (1993-1999) were restricted to the Alabama River below Millers Ferry and Claiborne Locks and Dams in Clarke, Monroe, and Wilcox Counties, Alabama (Mayden and Kuhajda 1996, Federal Register 2000). Last Alabama sturgeon captured (May 2000, P. Kilpatrick, pers. comm.) from Cahaba River was first specimen captured above Millers Ferry Lock and Dam since 1985.

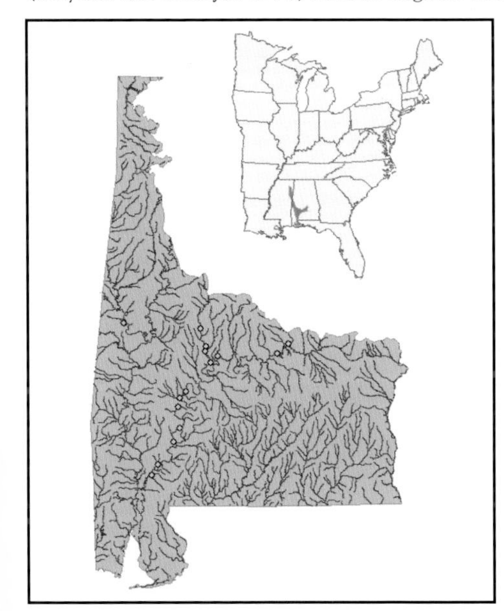

HABITAT. Restricted to main channels of large flowing rivers over relatively stable gravel and sand substrates, but also will occur over softer sediments.

LIFE HISTORY AND ECOLOGY. Spawning habitats and reproductive biology not well known, but captured specimens did produce gametes in March, and gravid females have been collected in March through May. Not sexually mature until five to seven years of age; spawns only once every two to three years. Spawning cues include water temperature and high river discharge; spawning likely occurs over hard-bottomed substrates in main-channel areas. Eggs (2.4-3 mm [0.09-0.12 in.] diameter) are adhesive, indicating a stable and silt-free substrate necessary for successful hatching. Larvae of other *Scaphirhynchus*, and likely those of Alabama sturgeon, require long stretches of flowing water in which to drift and develop successfully. Stomach contents include oligochaetes, fragments of mollusk shells, odonates, ephemerids, and heptageniids. Larval aquatic insect diet indicates foraging occurs in sandy depositional areas with little silt and slow to moderate flow (Williams and Clemmer 1991, Mayden and Kuhajda 1996).

BASIS FOR STATUS CLASSIFICATION. One of the rarest vertebrates in the world, with only 10 specimens collected since 1985, all within Alabama. Restricted to a fraction of former distribution due to loss and fragmentation of habitat resulting from navigational development, and perhaps historically from overfishing. Currently threatened by reduced distribution and small population numbers, blockage of upstream migratory routes by dams, loss of heterogeneic spawning habitat due to channel modifications, water quality degradation and sedimentation, and altered flow regimens affecting spawning cues and larval development (Federal Register 2000). **Listed as *endangered* by the U.S. Fish and Wildlife Service in 2000.**

***Prepared by*: Bernard R. Kuhajda**

PALEZONE SHINER
Notropis albizonatus Warren and Burr

OTHER NAMES. Paleband Shiner, White-Zone Shiner.

Photo—Patrick E. O'Neil

DESCRIPTION. A member of the swallow-tail shiner species group. A slender minnow (max. size = 65 mm [2.6 in.] SL), with relatively large scales (35-39 lateral-line scales), usually with seven anal rays, and with 0,4-4,0 pharyngeal teeth (Warren *et al.* 1994). Body straw-yellow dorsally and lighter yellow to white ventrally. Dorsal scales darkly outlined with melanophores, and a dark (reflective silver-colored in life) lateral stripe extends along body becoming more diffuse around snout. A yellowish lightly pigmented supralateral stripe about two scale rows deep extends along body above lateral stripe, contrasting with dark lateral stripe below and darkly outlined dorsal scales above; species named for this pale zone. A small, triangular, basicaudal spot present at end of lateral stripe, and dark streaks present dorsally and ventrally in caudal fin. Does not develop bright breeding colors. Similar species in Paint Rock River system include mimic and sawfin shiners, bigeye chub, and bluntnose minnow. All these minnows have deeper bodies than palezone shiner, lack lightly pigmented stripe above a dark lateral stripe, and typically have eight rather than seven anal rays.

DISTRIBUTION. Historically known from only four tributary systems: Little South Fork of Cumberland River, Kentucky; Paint Rock River system, Alabama; Marrowbone Creek, Cumberland River system, Kentucky; and Cove Creek, Clinch River system, Tennessee (Warren *et al.* 1994). Known based on records of only single specimens from both Marrowbone (collected 1947) and Cove Creeks (collected 1936). More recent collections in Marrowbone Creek have failed to produce additional specimens. Now considered extirpated in that system, possibly due to effects of cold-water discharges from Lake Cumberland in lower reaches, and deforestation of watershed. Cove Creek population now also considered extirpated due to impoundment by Norris and Cove Lake Dams, and impacts from surface mining. In Little South Fork of Cumberland River, most abundant in lower 10 kilometers (six miles) upstream of embayment of Lake Cumberland, although has been collected in a 48 kilometers (30 miles) reach of the river. Occurs in only about 27 kilometers (16.8 miles) of Paint Rock River system: 8.8 kilometers (5.5 miles) of upper Paint Rock River from river mile 54.5 upstream to origin at confluence of Hurricane Creek and Estill Fork; lower 6.4 kilometers (four miles) of Hurricane Creek; lower 9.3 kilometers (5.8 miles) of Estill Fork; and lower 2.4 kilometers (1.5 miles) of Larkin Fork (Shepard *et al.* 1997). Although sometimes collected in relatively high numbers at some localities, occurrence is sporadic, and apparently suitable habitat within distribution frequently fails to yield specimens.

HABITAT. Usually silt-free runs and flowing pools with sand, gravel, and bedrock substrates; also occasionally taken at base of riffles and in backwaters with little flow (Warren *et al.* 1994). Upper Paint Rock River and Little South Fork of Cumberland are small upland rivers with excellent habitat and water quality that drain relatively undisturbed forested watersheds. Judged by its extremely limited distribution, this species apparently requires high-quality habitats provided by such protected watersheds.

LIFE HISTORY AND ECOLOGY. Based on a life history study in Little South Fork of Cumberland River (Poly 1997) and observations in original description (Warren *et al.* 1994), palezone shiner usually lives to two years although a few individuals may survive to a third year. Diet consists mainly of nematoceran Diptera larvae. Males and females sexually mature at 35 to 40 millimeters (1.4 to 1.6 inches) SL. Spawning period probably from mid-May through late June, possibly into early July, based on presence of gravid females and tuberculate males. Spawning behavior unknown.

BASIS FOR STATUS CLASSIFICATION. Two of four known populations have been extirpated. Now occurs in a very limited distribution in only two disjunct tributary systems (Warren and Burr 1998). Sedimentation and toxic runoff from surface mines, logging, cattle watering, and poorly managed unpaved roads, as well as brine discharge and petroleum leakage from oil wells are concerns in Little South Fork system (Warren *et al.* 1994). In Paint Rock River system, agricultural activity, particularly cattle production, is impacting system. Numerous improvised cattle watering areas in Estill Fork, Larkin Fork, Hurricane Creek, and the Paint Rock River upstream of Princeton produce sediment problems through disruption of banks, and increased nutrient loads from cow manure being deposited directly into streams and as runoff (Shepard *et al.* 1997). Fine sediment deposits are a problem in some pools, and algal turbidity and accumulations of filamentous algae increase progressively in Paint Rock River downstream of junction of Estill Fork and Hurricane Creek. **Listed as *endangered* by the U.S. Fish and Wildlife Service in 1993.**

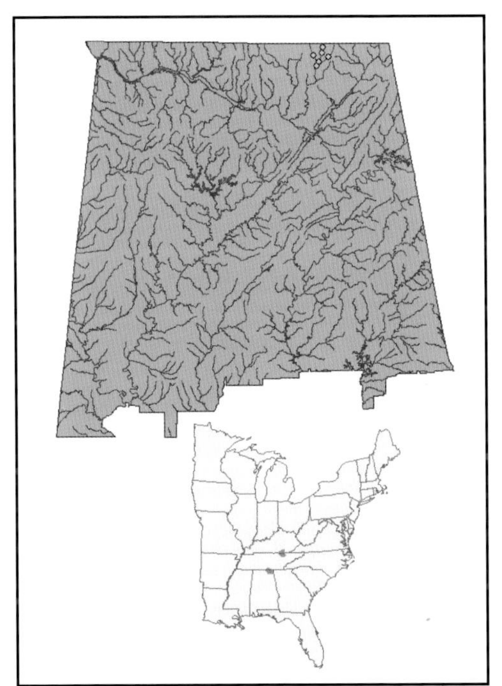

***Prepared by*: Thomas E. Shepard**

CAHABA SHINER

Notropis cahabae Mayden and Kuhajda

OTHER NAMES. None.

Photo—Robert Stiles

DESCRIPTION. A member of the *Notropis volucellus* species group; all are small shiners (to 77 mm [3 in.] TL) with anterior lateral-line scales much taller than wide, a subterminal mouth, eight anal-fin rays, and pharyngeal teeth 0,4-4,0. Distinguished from other members of the species group by possessing a lateral stripe on caudal peduncle with straight dorsal and ventral margins rather than an expanded ventral edge before the wedge-shaped caudal spot. Other characteristics that separate Cahaba Shiners from syntopic Mimic Shiners include a reduced or absent predorsal stripe, dorsal caudal peduncle scales uniformly pigmented and dusky gray rather than more heavily pigmented posteriorly and straw colored anteriorly, and breast scales present rather than absent. Nuptial males have small tubercles on head and along rays on pectoral fins, and pectoral and dorsal fin-rays more heavily pigmented than females (Mayden and Kuhajda 1989). Specimens from two disjunct populations (Cahaba River and Locust Fork of Black Warrior River) show only slight morphological differentiation, but phylogenetic analyses of mitochondrial DNA reveal that these two populations are genetically divergent and are Evolutionary Significant Units (ESUs).

DISTRIBUTION. Restricted to two river systems in Mobile River Basin in Alabama: main stem of Cahaba River and predominately in main channel of Locust Fork. Within Cahaba River, historic distribution includes 140 river kilometers (76 miles) from Helena, Shelby County, to Centreville, Bibb County, with a few waif individuals reported further downstream (below Fall Line) in Bibb and Perry Counties. Present stronghold reduced to only 28 river kilometers (15 miles) between Fall Line at Centreville upstream to Piper Bridge (Mayden and Kuhajda 1989, Shepard *et al.* 1995). Current distribution in Locust Fork system includes 118 river kilometers (73 miles), from first shoal upstream of Bankhead Reservoir embayment (just above Fivemile Creek) upstream to Alabama Highway 160 near Cleveland. Also found in lower eight kilometers (five miles) of Blackburn Fork (Shepard *et al.* 2002). Recently discovered in Locust Fork; therefore, no historic distribution is available, but may have been present in main stem of Black Warrior River as far downstream as Tuscaloosa, where an associate species, the coal darter, was collected in 1889 before impoundment (Shepard *et al.* 2002). Also may have occurred in main stem of Coosa River prior to extensive impoundment (Mayden and Kuhajda 1989).

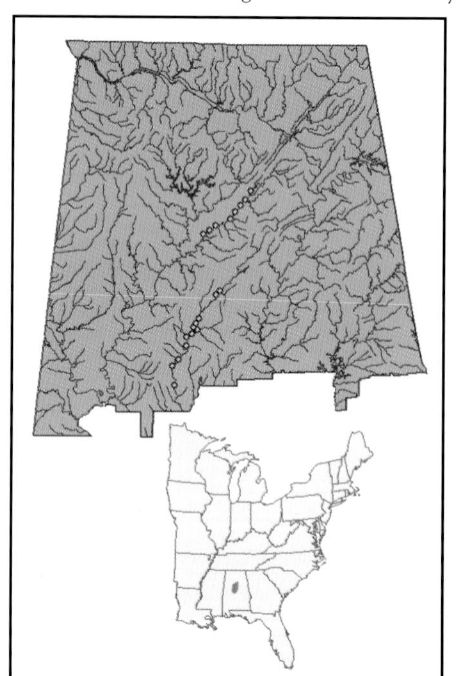

HABITAT. Principal habitat includes main channel of Cahaba River and Locust Fork, where shoal macrohabitats predominate. Although found in lower reaches of Blackburn Fork, species not associated with smaller tributaries except during times of high water, when individuals move into mouths of creeks and streams. Like other members of the species group, typically found in microhabitats of quiet shallow backwater just below, or adjacent to, riffles and runs primarily over clean sand or sand-gravel substrates (Mayden and Kuhajda 1989).

LIFE HISTORY AND ECOLOGY. In reproductive condition from May to July, are mature at one year of age, and may spawn a second year (Mayden and Kuhajda 1989). In aquaria, spawning occurred May through October, with eggs attached to yarn mops in shallow areas with little to no current. Eggs were large (nearly 3 mm diameter [0.12 in.]), adhesive, deposited in large gelatinous masses of 20 to 100, and hatched in less than three days at 20-22° C (68-72°F). Larvae were motionless on bottom of aquaria for several days before becoming active (Rakes and Shute 2001). Adults probably feed on small crustaceans, aquatic insect larvae, and perhaps some vegetation. Differences in abundance occur between two populations. In Cahaba River, the species is rare, being outnumbered by mimic shiners 38 to one. In Locust Fork, Cahaba shiners are almost five times more abundant than mimic shiners.

BASIS FOR STATUS CLASSIFICATION. As a main channel species, extremely vulnerable to acute and chronic pollution. Restricted to two river systems within Alabama, both with degraded water quality. In Cahaba River, area upstream of Helena is one of the most rapidly developing areas in Alabama. A mean discharge of 20 mgd of treated wastewater is leading to eutrophic conditions within distribution of the species, and removal of an average of 53 mgd of water at a pumping station upstream of U.S. Highway 280 magnifies water quality concerns, particularly in periods of low flow. Sedimentation related primarily to construction also a major concern (U. S. Fish and Wildlife Service 1992, Shepard *et al.* 1995). In Locust Fork system, eutrophication and sedimentation related to agricultural activities are impacting habitat. Coal surface mining also a source of sedimentation and water-quality degradation in sections of the species' distribution. A water-supply reservoir proposed for construction near Jefferson-Blount county line by Birmingham Water Works Board would probably eliminate this species from Locust Fork system through inundation of upstream habitat and impacts to downstream habitat related to construction and altered flow regimes. **Listed as *endangered* by the U.S. Fish and Wildlife Service in 1990.**

***Prepared by*: Bernard R. Kuhajda and Thomas E. Shepard**

IRONCOLOR SHINER

Notropis chalybaeus (Cope)

OTHER NAMES. None.

Photo—Patrick E. O'Neil

DESCRIPTION. A member of *texanus* species group; sister group relationship unknown (Swift 1970, Mayden 1989). A small (max. = approx. 60 mm [2.5 in.] TL) shiner, with a moderately deep and compressed body and a blunt snout distinctly less than diameter of eye. Mouth relatively small, oblique, and terminal. Origin of dorsal fin directly above insertion of pelvic fins. Lateral scale rows number 31 to 36, usually 33 or 34; however, lateral line incomplete, usually with 10 or more unpored scales. Predorsal scales varying from 12 to 20, usually 14 to 18; circumferential scales usually 24 to 28. Anal fin rays eight, rarely seven or nine; pectoral fin rays 11 to 13. Pharyngeal teeth 2,4-4,2. Tubercles well developed on lower jaw, occurring in a single or double row and projecting outward. Few tubercles on pre- and supraorbital regions, lower part of cheek, and lower edge of snout. Tubercles also on anterior half of body and lateral line. Pectoral fins with shagreen of tubercles on rays two to five or six. Sharply delimited midlateral dark stripe continuing around snout. Inconspicuous basicaudal spot no larger or darker than lateral stripe. Lips, oral valve, and floor and roof of mouth, black. Predorsal stripe faint, if present. Melanophores in anal fin virtually absent. Breeding males pale orange on body, dorsal, and caudal fins. Distinguishable from the coastal and weed shiners in its smaller size, less robust body, more intense lateral stripe, and usually eight (*vs.* usually seven) anal rays. The redeye chub differs in having a small barbel in corner of mouth and 0,4-4,0 pharyngeal teeth. The blackmouth shiner, which is known from Florida, Mississippi, and Alabama, has a black mouth but differs from the ironcolor shiner in having 10 or 11 anal rays and a very oblique (almost vertical) mouth. The blackmouth shiner is sympatric with several minnows (*e.g.*, dusky, coastal, and weed shiners; redeye chub; and pugnose minnow) with which it shares a midlateral stripe of varying intensity that continues forward around snout and a light narrow area between midlateral stripe and darker field of dorsolateral scales. It differs from all sympatric minnows in having unique combination of eight anal rays and black interior of mouth.

DISTRIBUTION. Lowland regions of Atlantic and Gulf seaboards, from lower Hudson River drainage in New York (Smith 1985) south to vicinity of Lake Okeechobee, Florida, and west to Sabine River drainage in Louisiana and Texas. Disjunct populations occur farther west to San Marcos River in Texas and Red River drainage in extreme southeastern Oklahoma and lowlands of Arkansas (Robison 1977). Ranges north into Mississippi River Valley to Wolf River in Wisconsin, and east in Illinois River system in Illinois (Smith 1979) and Indiana and to Lake Michigan drainage in southwestern Michigan. Usually occurs less frequently in western and northern parts of distribution, but sometimes locally common. Although widespread throughout Florida, excluding peninsula below Lake Okeechobee, conspicuously absent from certain streams such as Econfina and Bear Creek systems and upper Suwannee River drainage. Uncommon in Alabama; was known in all coastal streams in Florida from Chipola River west to Perdido River, as well as Mobile Delta area and lower Tombigbee and Escatawpa River systems.

HABITAT. Varies somewhat throughout its distribution. In Alabama, associated with small, sluggish but clear creeks with sand substrates, abundant aquatic vegetation, as well as flowing swamps with stained acidic waters typical of coastal areas.

LIFE HISTORY AND ECOLOGY. At our latitude the breeding period probably extends from March to August or September. Prespawning behavior consists of males chasing after females, which make quick retreats unless ready to spawn. Spawning probably takes place when nuptial pairs, vent to vent, dash across a pool. Fertilized as they are broadcast, eggs fall to substrate. Eggs 0.9 millimeters (0.03 inches) in diameter and artificially fertilized in laboratory and incubated at 16.5°C (61.7°F), hatch in 52 to 56 hours. Hatchlings are 2.3 millimeters (0.09 inch) TL. Larvae require 47 days to reach 7.4 millimeters (0.29 inch) SL (Mansueti and Hardy 1967, Jones *et al.* 1978). Growth rate seems uncommonly slow and suspected to be faster in nature. Ironcolor shiners are sight feeders, and their diet consists of insects and microcrustaceans. Plant materials (algae), thought to come from gut of ingested prey, pass undigested (Marshall 1947, McLane 1955, Sheldon and Meffe 1993).

BASIS FOR STATUS CLASSIFICATION. Rare, endangered, or extirpated in several states on the periphery of distribution (C. R. Gilbert and J. D. Williams, in lit.). Habitat degradation in Mississippi may be driving small populations to extinction (Albanese and Slack 1998). Because of limited and degraded habitat, have disappeared from historically known locations and may be extirpated in Alabama (Boschung and Mayden, in press).

***Prepared by*: Herbert Boschung and Richard Mayden**

ALABAMA CAVEFISH

Speoplatyrhinus poulsoni Cooper and Kuehne

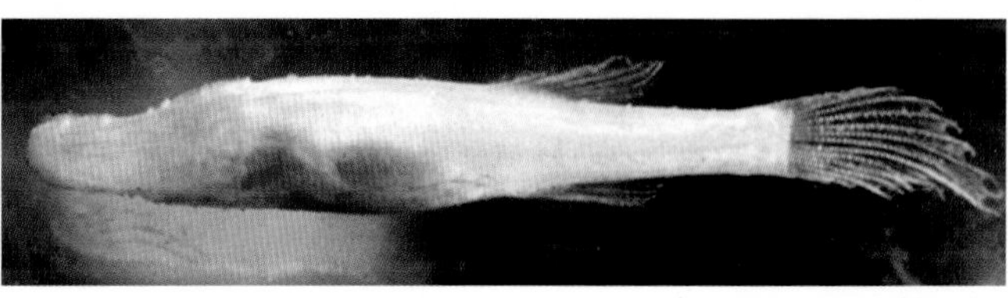
Photo—Richard Mayden

OTHER NAMES. None.

DESCRIPTION. All cavefish have small or rudimentary eyes, reduced pigmentation, no or very small pelvic fins, anus and urogenital opening positioned between gill membranes, and rows of sensory papillae on head, body, and caudal fin. The Alabama Cavefish is white, blind, lacks pelvic fins, and is distinguished from all other cavefish by its long, anteriorly depressed head; a laterally constricted and dorsally flattened snout, which gives it a bill-like appearance; absence of bifurcated fin rays; incised fin membranes; and larger and fewer caudal sensory papillae. No sexual dimorphism. Attains a total length of seven centimeters (2.8 inches) (Cooper and Kuehne 1974).

DISTRIBUTION. Restricted to a single cave (Key Cave) just north of Tennessee River in Lauderdale County, Alabama. Surveys of caves surrounding Key Cave in 1980s and in 1999 failed to find any new populations (Kuhajda and Mayden 2001). Water flow in Key Cave, and perhaps some historic habitat within Key Cave aquifer, has been altered with construction of Pickwick Reservoir. This impoundment of the Tennessee River has submerged and reduced the historic outflow of Key Cave (Collier Spring), thereby reducing water flow through Key Cave and possibly depressing population size through lower reproduction (see below) (Aley 1990). Current population extremely small, maximum number of specimens observed during single visit to cave is 10, and population estimates are less than 100 individuals (U. S. Fish and Wildlife Service 1990). One specimen of the widespread Southern Cavefish has been collected in Key Cave (Kuhajda and Mayden 2001).

HABITAT. A stygobitic fish restricted to Key Cave. Occupies deep subterranean pools (up to five m [16 ft.] deep) with silt and rock bottoms and little or no flow. Pools typically clear, but can become cloudy. A maternal colony of gray myotis (bats) present in Key Cave from April to October deposits guano that is an important nutrient for the aquatic environment of cave. Temperature is somewhat stable (15°C [59°F]), but water depth can fluctuate extensively (U.S. Fish and Wildlife Service 1990, Kuhajda and Mayden 2001).

LIFE HISTORY AND ECOLOGY. Seasonal flooding stimulates hormonal changes within other species of cavefish, which initiate growth and reproduction. A survey of Key Cave (1992-1997) indicated recruitment was occurring; three size classes were present (Kuhajda and Mayden 2001). A female with ova was collected in late May, indicating summer spawning. Specific reproduction and early life history lacking, but other species of cavefish mature at a late age, have very low fecundity, and do not spawn every year. Alabama cavefish probably feed on copepods, isopods, amphipods, and small conspecifics (Cooper and Kuehne 1974, U.S. Fish and Wildlife Service 1990).

BASIS FOR STATUS CLASSIFICATION. Restricted to a single cave in Alabama with a population of less than 100 individuals, making it one of the rarest vertebrates in the world. Disruption of the gray myotis colony could affect nutrient regime in Key Cave, and illegal collecting of individuals could impact population (U.S. Fish and Wildlife Service 1990). Even with establishment of Key Cave National Wildlife Refuge in a high recharge area of the Key Cave aquifer, threats to ground water continue from encroaching urbanization, specifically a proposed industrial park. Urbanization concerns include lowering of water table and a decrease in seasonal flooding due to impermeable surfaces. Runoff into the recharge area of Key Cave from a proposed industrial site or adjacent highways is a potential source of acute and chronic deterioration of water quality (Kuhajda and Mayden 2001). **Listed as *endangered* by the U.S. Fish and Wildlife Service in 1988.**

***Prepared by*: Bernard R. Kuhajda**

PYGMY SCULPIN

Cottus paulus Williams

Photo—Carol Johnston

OTHER NAMES. Originally described as *Cottus pygmaeus*. Subsequently *pygmaeus* was discovered to have been already used in genus *Cottus*. Williams (2000) proposed replacement name *Cottus paulus*.

DESCRIPTION. A diminutive species (rarely > 38 mm [1.5 in.] SL), distinguished from other Alabama cottids by presence of a connection between spiny and soft dorsal fins and two, rather than three, preopercular spines. Has flattened head and robust body. Body coloration usually consists of gray background with areas of white; usually three dark bands that circle, or partially circle, posterior half of body. Individuals can change color patterns. Courting males, or those engaged in agonistic encounters, may become almost totally black, while females being courted become light gray or white.

DISTRIBUTION. Found only in Coldwater Spring and spring run in community of Coldwater, Calhoun County, Alabama.

HABITAT. Coldwater Spring, one of largest springs in northern Alabama, has basin of approximately 0.41 hectare (1.24 acres); run encompasses another 0.16 hectare (0.50 acre). Spring discharges approximately 32 million gallons of water per day. Spring basin contains a variety of habitats including rocky areas, a central area of sand substrate and areas dominated by moss, water milfoil, and coontail. Spring run has a substrate of river weed covered rocks in its upper end, while rest of run has a substrate of sand

and gravel with scattered cobble. Vegetation in middle and lower portion of run consists of stands of bur-reed and some watercress.

LIFE HISTORY AND ECOLOGY. Population in spring run feeds primarily on gastropods, amphipods, isopods, and aquatic insects (McCaleb 1973). No feeding analysis available for population in spring basin but recently completed macroinvertebrate sampling reveals fauna dominated by isopods and amphipods, which are presumed to represent bulk of diet there. Nesting occurs under rocks, primarily in areas of shallow water and relatively high water velocity. Will nest readily under artificial shelters (*e.g.*, tiles) placed in appropriate depths, flows, and substrate composition. Average number of clutches per nest is 4.2 (SD = 2.0, n = 18) with an average clutch consisting of 25 eggs (SD = 5.5, n = 43) (Johnston 2001). A male guards nest site until eggs have hatched; however, males may be replaced by other males that then provide allopaternal care for eggs (Johnston 2000). Spawning may occur throughout year with peak spawning periods from April to July, and possibly as late as October. Thus, some spawning apparently occurs throughout year, but the majority takes place from March through October with a peak in July and a possible second peak in late September and October (McCaleb 1973, Johnston 2001, Stiles 2002). A recent population study (June 1997 to February 1999) found sculpin populations in spring basin varying between a high of 23,751 (February 1999) and a low of 8,499 (May 1998). Population in run averaged 2,799 individuals. Following hatching, post-larval pygmy sculpin congregate in mats of moss with some moving into areas of water milfoil and coontail (Stiles 2002). Thus, vegetated portions of spring act as nursery areas for young sculpin, as well as, additional habitat for adults.

BASIS FOR STATUS CLASSIFICATION. Found only in Coldwater Spring, which is owned by the Anniston Water Works and Sewer Board and is major water source for much of Anniston and surrounding portions of Calhoun County. The water works withdraws an average of approximately 12 million gallons per day. An agreement between the water works and U.S. Fish and Wildlife Service provides protection for spring. Water works can withdraw all but 1.9 million gallons per day; however, it has never withdrawn more than 18 million gallons per day. Under agreement, an emergency plan is in place to rescue a sample of the pygmy sculpin population in case of a catastrophic event at the spring. There has been recent concern that toxic compounds, in particular trichloroethylene, might be entering Coldwater Spring aquifer from Anniston Army Depot. Trichloroethylene has been detected regularly in Coldwater Spring since 1981; however, amounts have never exceeded maximum contaminant limits for safe public drinking water. Nonetheless, with a recharge area that may be as large as 145 square kilometers (90 square miles), there is always threat of major contamination of spring. **Listed as *threatened* by the U.S. Fish and Wildlife Service in 1989.**

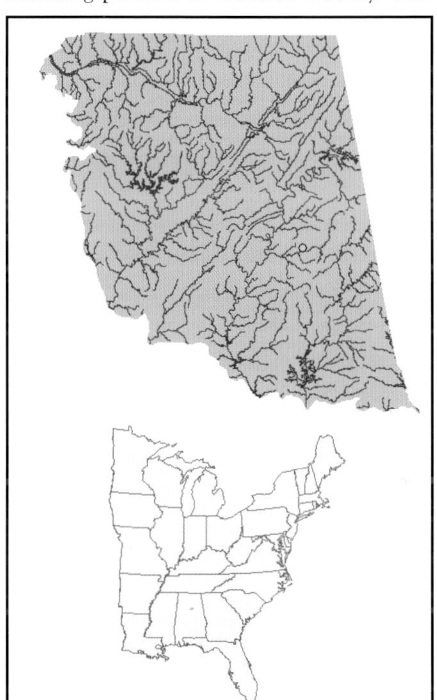

***Prepared by*: Robert A. Stiles and Melvin L. Warren, Jr.**

SPRING PYGMY SUNFISH

Elassoma alabamae Mayden

Photo—*Patrick E. O'Neil*

OTHER NAMES. Languished as a taxonomically undescribed, but well known, species for about 50 years (Mayden 1993), during which it was referred to as *Elassoma* sp. or *Elassoma* species.

DESCRIPTION. Smallest species in genus *Elassoma* (average adult body size = 17.4 mm [0.7 in.] SL; maximum body size = 25 mm [1 in.] SL; (Mayden 1993). Male distinguished from congeners by clear or white windows present on dorsal and anal fins (vs. absent in other *Elassoma*); broad, lateral vertical bars (vs. narrow or indistinct), and indistinct posterior borders on basicaudal spots (vs. bordered posteriorly). Breeding males brown to black with five to eight (usually six to seven) broad, black, or dark olive bars along flanks that are separated by four to eight (usually five or six) narrow, iridescent blue-green or cream-colored bars. Female coloration drab, mottled brown with barring less distinct than in male (Mayden 1993).

DISTRIBUTION. Endemic to Alabama. Known historically only from three springs and associated habitats in Tennessee River drainage in northern Alabama (Mayden 1993): Cave Spring, Lauderdale County, Alabama (population extirpated by closure of Pickwick Dam in 1938); Pryor Spring, Limestone County, Alabama, (natural population extirpated in 1940s) (Jandebeur 1979, 1982; Mayden 1993); and Beaverdam Spring and swamp complex (inclusive of Moss Spring and Lowe's Ditch), Limestone County, Alabama (only extant natural population). In 1984 and 1987, individuals from Moss Spring transferred to Pryor Springs locality resulting in species being reestablished in Pryor Spring system (Mettee and Pulliam 1986, Mayden 1993, Mettee *et al.* 1996). Extant populations restricted to Beaverdam Creek watershed (downstream to near confluence with impounded waters of Wheeler Reservoir) and Pryor Spring system (Mayden 1993, Mettee *et al.* 1996).

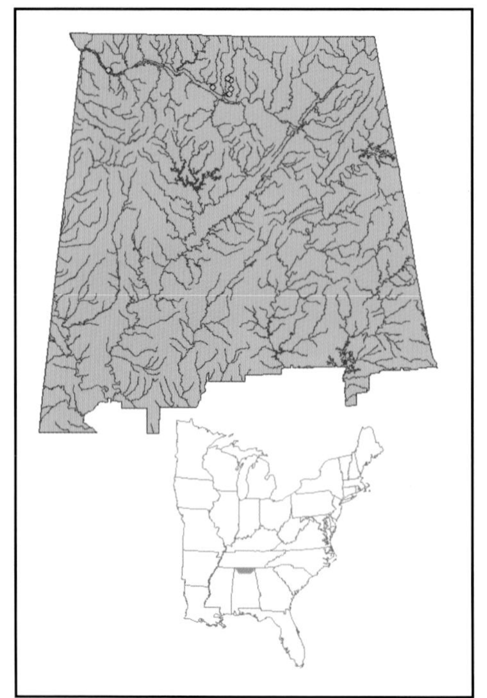

HABITAT. Clear to slightly stained springs, spring runs, and associated spring-fed wetland habitats. Occurs in water column associated with rooted, submergent vegetation (e.g., hornwort, water milfoil, bladderwort, and water weed). May use different spring and wetland macrohabitats at different times of year (Jandebeur 1979, Darr and Hooper 1991, Mayden 1993).

LIFE HISTORY AND ECOLOGY. Associated consistently with dense submergent vegetation in shallow water (15 to 60 cm [0.5 to 2 ft.]) (Mettee and Ramsey 1986, Mayden 1993). Most intense spawning period from March to April (Darr and Hooper 1991, Mayden 1993), with females producing up to 65 eggs per spawning (Mettee and Ramsey 1986). Courtship and spawning behaviors of *Elassoma* are complex (Mettee 1974, Walsh and Burr 1984), but no nests are constructed and eggs are usually attached to aquatic vegetation above substrate. Investigations of reproductive biology and demographics suggest spring pygmy sunfish is an annual species (Mettee 1974, Mettee and Ramsey 1986, Darr and Hooper 1991). Adults apparently spawn at one year of age and die during late spring or early summer. By fall, population consists entirely of young-of-the-year.

BASIS FOR STATUS CLASSIFICATION. Considered endangered (Warren *et al.* 2000) due to extremely small geographic distribution, short life span, and known sensitivity to habitat alterations. Extremely vulnerable to extirpation (Mayden 1993). Pryor Spring Branch native population likely extirpated in 1940s by herbicide treatments to control spring vegetation, channelization, agricultural pollutants, or a combination of these factors (Jandebeur 1979, Mayden 1993). Only extant native population resides in one watershed crisscrossed by roads and subject to intensive agricultural activities (*e.g.*, aerial application of herbicides and pesticides). A careless mistake (*e.g.*, pesticide spill) could severely reduce or eliminate this native population. The U.S. Fish and Wildlife Service removed spring pygmy sunfish as a candidate for recognition under Endangered Species Act in 1996 because of success of introduced Pryor Spring population and discovery of "additional populations" on Wheeler National Wildlife Refuge (Federal Register 1996). However, "additional populations" consisted of capture of a few scattered individuals from near Wheeler Reservoir and upstream in Beaverdam Creek (P. W. Shute, pers. comm.). Discovery of individuals near confluence of Beaverdam Swamp and Wheeler Reservoir constitute downstream distribution extensions, but were within Beaverdam Creek and Swamp complex, a system long known to harbor the species. The basis for considering a distribution extension within a single watershed as an "additional population" was not specified (Federal Register 1996) and is questionable.

***Prepared by:* Melvin L. Warren, Jr.**

SLACKWATER DARTER

Etheostoma boschungi Wall and Williams

Photo—Catherine Phillips

OTHER NAMES. None.

DESCRIPTION. Subgenus *Ozarka*. Max. size = 65 millimeters (2.6 inches) SL. Three dark saddles on dorsum; dark suborbital bar; females brown with blotches; breeding males with bright orange belly, cheeks, and lower jaw and orange submarginal band in first dorsal fin. Males develop breeding tubercles on posterior belly scales and rays of anal and pelvic fins. Has interrupted supratemporal canal, modally seven infraorbital pores, an incomplete lateral line, and two anal spines. Belly scaled, breast and cheeks unscaled, nape partly to fully scaled, opercle unscaled to partly scaled (Page 1983, Etnier and Starnes 1993).

DISTRIBUTION. Restricted to tributaries of Buffalo River, Flint River, Cypress Creek, Swan Creek, and Shoal Creek, all within Highland Rim region of Tennessee River drainage in Tennessee and Alabama. In 1992-1994, known to occur in 22 locations throughout distribution, including 13 sites in Alabama (McGregor and Shepard 1995); 15 of these localities were known breeding sites. Unfortunately, recent efforts to collect species during breeding and nonbreeding seasons in Alabama have failed, although species still present in Tennessee (Johnston and Hartup 2002).

HABITAT. Members of subgenus *Ozarka* unusual in having distinct breeding and nonbreeding habitats. During nonbreeding season, slackwater darters found in sluggish parts of small to medium-sized streams, often in association with packs of leaves and other debris (Boschung and Nieland 1986). During late winter, adults migrate up to four kilometers (2.5 miles) upstream to flooded seepage areas to spawn (Boschung and Nieland 1986).

LIFE HISTORY AND ECOLOGY. Egg attachers, laying eggs on submerged vegetation without parental care (Boschung and Nieland 1986). Spawning takes place in flooded seepage areas in early spring, when water temperature reaches about 14°C (57.2°F) (typically late February-March). Adults leave spawning areas in April and return to non-breeding habitat. Longevity estimated about three years. Diet consists of aquatic isopods, amphiods, mayflies, midge larvae, and limpets.

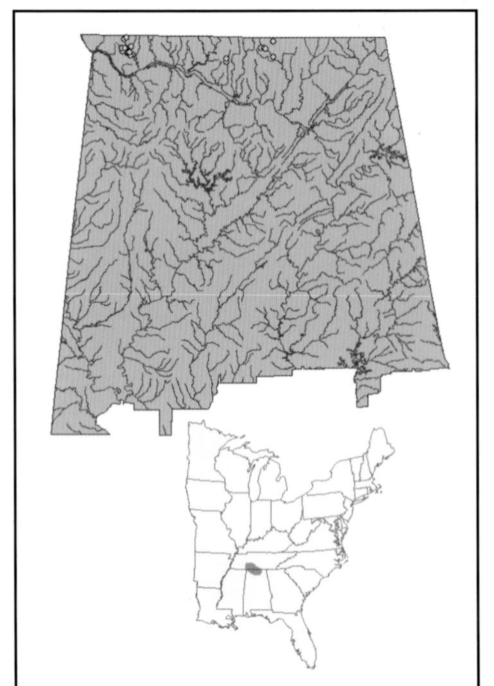

BASIS FOR STATUS CLASSIFICATION. Although nonbreeding habitat appears intact in many places, species has not been found recently in Alabama. Many previous breeding sites have been destroyed, or altered by human activity. Barriers to migration also a concern. Considered endangered in Alabama (Johnston and Hartup 2002). **Listed as *threatened* by the U.S. Fish and Wildlife Service in 1977.**

***Prepared by*: Carol E. Johnston**

HOLIDAY DARTER

Etheostoma brevirostrum Suttkus and Etnier

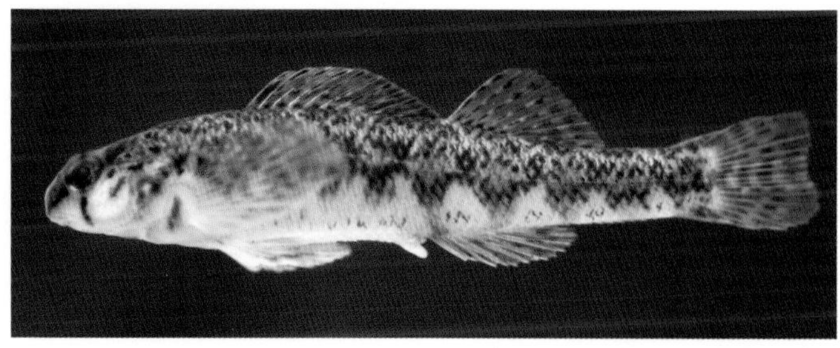

Photo—Malcolm Pierson

OTHER NAMES. Upland Snubnose Darter (Sizemore and Howell 1990).

DESCRIPTION. A small percid (approx. 50 mm [2 in.] TL). Breeding male has median red band in blue-green anal fin, a red ocellus in first membrane of spiny dorsal fin, and a narrow red band in soft dorsal fin (Mettee *et al.* 1996). Green blotches occur on back and sides.

DISTRIBUTION. Restricted to Coosa River system above Fall Line. Occurs in Ridge and Valley, Piedmont, and Blue Ridge provinces in Alabama, Georgia, and Tennessee (Suttkus and Etnier 1991). In Alabama, known from Shoal Creek (Cleburne County) and a few springs in Choccolocco Creek system.

HABITAT. Prefers medium to large, clear streams with moderate to fast current and a variety of substrates that include boulders, rubble, gravel, sand, and river weed.

LIFE HISTORY AND ECOLOGY. Based on field observations and a few available specimens, spawning occurs in April and May. Diet consists of aquatic insect larvae and microcrustaceans (Mettee *et al.* 1996).

BASIS FOR STATUS CLASSIFICATION. In Alabama, restricted to upper reaches of Shoal Creek in Talledega National Forest and three springs in upper Choccolocco Creek system. Riverine habitat has been reduced by construction of small water-control impoundments. Species absent during recent sampling in lower reaches of Shoal Creek below Whitesides Mill Dam. Any further decline in distribution, or reduction in population size, would make species a likely candidate for some level of federal protection.

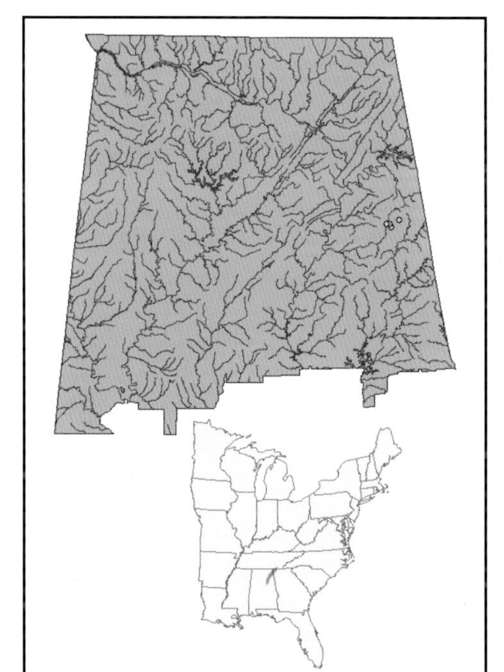

Prepared by: **J. Malcolm Pierson**

VERMILION DARTER

Etheostoma chermocki (Boschung, Mayden, and Tommelleri)

Photo—L. J. Davenport

OTHER NAMES. None.

DESCRIPTION. A relatively large (max. SL about 60 mm [2.4 in.]) snubnose darter of *Etheostoma* (Boschung and Mayden 1991). Differentiated from warrior darter by stockier body, shorter caudal peduncle, taller dorsal fins, and more extensive vermilion coloration along venter. No significant meristic differences present in sexes, but significant morphometric differences (Boschung and Mayden 1991). Males possess longer head, greater body depth, greater snout length, longer spinous dorsal fin base, longer dorsal fin spines, soft dorsal spin rays, greater caudal peduncle fin length, and wider trans-pelvic base. Lateral line complete and virtually straight from upper margin of gill opening to base of caudal fin. Sexual dichomatism conspicuous. Marked variation in colors of both sexes occur from spring to summer, with males being more brightly colored throughout year, especially during spring. Nuptial males have pale olive to cream-colored body, with back crossed by eight dark olive saddles (Boschung and Mayden 1991). Orange-red spots with olivaceous patches present along lateral line. Lower sides, venter, and lower caudal perduncle area brilliant vermilion. Females lack bright coloration. Dorsum of body and head dark olive to cream, with lateral line less distinctly colored. Some small red orange scale patches can be present above lateral line (Boschung and Mayden 1991).

DISTRIBUTION. Currently 11.6 kilometers (7.2 miles) of mainstream of Turkey Creek and lower reaches (0.8 km [0.5 mi.] total) of Dry Creek to upper range of Beaver Creek, both of which intersect Turkey Creek, a tributary of Locust Fork of Black Warrior River, northeastern Jefferson County (Mettee and O'Neil 1990, Stiles and Blanchard 2003). Populations seem clustered at various locations throughout distribution (Boschung and Mayden 1991).

HABITAT. Small to medium (3-20 m [10-66 ft.] wide) gravel-bottom streams of depths of 0.1 meter to more than 0.5 meters (4 inches to more than 1.6 feet) with pools of moderate current alternating with riffles of moderately swift current. Riffles are of cobble or coarse gravel or smaller stones, whereas pools are rock, sand, or silt. Most favorable habitat seems to be zones of transition between a run/riffle and pools, especially where some vegetation, including watercress or pondweed present (Boschung and Mayden 1991).

LIFE HISTORY AND ECOLOGY. Little known. Underwater, direct observation of spawning pairs at confluence of Turkey Creek and

Tapawengo Springs revealed use of clean, bare cobble and rocks dispersed on a fine silt and pebble bottom as substrate for egg deposition (R. A. Stiles, pers. obs.). Typically live two to three years and feed primarily on snails and aquatic insects. Recent population estimate was 3,300 individuals. A survey conducted in the summer of 2003 indicates that the population has declined substantially from this earlier estimate (Stiles and Blanchard 2003).

BASIS FOR STATUS CLASSIFICATION. Geographic isolation, coupled with existence in a stream habitat that flows through both floodplains and steep-sided embankments, make species particularly susceptible to impact of siltation and nutrification events. Small population and its arrangement into clustered segments throughout distribution suggests a toxic event (such as a chemical release) could be catastrophic (Federal Register 2001). Nonpoint source pollution continues to escalate in this rapidly-developing area of Jefferson County and will likely provide an on-going challenge to survival. **Listed as *endangered* by the U.S. Fish and Wildlife Service in 2001.**

***Prepared by*: Paul D. Blanchard and Daniel Drennen**

BRIGHTEYE DARTER

Etheostoma lynceum Hay

OTHER NAMES. None.

Photo—Patrick E. O'Neil

DESCRIPTION. Once considered a subspecies of banded darter, but subsequently elevated to distinct species. Short (adults in Alabama up to 45 mm [1.8 in.] SL) and robust with a wide frenum on upper lip, enlarged pectoral fins, and distinct pre-, sub-, and postorbital bars. Six dark dorsal saddles extend over back and become confluent with six to nine yellowish-green lateral blotches. Body light yellow to grayish yellow, although body of breeding male takes on a green-gray or spruce-gray hue. Breast region, anal fin, most of pectoral and pelvic fins, and lower and upper parts of caudal fin a light green to bluish-green color. Two light yellow to cream-colored spots found at base of caudal fin. Spiny dorsal fin has dark base, followed by a lighter reddish-brown band, then a wide green band marginally; soft dorsal fin similarly colored but with an additional light band distally. Meristic counts include 36-47 lateral line scales, seven to 12 dorsal fin spines and nine to 13 rays, two anal fin spines and seven to 10 rays, and 12-16 pectoral fin rays (Etnier and Starnes 1993). Breast and anterior part of belly generally naked; nape, cheeks, opercles, and prepectoral areas scaled.

DISTRIBUTION. Occurs in streams of upper Coastal Plain generally east of Mississippi River from Obion River system in Kentucky and Tennessee, south to Pascagoula River system in Mississippi and Alabama, and west to Lake Ponchartrain system in Louisiana. In Alabama, found only in Puppy Creek and Big Creek systems in Mobile County (Mettee *et al.* 1996).

HABITAT. Found over sand and mud in sluggish pools and over gravel and pebble substrates in swift runs and channels in small to medium-sized Coastal Plain streams. Preferred habitat in Puppy Creek was a swift run where stream had carved up to one-meter (three-foot) deep grooves and channels into soft clay substrate. Preferred extensive patches of rooted aquatic vegetation and detrital snags associated with channel margins (O'Neil *et al.* 1984).

LIFE HISTORY AND ECOLOGY. Little known in Alabama. Populations in Homochitto River, Mississippi, spawn from late February-early March to mid-May (Bell and Timmons 1991). June collections of young-of-year indicate spawning in late April and May in Puppy Creek. In Kentucky, feed on aquatic insect larvae. Oldest individual examined by authors three to four years of age.

BASIS FOR STATUS CLASSIFICATION. Fairly common outside of Alabama. Several records exist in Tennessee and Mississippi (Etnier and Starnes 1993, Ross 2001). However, in Alabama, species is at its southeastern distributional limit in the Escatawpa River, and is known only from Puppy and Big Creeks, Mobile County. Puppy Creek from Alabama Highway 217 to its source has been listed on the 303(d) list of impaired waters by Alabama Department of Environmental Management for pathogens from urban and storm sewer runoff. As such, this population appears vulnerable over time to effects of nonpoint source pollution due to both urban expansion of Citronelle community and potential runoff from Citronelle oil field.

***Prepared by*: Patrick E. O'Neil**

LOLLIPOP DARTER

Etheostoma neopterum Howell and Dingerkus

OTHER NAMES. None.

Photo—J. R. Shute

DESCRIPTION. A relatively large (max. length > 75 mm [3 in.]) member of the spottail darter (*Etheostoma squamiceps*) group of subgenus *Catonotus* (Howell and Dingerkus 1978). This group typified by an incomplete lateral line and usually incomplete head canals, a laterally flattened body, and reduced number of dorsal spines. Page *et al.* (1992) revised the *Etheostoma squamiceps* group and redescribed the species as follows: females and nonreproductive males drab olive to brown. Dorsal saddles may be present, and sides of bodies may be blotched or somewhat mottled. A vertical row of three dark spots present at base of caudal fin, and caudal fin has five to 12 alternating clear or yellow and brown stripes. Bodies of breeding males almost uniformly dark brown or gray, and heads are blackened.

With exception of caudal fin, all other fins on breeding males are black. First dorsal fin dark with a clear band above base, and with a small yellow knob on tip of each spine. Second dorsal fin sometimes has a clear basal band and a series of clear dots or stripes. Each ray tipped with large yellow knob. There are 41 to 52 total lateral scale rows; pored scales 19 to 33; dorsal fin usually with eight or nine spines (eight to 10) and 10 to 12 soft rays.

DISTRIBUTION. Known only from the Factory, Holly, Butler Creek systems in the Shoal Creek watershed of Tennessee drainage, Lauderdale County, Alabama, and Lawrence and Wayne Counties, Tennessee (Ceas and Page 1995). Although historical information on distribution is scant compared with present day data, present and historic distributions are probably similar (Ceas and Page 1995).

HABITAT. Gently flowing shallow pools and eddies in headwater streams or springs where substrate is gravel, cobble, and sand. Usually found beneath larger rocks, woody debris, aquatic vegetation, or undercut banks, like most other members of subgenus *Catonotus*.

LIFE HISTORY AND ECOLOGY. Details of feeding habits and reproductive activity are not published, but Ceas and Page (1995) provided some life history information on other members of species group. Like other members of group, primary food is probably aquatic insect larvae. Spawning season may be from late March through May, depending upon water conditions. Males guard nests beneath slabrocks or other larger cover objects, where females attach eggs on underside. Because males may spawn with many females, a single male may eventually collect more than 1,000 eggs. Young mature rapidly; females can become sexually mature at one year; males, not until two years of age. After hatching, young disperse, mostly downstream, and return upstream after about a year.

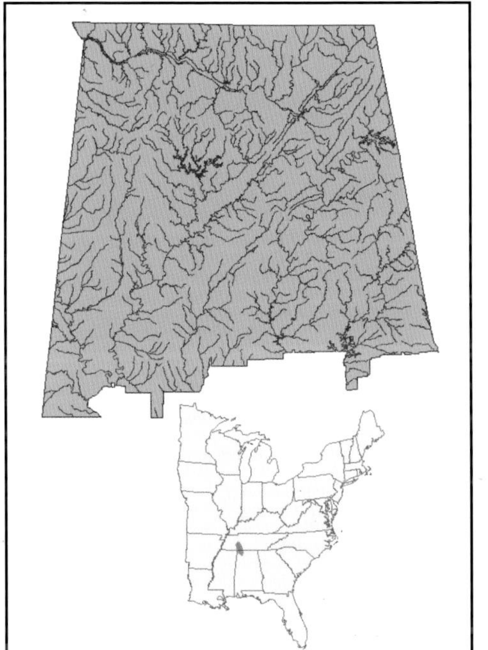

BASIS FOR STATUS CLASSIFICATION. Only 13 known breeding sites documented (Ceas and Page 1995), and global distribution is restricted to small, headwater streams in Wayne County, Tennessee, and Lawrence County, Alabama. Many activities that modify or decrease flow (channelization, diversion, water withdrawal for irrigation, small impoundments), or could affect water and habitat quality in these areas (such as removal of riparian buffer and stream bank destabilization from grazing), are threats because of highly restricted distribution.

Prepared by: **Peggy Shute**

WATERCRESS DARTER

Etheostoma nuchale Howell and Caldwell

Photo—W. Michael Howell

OTHER NAMES. None

DESCRIPTION. A small (max. total length = 56 mm [2.2 in.]) (Howell 1986) rather robust member of subgenus *Oligocephalus*. Has nine or so saddles on back and usually a pale stripe across dorsal midline of nape. Sides bear concentrations of dark spots that tend to form horizontal streaks anteriorly and vertical bars posteriorly. Nuptial males have reddish orange blotches on sides, particularly between posterior bars, and orange on venter. Spiny dorsal and soft dorsal fins have four alternating bands of blue and red beginning with a blue band on outer margin. Anal fin bright blue with a few flecks of red or orange. At some sites watercress darter sympatric with redspot darter, another member of subgenus. Redspot darter larger and differs from watercress darter in having moderately to broadly connected gill membranes and more scales in lateral line (48-55 vs. 36-39 in watercress darter). In addition, nuptial males have distinct red spots on sides, a broad red band in basal portion of anal fin, and a red band in posterior portion of caudal fin.

DISTRIBUTION. Currently inhabits four springs and their runs, two spring-fed ponds, and a small spring-fed creek all located in the Black Warrior River system, Jefferson County, Alabama. One spring (Glen Spring) and two spring-fed ponds are located in southern portion of City of Bessemer, and are tributaries to Halls Creek. A second spring (Seven Springs) is located in the Birmingham community of Powderly and is a tributary to Nabors Branch, whereas a third (Roebuck Spring), a tributary to Village Creek, begins on property of the Alabama Department of Youth Services, Vacca Campus, in the Roebuck section of Birmingham. Final populations inhabit Penny Spring, the spring run, and an adjacent spring-fed creek, all tributary to Turkey Creek in Pinson. These final populations result from successful transplants from Roebuck Spring.

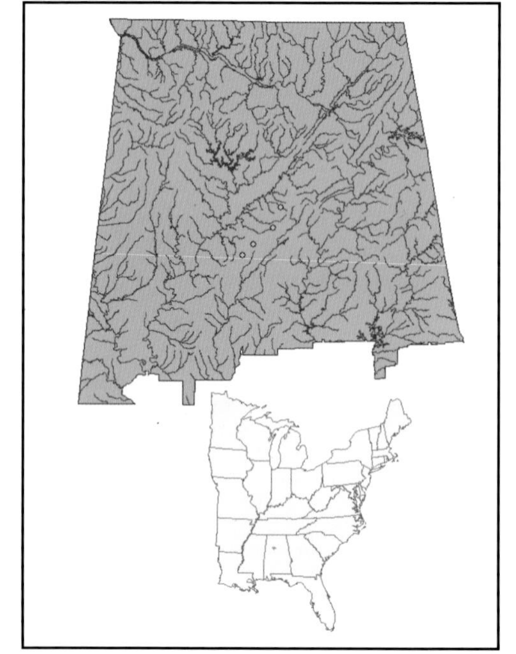

HABITAT. Restricted to springs, spring runs, and spring-fed ponds. Inevitably associated with stands of aquatic vegetation including watercress, stonewort, water milfoil, bur-reed, mosses, and algae. Apparently prefers areas with little or no current and avoids areas of spring runs and creeks with swift current.

LIFE HISTORY AND ECOLOGY. Little known. Apparently feeds on small snails, crustaceans, and insect larva (Howell 1986). Spawning occurs from March to July. Eggs deposited on aquatic vegetation (R.A. Stiles, pers. obs.), but no information available on clutch size or fecundity.

BASIS FOR STATUS CLASSIFICATION. Two spring fed ponds in Bessemer are incorporated into Watercress Darter National Wildlife Refuge. Penny Spring and adjacent creek are on land recently purchased by Black Warrior-Cahaba Land Trust, which is also in the process of purchasing Seven Springs and its run. Thus, these populations have, or will have, a considerable measure of protection. Most vulnerable populations are those at Glen Spring and in Roebuck. Over years, population in Glen Spring appears to have declined apparently due to decreased water flow (Howell 1989; R. A. Stiles, pers. obs.) probably resulting from increased urban development in Bessemer and its effects on recharge of aquifer. Run from spring at Roebuck flows through a City of Birmingham recreational complex. While there is an agreement between Birmingham Parks and Recreation Board and U.S. Fish and Wildlife Service to protect this spring and its run, city has not always been sensitive to needs of fish. **Listed as *endangered* by the U.S. Fish and Wildlife Service in 1970.**

***Prepared by*: Robert A. Stiles**

RUSH DARTER

Ethesotoma phytophilum Bart and Taylor

OTHER NAMES. None.

Photo—Henry L. Bart

DESCRIPTION. A recently described member of subgenus *Fuscatelum*, endemic to upper Black Warrior River system in Alabama (Bart and Taylor 1999). An upland species, closely related to mostly lowland-distributed goldstripe darter. Khaki-colored in life, with darker brown pigment forming dashes and saddles over top and sides of body. Distinctive "gold" (depigmented) stripe, typical of goldstripe darter, not well developed. Occurs sympatrically with, and superficially resembles redspot darter, when latter not in breeding color. Rush darter differs in having a fully scaled nape and cheek (*vs.* unscaled to only partly scaled in redspot darter) and an interrupted supratemporal canal (*vs.* uninterrupted in redspot darter). Species also resembles nonbreeding watercress darter, which was introduced into area (Penny Spring) historically inhabited by rush darter in 1988 (Mettee *et al.* 1989). Species differs from watercress darter in having a fully scaled nape and cheek.

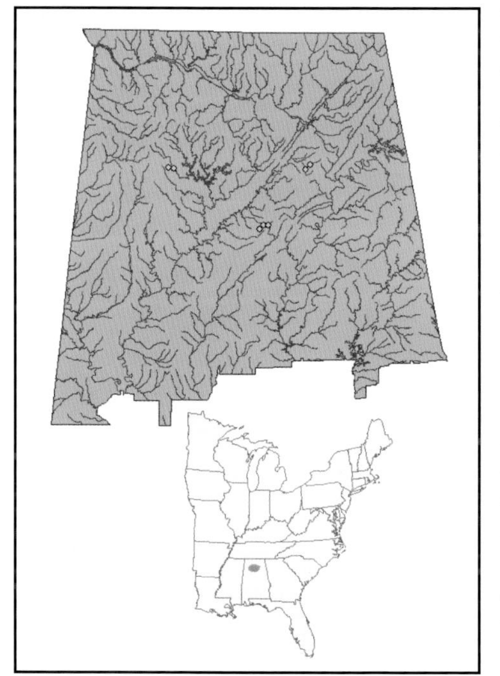

DISTRIBUTION. Endemic to upland portions of Black Warrior River system in Alabama (Appalachian Plateau and Valley and Ridge Provinces). Only three disjunct populations known: one in tributaries of Clear Creek system, a tributary to Sipsey Fork of Black Warrior River in Bankhead National Forest; one in spring-fed tributaries of Turkey Creek near Birmingham, Alabama; and one in a spring-run tributary to Little Cove Creek near Gadsden, Alabama (the latter two populations in Locust Fork of the Black Warrior).

HABITAT. Favors shallow, flowing water along margin of small streams and spring runs. Almost always associated with root masses of emergent vegetation (rushes, dock, and bur-reed) along stream margin. Never taken in watercress or submerged aquatic vegetation at mid channel in spring runs.

LIFE HISTORY AND ECOLOGY. Males taken in March and April are tuberculate and exhibit dusky breeding pigmentation. Females collected in March and early April have enlarged ovaries with large mature oocytes. Other aspects of life history unknown.

BASIS FOR STATUS CLASSIFICATION. An uncommon, difficult-to-collect, species. Known from less than 100 specimens. Collected recently in streams in both Turkey Creek and Clear Creek systems, but sporadic in both. Type locality in Turkey Creek system is a spring run that forms part of roadside drainage system along Alabama Highway 79, just southwest of Pinson, Alabama. Stream has been modified extensively and receives runoff from several businesses. An accidental spill of hazardous material from a business, or from trucks traveling along highway, would put this population in jeopardy. A single specimen was taken in Penny's Spring northwest of Pinson in 1979. The watercress darter was introduced into this spring in 1988 to reduce the then-looming threat of extinction from its native habitat (Mettee *et al.* 1989). Watercress darter now appears to be only species in Penny's Spring. Cove Creek population represented by a single specimen collected in 1975. Repeated surveys of area over past decade have failed to produce additional specimens. Population in Clear Creek system may receive some measure of protection from location in Bankhead National Forest. Regarded as endangered by several southeastern ichthyologists (Warren *et al.* 2000). **A *candidate for listing* by the U.S. Fish and Wildlife Service.**

***Prepared by*: Henry L. Bart, Jr.**

BOULDER DARTER

Etheostoma wapiti Etnier and Williams

OTHER NAMES. None.

DESCRIPTION. A medium-sized (max. length = 75 mm [3 in.]) darter in subgenus *Nothonotus*. Adult males uniformly gray, and marked with 10-14 dark horizontal lines on sides and black margins on median fins. All fins, except pectorals, green, as are distal portions of both dorsal fins and anal fin. Green on pelvic fins mostly confined to interradial membranes. Dorsal, anal, and caudal fins have pale yellow-green submarginal bands. Bases of both dorsal fins black. Throat and belly also bright emerald green. Red pigment absent on males. Both males and females have dark suborbital bar, humeral spot, and a pair of discrete spots at base of caudal fin. Females and juveniles olivaceous, and may have eight or nine dorsal saddles, and 10-11 blotches along side. Adult females may have some red pigment near margins of first membranes of first (spinous) dorsal fin. Young males may have metallic bronze, scale-sized, flecks on body. Boulder darter's sharper snout, suborbital bar, presence of cheek scales, and lack of bright red pigments in males distinguishes it from sympatric and similar bluebreast darter, although juveniles and females can be difficult to differentiate (Etnier and Williams 1989).

Photo—J. R. Shute

DISTRIBUTION. Probably once an inhabitant of the main channel of the Tennessee River and some larger tributaries in southern bend area in northern Alabama (Biggins 1989). Historically, found in Shoal Creek, which enters the Tennessee River in northern Alabama (into Wilson Reservoir) 30.5 kilometers (19 miles) downstream from mouth of the Elk River, but this population extirpated (Etnier and Williams 1989, Pierson 1990). Presently known from about 96.5 kilometers (60 miles) of mainstem Elk River. Upstream limit in the Elk River, near Fayetteville, may be associated with inability to tolerate fluctuating water levels and temperatures associated with operation of Tims Ford Dam (Burkhead and

Williams 1992). Downstream in the Elk River, found in appropriate habitats to just above influence of impoundment of Wheeler Reservoir in Limestone County, Alabama. They do not occur contiguously throughout 60-mile reach of Elk River, but are restricted to a few sites with appropriate habitat (O'Bara and Etnier 1987, Biggins 1988). In addition to the mainstem Elk River, also known from lower ends of Indian and Richland Creeks that enter Elk River within 60-mile reach inhabited by the species.

HABITAT. Adults prefer deep runs or flowing pools with moderate to swift current and are found beneath large boulder substrates. Often found in association with limestone rubble from bridges or mill dams as well as "natural" rubble and boulders from adjacent bluffs (O'Bara and Etnier 1987). Preferred juvenile habitat unknown, but likely somewhat different from that of adults (Rakes *et al.* 1999, P. L. Rakes and J. R. Shute, pers. comm.).

LIFE HISTORY AND ECOLOGY. Limited information, but comparisons with closely related darters and observations of boulder darters held in laboratory aquaria suggest that they choose as nesting sites a large rock (slabrock or boulder) with an opening downstream, or away from, current. Males guard nest site, and may mate with many females, who attach eggs to underside of rock in a wedge-shaped cavity between boulders (Burkhead and Williams 1992). Unlike many darters, boulder darter eggs may be placed on top of other eggs in a gelatinous mass (J. R. Shute and P. L. Rakes, pers. comm.). Hatchlings reared in captivity relatively small, possess a small yolk-sac, and are pelagic for several weeks before becoming benthic (Rakes *et al.* 1999), which suggests their larvae may be carried by current considerable distances downstream before settling out to benthic existence typical of adults. Because of their relatively small size, young require very small organisms for food. In captivity, they were fed copepods, ostracods, and other fine planktonic foods (Rakes *et al.* 1999). Like most other darters, adults probably feed mostly on aquatic insect larvae. Lifespan is probably three or four years.

BASIS FOR STATUS CLASSIFICATION. Sizes of localized populations surveyed in 1986 increased as one moved downstream in Elk River (O'Bara and Etnier 1987). Fluctuating water temperatures and flows resulting from previous operations at Tims Ford Dam may leave nests stranded on dry land, force males to abandon nests whose habitat parameters had changed as a result of fluctuating flows, or result in egg mortality because of dissolved oxygen and temperature fluctuations (Burkhead and Williams 1992). If larvae do float with current, present rarity may be explained in part by low probability of juveniles finding appropriate habitat in localities where they "settle" and become benthic. TVA recently made changes in operation of Tims Ford Dam to improve water quality for aquatic ecosystem downstream, including the boulder darter (Anonymous 1995). In addition, because of water pollution controls and an upgraded treatment facility in Lawrenceburg, Tennessee water quality of Shoal Creek has improved. Techniques have been developed to raise the species as part of a proposal to reintroduce it and other rare aquatic species into Shoal Creek. Also, a multi-agency project to increase habitat for Boulder Darters has resulted in several tons of rock being added at several localities to improve adult habitat, which has been described as one of most important factors limiting the Elk River population. If larvae do drift with current, this habitat improvement may result in an increase in number of sites where darters occur throughout 96.5-kilometer (60-mile) reach.

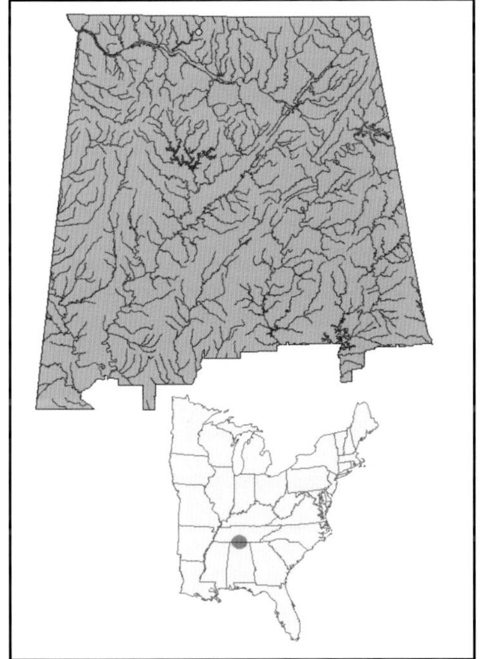

Prepared by: **Peggy Shute**

GOLDLINE DARTER

Percina aurolineata Suttkus and Ramsey

Photo—Patrick E. O'Neil

OTHER NAMES. None.

DESCRIPTION. A medium-sized (adult max. length = 74 mm [2.9 in.]) member of subgenus *Hadropterus* (Mettee *et al.* 1996). Has series of eight or nine large ovoid, dark mid-lateral blotches, usually connected. Common name derived from pale amber to gold area just dorsal to these blotches. Between this "gold stripe" and dorsal fin base is a series of rust-red to brown streaks that may join to form a continuous stripe. Median fins dusky, with three distinct dark spots at base of caudal fin. In Cahaba River, often associated with blackbanded darter, which differs from goldline darter in having vertically elongated, rather than ovoid, mid-lateral blotches; also lacks series of rust-red or brown streaks between blotches and dorsal fin bases.

DISTRIBUTION. In Alabama, inhabits middle portion of Cahaba River and two of its larger tributaries, Little Cahaba River and Schultz Creek. In main channel of Cahaba, historically collected as far upriver as Shelby County Highway 52 bridge just west of Helena; however, has been extirpated from upper portion of distribution. Recent surveys located fish from vicinity of Blue Girth Creek (Shepard *et al.* 1996) upstream to slab bridge located approximately two air miles northwest of community of Marvel. In Little Cahaba River, Bibb County, recently located as far upstream as two miles above Bulldog Bend, and in Schultz Creek, it occurs at least to vicinity of State Highway 5 bridge north of Brent. Also occurs in Coosawattee River system of northern Georgia.

HABITAT. Riffles and runs of medium to large streams. Usually found in swift to moderate current over a substrate of cobble or small boulders interspersed with sand, gravel, and pebbles. Riffles may have growths of river weed on rocks and be bordered by stands of water willow. Both goldline and blackbanded darters often found in stands of water willow apparently seeking shelter and perhaps food; however, also present in riffles and other areas where plants not present.

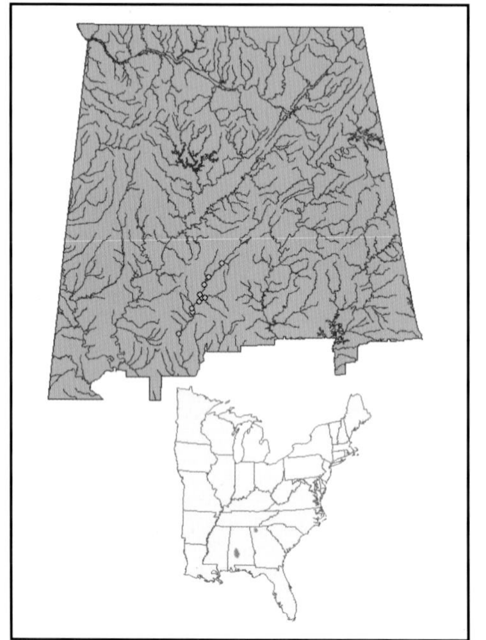

LIFE HISTORY AND ECOLOGY. Little known. A benthic feeder taking insects and possibly other macroinvertebrates from rocks (R.A. Stiles, pers. obs.). Apparently spawns from late March to early June. Eggs buried in sand or fine gravel in eddy areas just downstream of, or between, rocks.

BASIS FOR STATUS CLASSIFICATION. While settlement of lawsuit involving EPA and Jefferson County has forced latter to improve its waste water treatment plant discharges into the Cahaba River, massive urban development in Jefferson-Shelby County portion of drainage threatens main channel populations with increased siltation and nonpoint source urban runoff. In addition, portion of river in southern Shelby and Bibb Counties continues to experience siltation from clear cuts, strip mines, coal bed methane sites, and associated roads. Little Cahaba populations imperiled by toxic runoff from wood treatment plants and siltation from limestone and dolomite quarries. **Listed as *threatened* by the U.S. Fish and Wildlife Service in 1992.**

***Prepared by*: Robert Stiles**

BLOTCHSIDE LOGPERCH
Percina burtoni Fowler

OTHER NAMES. None.

Photo—J. R. Shute

DESCRIPTION. Among the largest (max. length approaching 180 mm [7 in.]) darters, yellowish, with numerous narrow dorsal saddles that continue onto side of body as bars. Snout projects somewhat "pig-like." Differs from widespread and often sympatric logperch in having a naked nape, a persistent red band in spiny dorsal fin, and lateral blotches fused to form a black lateral band. Lateral line complete with 79 to 94 scales, dorsal fin with 15 to 18 spines and 14 to 16 soft rays (Etnier and Starnes 1993).

DISTRIBUTION. Cumberland and Tennessee River drainages in Kentucky, Virginia, Tennessee, North Carolina, and Alabama (Etnier and Starnes 1993, Mettee *et al.* 1996).

HABITAT. Medium-sized streams and small rivers. Occurs over variety of substrate types in about 0.5-meter (1.5-foot) deep water in riffles and runs (Leftwich and Angermeier 1993). Typically found in swift to moderate currents.

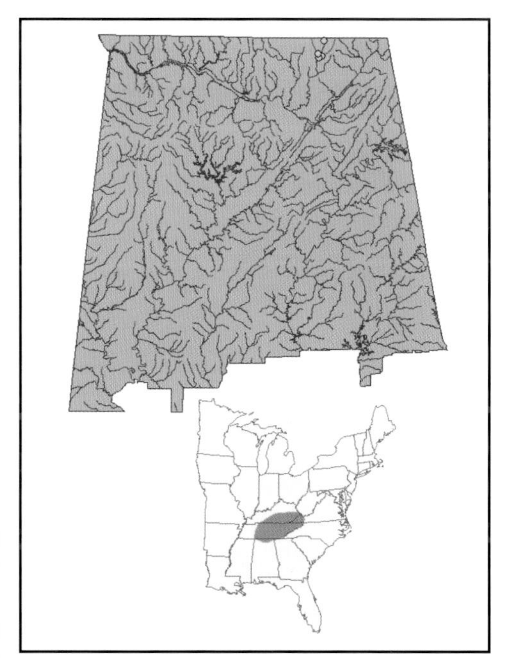

LIFE HISTORY AND ECOLOGY. Details of reproduction and growth virtually unknown. Spawning season is April and May, and reproductive behavior assumed similar to other logperches (Etnier and Starnes 1993). Sexual maturity may not be reached until individuals are three years old. Lifespan is probably at least four years. Pig-like snouts used to flip stones to eat insect larva exposed on undersides.

BASIS FOR STATUS CLASSIFICATION. Present distribution is much reduced over historical distribution, which was relatively widespread. Where it is presently known, including Paint Rock River headwaters (Estill Fork), Alabama, found in low numbers. This Alabama population was discovered relatively recently (P. Shute, pers. obs.). May also occur in the Shoal Creek system (Etnier and Starnes 1993, Mettee *et al.* 1996).

***Prepared by*: Peggy Shute**

SLENDERHEAD DARTER

Percina phoxocephala (Nelson)

Photo—Patrick E. O'Neil

OTHER NAMES. None.

DESCRIPTION. A member of subgenus *Swainia* characterized by 10-16 round, light brown blotches along sides, a moderately pointed snout, a wide maxillary frenum, and a single black spot at caudal fin base. Adult size varies up to 80 millimeters (3.2 inches) SL. Back usually yellowish brown with a darker brown vermiculated pattern. Bright coloration absent except for an intense orange band along outer margin of spiny dorsal fin. A distinct black preorbital bar present, and a dusky to dark suborbital bar sometimes present. Spiny dorsal fin of males brown basally, orange submarginally, and occasionally with a dark band marginally. Meristic counts include 58 to 80 lateral-line scales; 11 to 14 dorsal fin spines and 11 to 15 rays; two anal fin spines and seven to 10 rays; and 13 to 15 pectoral fin rays (Mettee *et al.* 1996). Cheeks, opercles, and nape scaled; anterior portion of belly naked, and belly has a row of well-developed midventral scales.

DISTRIBUTION. Occurs in Mississippi River Basin from Ohio west to central Minnesota and northeastern South Dakota, and south to Oklahoma and Alabama (Thompson 1979). Records in Alabama limited to Bear Creek system in Tennessee River drainage (Mettee *et al.* 1996).

HABITAT. Reportedly prefers a variety of habitat types, varying from small-stream riffles, to riffles of cobble and rubble, and sandy pools of larger rivers. Gravel raceways with moderate to swift current reportedly preferred (Thomas 1970). In Alabama, have been collected in knee- to waist-deep pools over sand and silt mixed with detritus in Cedar Creek, a tributary to Bear Creek in northwestern Alabama.

LIFE HISTORY AND ECOLOGY. Little known in Alabama. Peak spawning in gravel riffles has been reported from late May to June in Illinois (Page and Smith 1971). A second report from Illinois suggested a longer spawning season extending into August and a maximum life span of four years. Predominant food items include mayflies, midges, and caddisflies.

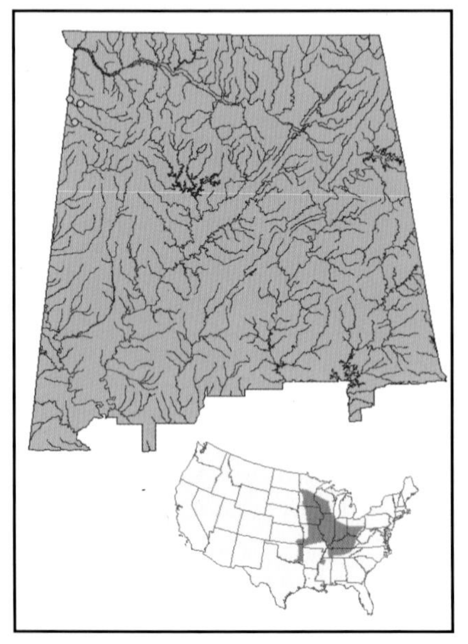

BASIS FOR STATUS CLASSIFICATION. Never common in Alabama streams because Bear Creek at very margin of its distribution. Individuals were found at several localities in Bear Creek system in 1960s and 1970s, based on collection data from the University of Alabama Ichthyological Collection and Auburn University Museum, but only from one to four individuals taken at those sites. Three sites known in Bear Creek main channel, three in Cedar Creek, two in Little Bear Creek, and two in Bear Creek, Tishomingo County, Mississippi. Last taken in Mississippi in 1974 and may be extirpated from that state (Ross 2001). Recent sampling by Alabama Geological Survey biologists located species in Bear Creek (near Natchez Trace), and two locations in Cedar Creek. Construction of TVA impoundments on tributary main channels, increased sedimentation of their preferred main channel benthic habitat, and scarcity of recent collections form basis for priority designation.

***Prepared by*: Patrick E. O'Neil**

SNAIL DARTER

Percina tanasi Etnier

Photo—J. R. Shute

OTHER NAMES. None.

DESCRIPTION. A medium-sized darter (max. size = approx. 75 mm [3 in.]). Dorsum brown to bronze-green with four dark saddles usually evident: the first is beneath front of spinous dorsal fin origin, second between spinous and soft dorsal fins, third beneath middle of soft dorsal, and fourth on smallest part of caudal peduncle. Saddles may be indistinct at times. Anal fin of males elongated, and may reach to middle of caudal fin in mature individual. Nuptial tubercles appear on anal fin, lower caudal fin, pelvic fins, breast, and cheeks of males (Etnier 1975, Etnier and Starnes 1993).

DISTRIBUTION. Historically, may have been found in the main channel of middle portion of the Tennessee River from northeastern Alabama, and possibly lower reaches of major tributaries (Etnier and Starnes 1993). However, until very recently the only known naturally occurring population persisted in lower 24 kilometers (15 miles) of Little Tennessee River and a small portion of adjacent Tennessee River, upper Watts Bar Reservoir, in Loudon County, Tennessee (Etnier 1975, U.S. Fish and Wildlife Service 1975, Starnes 1977). This population disappeared after the Tellico Reservoir was created in 1979 and has not been seen in subsequent surveys there. In 1975, prior to closure of Tellico Reservoir, several hundred individuals were moved from the Little Tennessee River into the lower Hiwassee River, Polk County, Tennessee, and in 1978 the species was introduced into the lower Holston River, Knox County, Tennessee (Etnier and Starnes 1993, Biggins and Eager 1983, Biggins 1984). By 1984, found in small numbers in four other Tennessee River tributaries and a mainstem section of the Tennessee River. Presently, relatively abundant in lower French Broad, Holston, and Little Rivers near Knoxville, and in the Hiwassee River. Somewhat less abundant, but still found in, Sewee Creek, South Chickamauga Creek, the Sequatchie River, and in the Paint Rock River. Last collected in 2002. Probably also occurs in mainstem impoundments near mouths of these streams.

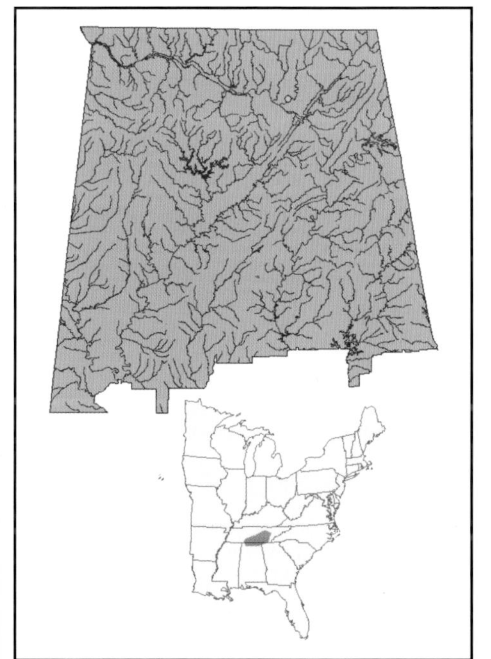

HABITAT. Large, free-flowing rivers with extensive areas of clean-swept, sand-gravel shoals for reproduction and feeding.

LIFE HISTORY AND ECOLOGY. Spawns on swift gravel shoal areas from February to April. Adhesive eggs hatch in 15 to 20 days, and larvae drift to deeper areas downstream. After four to seven months, juveniles migrate back upstream to shoal areas. Lifespan is three or four years. Feeds primarily on immature river snails, which compose about 60 percent of annual diet. Larvae drift into deeper downstream areas, which may include reservoirs. For life cycle to be completed, nursery areas should be relatively free of pollution, and there should be no obstructions to upstream migration. Probably most successful when abundant populations of primary food, pleurocerid river snails, are present in shoal areas (Starnes 1977, Hickman and Fitz 1978).

BASIS FOR STATUS CLASSIFICATION: Paint Rock River population, the only known occurrence in Alabama (Pierson 1990), is on periphery of species' global distribution, and has always been relatively small, in comparison with other populations. Potential for extirpation in Alabama is great because of small population size, limited distribution within state, and its unique life history characteristics (larval drift). **Status downgraded from *endangered* to *threatened* by the U.S. Fish and Wildlife Service in 1984** (Biggins 1984).

***Prepared by*: Peggy Shute, with input from Charles F. Saylor**

WARRIOR BRIDLED DARTER
Percina sp. cf. *macrocephala*

OTHER NAMES. Muscadine Darter.

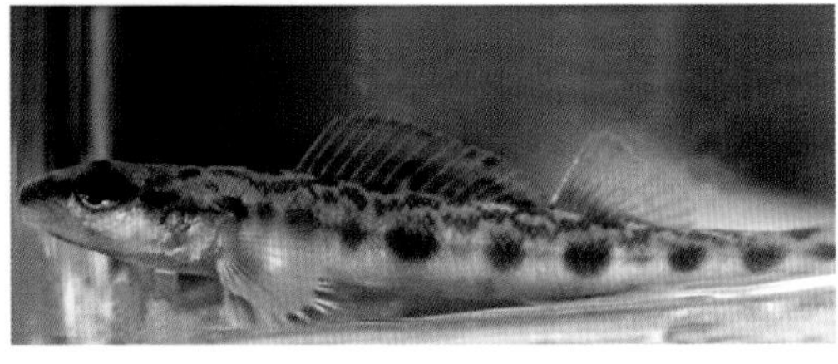

Photo—Joseph Tomelleri

DESCRIPTION. A member of subgenus *Alvordius*; all members have a midlateral row of dark blotches with or without a prominent dark stripe, a premaxillary frenum well developed, and lack breeding tubercles and bright colors. Three similar, undescribed species within this subgenus, including the warrior bridled darter, endemic to Mobile Basin; all are thin and elongate darters with seven to nine oval to quadrate midlateral blotches connected with a narrow to wide stripe, have weakly developed dorsal saddles, and lack a suborbital bar. The warrior bridled darter is distinguished from the other two species by having less than 50 percent of nape scaled compared to 80 percent or greater. Adults typically have a moderate lateral stripe connecting blotches, but stripe may be nonexistent in juveniles. No sexual dimorphism. Attains a total length of seven centimeters (2.8 inches).

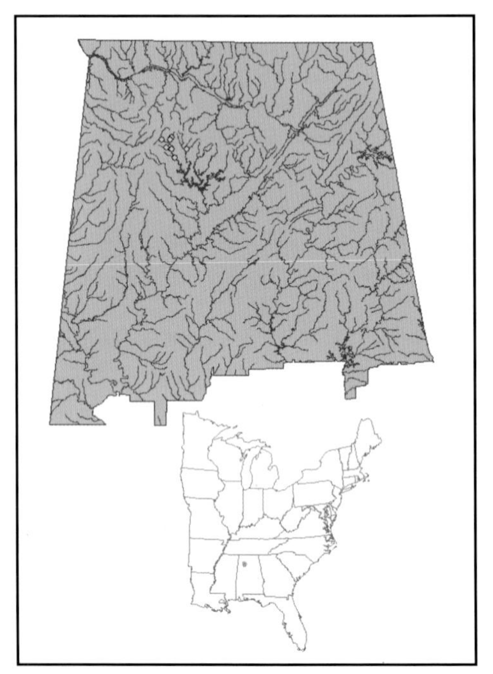

DISTRIBUTION. Restricted to Sipsey Fork System in Winston and Lawrence Counties, Alabama. Most records from main stem of Sipsey Fork from just above Lewis Smith Reservoir embayment to its origin. Also occurs in Borden and Caney Creeks, tributaries of upper Sipsey Fork. Does not occur in Thompson or Hubbard Creeks. Although there are no records from impounded section of Sipsey Fork, species likely occurred there before impoundment. Also known from one site in Brushy Creek, a major fork of Sipsey Fork System in Winston County (Kuhajda and Mayden 2002*b*, Powers *et al.* 2003).

HABITAT. Large upland streams 15-25 meters (49-82 feet) wide with substrates of sand, gravel, cobble, boulders, and bedrock. Typically occupy areas with slow to moderate current over larger substrates of cobble, boulders, and bedrock interspersed with finer sediments.

LIFE HISTORY AND ECOLOGY. Rarely abundant; few collections contain 10 or more specimens. Biology unknown, but its life history probably similar to the undescribed muscadine darter endemic to Tallapoosa River system. Muscadine darters spawn below 20°C (68°F) from late March to May, have relatively low fecundity (22-84 ova), relatively large ovulated ova (average = 1.63 mm diameter), only live two years, and have relatively restricted diets of aquatic invertebrate larvae (Wieland and Ramsey 1987).

BASIS FOR STATUS CLASSIFICATION. Currently only known from upper Sipsey Fork, two of its tributaries, and one site on Brushy Creek; populations presumed present in lower Sipsey Fork extirpated by Lewis Smith Reservoir. Collections typically produce few individuals, which may be reflective of presumed low fecundity and short life span. Some sedimentation associated with poor forestry management (clear cuts without appropriately wide streamside management zones) prevalent, especially in tributaries (Kuhajda and Mayden 2002*b*, Powers *et al.* 2003).

***Prepared by:* Bernard R. Kuhajda**

HALLOWEEN DARTER

Percina sp.

OTHER NAMES. None.

DESCRIPTION. A medium-sized (to 100 mm [4 in.] SL), undescribed darter not readily attributable to any existing subgenus within *Percina*. Possesses narrowly connected branchiostegal membranes, seven closely spaced rectangular dorsal saddles, a broad subocular bar, and an orange submarginal band above a dark basal band in the spinous dorsal fin of breeding males and females. Breeding males lack distinct tubercles, although they develop tubercluar ridges on anal and pelvic fin rays (Freeman and Freeman 1992).

Photo—Byron J. Freeman

DISTRIBUTION. Has a fragmented distribution restricted to four isolated stream reaches in Apalachicola River drainage including extreme upper Chattahoochee River and tributaries in northeastern Georgia, Uchee Creek within the middle Chattahoochee system in east-central Alabama, upper Flint River and tributaries in west-central Georgia, and lower Flint River tributaries in southwestern Georgia. In Alabama, species restricted to lower reaches of Uchee and Little Uchee Creeks, Russell County (Freeman *et al.* 2002).

HABITAT. High gradient, medium to large streams in Blue Ridge, Piedmont, and Coastal Plain with typical substrates of sand, gravel, cobble, boulders, and bedrock. Both adults and juveniles found almost exclusively in riffle and run habitats in shallow, relatively fast water over rocky substrate, often associated with river weed (Hill 1996).

LIFE HISTORY AND ECOLOGY. In Flint River system of the Piedmont region, spawning occurs from May to early June at 20-24°C (68-75°F), but may extend into summer months. Spawning behavior unknown. Reach more than 50 percent of maximum size during first summer and attain sexual maturity at one year. Females produce multiple clutches, and the largest average monthly mean clutch (116) low compared with other species of *Percina*. Maximum longevity three years. Diet composed of immature aquatic insects, with simuliids dominating from May to October and chironomids rest of year (Hill 1996, Freeman *et al.* 2002).

BASIS FOR STATUS CLASSIFICATION. Distribution significantly fragmented and restricted to four isolated stream reaches in Apalachicola River drainage. Assuming a continuous presettlement distribution between these four isolated populations, this species has disappeared from vast majority of former distribution primarily due to construction of impoundments. Current threats to remaining populations include water withdrawal, pollution, sedimentation, and numerous proposed impoundments. In Alabama, declining habitat quality in Uchee Creek system, especially due to pollution and sedimentation, is a significant threat (Freeman *et al.* 2002).

***Prepared by*: Carol E. Johnston and Bernard R. Kuhajda**

FISHES

PRIORITY 2
HIGH CONSERVATION CONCERN

Taxa imperiled because of three of four of following: rarity; very limited, disjunct, or peripheral distribution; decreasing population trend/population viability problems; specialized habitat needs/habitat vulnerability due to natural/human-caused factors. Timely research and/or conservation action needed.

Class Actinopterygii
Order Acipenseriformes
Family Acipenseridae

GULF STURGEON *Acipenser oxyrinchus desotoi*

Order Clupeiformes
Family Clupeidae

ALABAMA SHAD *Alosa alabamae*

Order Cypriniformes
Family Cyprinidae

BLUE SHINER *Cyprinella caerulea*
STREAMLINE CHUB *Erimystax dissimilis*
SHOAL CHUB *Macrhybopsis aestivalis hyostoma*
GHOST SHINER *Notropis buchanani*
DUSKY SHINER *Notropis cummingsae*
SUCKERMOUTH MINNOW *Phenacobius mirabilis*
STARGAZING MINNOW *Phenacobius uranops*
BROADSTRIPE SHINER *Pteronotropis euryzonus*
BLUENOSE SHINER *Pteronotropis welaka*

Order Siluriformes
Family Ictaluridae

MOUNTAIN MADTOM *Noturus eleutherus*
BRINDLED MADTOM *Noturus miurus*
FRECKLEBELLY MADTOM *Noturus munitus*
HIGHLANDS STONECAT *Noturus* sp cf. *flavus*

Order Perciformes
Family Centrarchidae

SHOAL BASS *Micropterus cataractae*

Family Percidae

LOCUST FORK DARTER *Etheostoma* sp. cf. *bellator*
SIPSEY DARTER *Etheostoma* sp. cf. *bellator*
BLUEBREAST DARTER *Etheostoma camurum*
LIPSTICK DARTER *Etheostoma chuckwachatte*
COLDWATER DARTER *Etheostoma ditrema*
TUSCUMBIA DARTER *Etheostoma tuscumbia*
BANDFIN DARTER *Etheostoma zonistium*
BLUEFACE DARTER *Etheostoma* sp. cf. *zonistium*
COAL DARTER *Percina brevicauda*
GILT DARTER *Percina evides*

GULF STURGEON

Acipenser oxyrinchus desotoi (Vladykov)

Photo—Robert Hastings

OTHER NAMES. Gulf of Mexico Sturgeon. Specific name has been incorrectly spelled "*oxyrhynchus*."

DESCRIPTION. A subspecies of *Acipenser oxyrinchus*, the Atlantic, common, or sea sturgeon. A large (reaching lengths of 3 m [9.5 ft.]) light to dark brown fish with a white belly, an extended flattened snout, and a heterocercal tail. Body elongated and somewhat rounded in cross-section. Five rows of large bony scutes are middorsal, midlateral, and ventrolateral in position; remainder of body scaleless except for scattered bony dermal ossifications. Internal skeleton composed mostly of cartilage. Mouth ventral and protractile, and preceded by four fleshy barbels. Largest and one of most long-lived (reaching ages of at least 42 years) fish occurring in Alabama fresh waters. Differs from Atlantic sturgeon (*A. o. oxyrinchus*) in having squarish (length much shorter than width) rather than oval bony shields, two conspicuous hooks on keel of dorsal shields (hooks absent in *A. o. oxyrinchus*), and a much longer spleen (Vladykov and Greeley 1963).

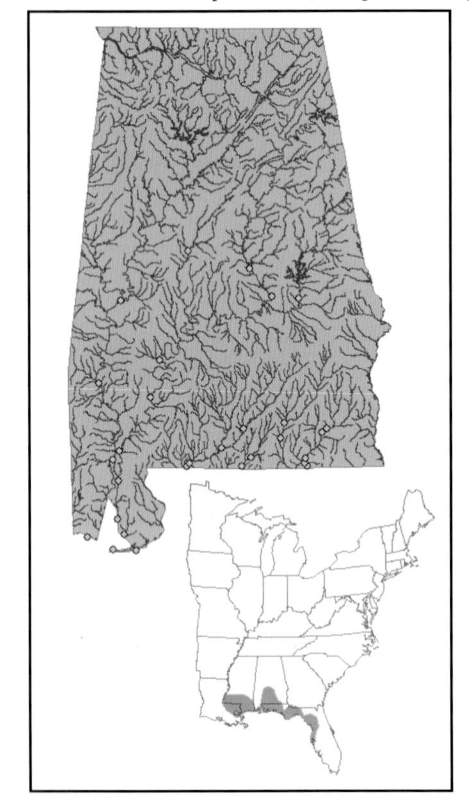

DISTRIBUTION. Coastal Gulf of Mexico and adjacent freshwater rivers between Suwannee River in Florida to Lake Pontchartrain in Louisiana, with sporadic occurrences south to Florida Bay and west to Rio Grande River. An anadromous subspecies, with spawning populations in the Suwannee, Apalachicola, Choctawhatchee, Yellow/Blackwater, Escambia, Pascagoula, and Pearl Rivers of Florida, Alabama, Mississippi, and Louisiana. Former spawning populations documented from Mobile/ Alabama River, as well as Ochlockonee of Florida and Tchefuncte of Louisiana. Ramsey (1976) reported historical records from Cahaba River at Centerville, Tallapoosa River at Tallassee, Coosa River near Wetumpka, Alabama River just above Tombigbee River, Tombigbee River near Demopolis, Choctawhatchee River in Geneva County, and Mobile River near the bay. Mettee *et al.*

(1996) reported a three-meter (9.5-foot) sturgeon landed from Coosa River near Coopers in 1924, and thousands of pounds of sturgeon harvested from the Mobile River in early 1900s. Now excluded from upper Alabama and Tombigbee Rivers upstream of dams at Claiborne and Coffeeville, respectively. Recent (since 1991) collection sites in Alabama include the Choctawhatchee River near Geneva; Pea River below dam at Elba; Yellow River above and below Alabama Highway 55; Conecuh River near Brewton; Alabama River below Claiborne Lock and Dam; Tombigbee River below Coffeeville; Tensaw and Blakeley Rivers in Mobile Delta; Fish River near Mobile Bay; Mobile Bay at Fairhope, Fort Morgan, and Dauphin Island; and in nearshore Gulf of Mexico near Gulf Shores and Bayou LaBatre. Number in Alabama river systems largely unknown. Recent (1999-2001) Choctawhatchee and Yellow River studies estimated population of adults and subadults as under 3,000 and 550, respectively.

HABITAT. Anadromous, inhabiting estuaries, bays, and nearshore waters of Gulf of Mexico during winter, mostly in waters less than 10 meters (33 feet) deep. Begin migrations into coastal rivers in early spring (March through May) to spawn when river water temperatures range from 16 to 23°C (60.8 to 73.4 °F). Remain in river system entire summer.

LIFE HISTORY AND ECOLOGY. Females reach sexual maturity at eight to 17 years; males at seven to 21 years (Huff 1975; Sulak and Clugston 1999). Spawning occurs in upper river reaches characterized by limestone banks, bluffs, and outcroppings with substrates consisting of limestone, gravel, or cobble, which provide adhesion sites and protection for demersal eggs and yolk sac larvae. Mature females produce average of 400,000 eggs at each spawning, which occur at intervals of three to five years. Males thought to spawn annually. Young-of-year remain in river through early February, where they feed upon detritus and aquatic invertebrates, including mayflies and caddisflies, oligochaete worms, and bivalve mollusks. Adults and subadults apparently do not feed in fresh waters (six to eight months), but rely upon winter feeding in marine habitats on benthic invertebrates including amphipods, lancelets, polychaetes, gastropods, shrimp, isopods, brachiopods, and ghost shrimp. About 70 kilometers (44 miles) of suitable spawning habitat identified in Conecuh and Pea Rivers, and in Alabama portion of Yellow and Choctawhatchee Rivers. Spawning documented with collection of fertilized eggs in Conecuh River, below and above mouth of Sepulga River, and in Pea and Choctawhatchee Rivers, above and below Geneva (Fox *et al.* 2000, Parauka and Giorgianni 2002). Exhibits a high degree of river-specific fidelity with very few fish documented as making inter-river movements from natal rivers. Significant genetic differences identified in five regional or river specific stocks. These include Lake Pontchartrain and Pearl River; Pascagoula River; Escambia and Yellow Rivers; Choctawhatchee River; and Apalachicola, Ocklockonee, and Suwannee Rivers.

BASIS FOR STATUS CLASSIFICATION. Once abundant in most rivers of Gulf Coast, numbers declined drastically during 1900s due to over-fishing and loss of river habitat blocked by dams (Lorio 2000). Other threats and potential threats include modifications to habitat associated with dredged material disposal, desnagging, and other navigation maintenance activities; incidental take by commercial fishermen; poor water quality associated with contamination by pesticides, heavy metals, and industrial contaminants; and aquaculture and incidental or accidental introductions. Also, life history characteristics, late maturation, and spawning periodicity may protract recovery efforts. **Listed as *threatened* by the U.S. Fish and Wildlife Service in 1991.**

Prepared by: **Robert W. Hastings and Frank M. Parauka**

ALABAMA SHAD

Alosa alabamae Jordan and Evermann

Photo—Patrick E. O'Neil

OTHER NAMES. White Shad.

DESCRIPTION. Adult females longer (373-467 mm [4.7-18.4 in.] TL) and heavier (360-1,265 g [0.8 -2.8 lb.] TW) than males (360-415 mm [14.2-16.3 in.] TL and 48-707 g [0.1-1.6 lb.] TW). Live individuals silvery overall; back has greenish to purplish metallic sheen that fades shortly after death; venter white. Lateral line has 55 to 60 scales. Dorsal fin has 15 to 17 rays and anal fin 18 to 19 rays; both fins have a distinct darkened margin. Has been confused with skipjack herring, but the two can be readily differentiated. Upper and lower jaw lengths on Alabama shad essentially equal with mouth closed and viewed from side, the front edge of lower jaw does not protrude beyond upper jaw. In contrast, lower jaw on Skipjack herring is distinctly longer and protrudes well beyond upper jaw when mouth closed and viewed from side. Alabama shad have 42 to 48 closely spaced gill rakers on the lower limb of first gill arch versus 20 to 24 on Skipjack herring (Hildebrand 1963). Individuals collected in Ohio River in early 1900s described as Ohio shad (Evermann 1902), but this species was subsequently synonymized with Alabama shad because of insignificant morphological differences (Berry 1964).

DISTRIBUTION. Reported from several major tributaries to the Mississippi River and east in larger Gulf Coast river systems to Suwannee River in northern Florida (Laurence and Yerger 1966; Mills 1972). Indi-viduals previously collected in upper and lower Tombig-bee, Black Warrior, Cahaba, Coosa, and Alabama Rivers within Mobile Basin as well as the Choctawhatchee and Conecuh Rivers in Alabama (Mettee *et al.* 1996).

HABITAT. An anadromous species. Adults live in marine and estuarine environments most of year and migrate into free-flowing rivers to spawn in spring.

LIFE HISTORY AND ECOLOGY. Adults in pre-spawning condition usually enter freshwater rivers in late February and March. Very young males and a few older males and females appear during middle to latter part of the spawning season. Actual spawning sites have not been documented in Mobile Basin rivers, but many young-of-the-year were collected along sand bars in upper Alabama and Cahaba Rivers prior to construction of Claiborne and Millers Ferry Locks and Dams in 1960s and 1970s (Mettee *et al.* 1996). Spawning occurs along sand bars and possibly over limestone outcrops in the Choctawhatchee and Conecuh Rivers. Most spawning activity completed from late March to early May when water temperatures reach 19-23°C (62.2-73.4°F)

(O'Neil *et al.* 2000). Most abundant age groups in Choctawhatchee River were three- to five-year-old males and three- to six-year-old females. Males two to three, and females two to four years old, were the most common spawning groups in Choctawhatchee River. Females produce from 100,000 to 250,000-plus eggs depending on age. Two spawning periods may occur in one season. Surviving individuals return to salt water in late April and May. Young-of-the-year remain in the spawning area for several weeks and reach 70 to 100 millimeters (2.8 to 3.9 inches) TL before moving downstream to salt water.

BASIS FOR STATUS CLASSIFICATION. High-lift navigational and hydroelectric dams have blocked upstream migrations to inland spawning areas, whereas dredging activities have eliminated sections of their spawning habitat (Coker 1930*b*). As a result, populations have declined throughout much of distribution. Species may be extirpated from the upper Tombigbee, Cahaba, Coosa, and upper Alabama Rivers in Alabama. Only one individual has been collected in the Black Warrior River since 1896, when this species was described from specimens collected at Tuscaloosa (Evermann 1896). Only five adults have been collected below Millers Ferry Lock and Dam in Alabama River in the past 30 years. All were collected following spring floods that inundated Claiborne Lock and Dam. Only known self-sustaining populations in Alabama occur in the Choctawhatchee and Conecuh Rivers. Major threats to these populations include increased sedimentation, herbicide and pesticide runoff from agricultural operations, prolonged drought, and possible reservoir construction for water supply on major tributaries.

***Prepared by:* Maurice F. Mettee**

BLUE SHINER

Cyprinella caerulea (Jordan 1877)

OTHER NAMES. None.

DESCRIPTION. A medium-sized (about 100 mm [4 in.] TL) minnow with a narrow and elongate body, and head relatively small and triangular in shape. Females and nonbreeding males dusky blue to olive above a conspicuous lateral stripe, and white to silver below lateral stripe. Breeding males develop bright metallic blue color with an intense blue-green lateral stripe. Fins yellow or orange edged in white, and dorsal fin membranes dark with streaks of white. Some features that distinguish this shiner from similar forms in Mobile Basin are: eight anal rays; pharyngeal teeth 1,4-4,1; and predorsal circumferential scale rows, 11 above lateral line and nine below.

Photo—Patrick E. O'Neil

DISTRIBUTION. Formerly limited to Cahaba and Coosa River systems of Mobile Bay drainage in Alabama, Georgia, and Tennessee. None observed in Cahaba River system since 1973. Endemic to Mobile Basin above Fall Line in upper Coosa River system. Good populations found in 8.4-kilometer (5.2-mile) segment of Little River in Cherokee County, Alabama. Recently an additional population found in Spring Creek, a tributary to Little River. Fair to good populations found in 29.6 kilometers (18.4 miles) of main channel Choccolocco Creek from mouth of Egoniaga Creek upstream to just above confluence with Jones Branch, Calhoun County, Alabama. Small populations reported in Jones Branch and an unnamed tributary to Choccolocco Creek near upstream limits of its distribution within Choccolocco Creek watershed (Howell and Linton 1996). In Weogufka Creek, found in Coosa Wildlife Management Area at a concrete slab ford near Horse Stomp Campground upstream to Weogufka community, Coosa County, a distance of 36 stream-kilometers (22.4 stream-miles).

HABITAT. Prefer clear, medium, or large streams and usually found in shallow pools or eddies with slow to moderate currents, or in quiet backwaters over sand, gravel, or a sand-gravel combination.

LIFE HISTORY AND ECOLOGY. Spawning in upper Coosa system occurs from late April to late July. Observed spawning in relatively deep water with moderate flow in crevices of logs and other woody debris in Conasauga River, Georgia (Johnston and Shute 1997). Length-frequency analysis revealed species lives for three years, and most spawning adults were two years old. Appears intolerant of high turbidity, and likely a surface and mid-depth feeder competively dependent on high visibility. Feeds on terrestrial insects and, to a lesser extent, immature aquatic insects.

BASIS FOR STATUS CLASSIFICATION. Intensive sampling by several collectors has not resulted in a rediscovery of species from Cahaba River system. Appears with more certainty it has been extirpated from Cahaba, where once it was fairly common. Last collection from Cahaba was in February 1973. Additionally, no populations (last collected in 1958) located from recent collections in Big Wills Creek and its tributaries in DeKalb and Etowah Counties. The Little River flows through Little River Canyon National Preserve, which is managed by the U.S. National Park Service. This population should remain stable. Upper Choccolocco Creek flows from Talladega National Forest, but soon enters a watershed that is 66 percent forest, 20 percent pasture, and 13 percent agriculture. The cities of Oxford and Anniston contribute a high nutrient load to Choccolocco Creek and, consequently, no Blue Shiners have been found in recent years downstream of these cities. Weogufka Creek's watershed primarily forested; however, forestry practices by former owner (Kimberly Clark Corporation) resulted in several large clearcuts in Weogufka watershed. The species' distribution has been reduced and fragmented by dams, loss of habitat, and water quality degradation. These isolated populations are vulnerable to habitat degradation and decreased genetic diversity. **Listed as *threatened* by the U.S. Fish and Wildlife Service in 1992.**

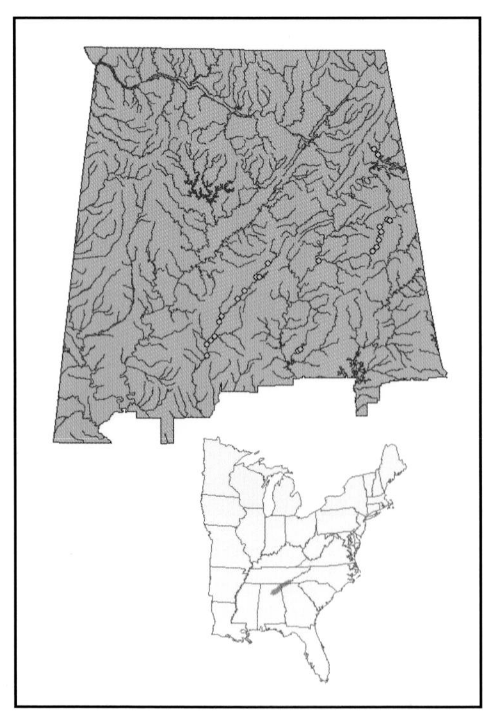

***Prepared by*: J. Malcolm Pierson**

STREAMLINE CHUB
Erimystax dissimilis (Kirtland)

Photo—Patrick E. O'Neil

OTHER NAMES. None.

DESCRIPTION. An elongated (max. size in Alabama = 115 mm [4.5 in.] SL) terete-shaped minnow with a small, horizontal, inferior mouth. Live specimens greenish olive to straw colored dorsally and white ventrally. Nine to 11 small, oval-shaped spots occur along silvery sides near the complete lateral line. Dorsal fin origin anterior to pelvic fin origin (Mettee *et al.* 1996). Although similar in appearance and sometimes confused, the streamline and blotched chubs are distinguished by lateral line scale counts (≥46 for streamline chub *vs.* ≤43 for blotched chub) and numbers of predorsal scale rows (19-25 for streamline chub *vs.* 16-19 for blotched chub).

DISTRIBUTION. From New York and Pennsylvania southward through Ohio Basin into Tennessee and Cumberland drainage. A previously recognized, disjunct population in White River, Arkansas, (Robison and Buchanan 1988) was described as *E. harryi*. This species and the closely related blotched chub occur syntopically throughout much of their respective distributions in Tennessee and Alabama (Etnier and Starnes 1993). Reaches southern limit of distribution in northern Alabama, where it is known from nine locations in Shoal Creek, and Elk and Paint Rock Rivers (Mettee *et al.* 2002).

HABITAT. Medium to large, clear streams having a moderate to swift flow and clean cobble and gravel substrates. Researchers in Arkansas and Ohio suggest species is very silt intolerant (Robison and Buchanan 1988).

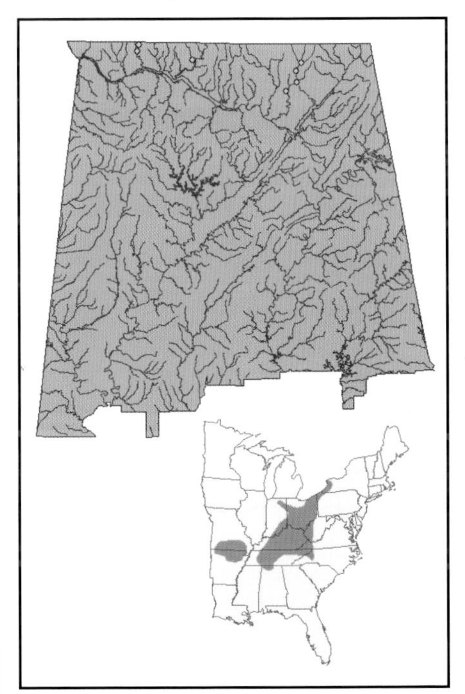

LIFE HISTORY AND ECOLOGY. Unknown in Alabama; poorly known elsewhere. Spawning occurs from mid-April into May in Tennessee and from late April to late May in Virginia (Etnier and Starnes 1993). May be triggered by water temperatures approaching 15°C (59°F) and increased stream discharge. Females produce up to 1,200 eggs per year. Life span slightly more than two years ,with a few females reaching age three. Periphyton, small bits of plant material, and aquatic insect larvae, primarily midges, mayflies, and caddisflies important dietary items in Tennessee (Harris 1986).

BASIS FOR STATUS CLASSIFICATION. Known distribution in the Tennessee drainage in Alabama currently limited to nine locations. Has been extirpated in parts of its original distribution by impoundments and increased siltation, leaving disjunct populations whose continued existence are threatened by habitat destruction and declining surface water quality (Etnier and Starnes 1993, Jenkins and Burkhead 1993).

Prepared by: **Maurice F. Mettee**

SHOAL CHUB

Macrhybopsis aestivalis hyostoma (Gilbert)

Photo—Patrick E. O'Neil

OTHER NAMES. Speckled Chub for the *Macrhybopsis aestivalis* species complex.

DESCRIPTION. Small (max. TL = males 68 mm [2.7 in.]; females 74 mm [3 in.]) chub with relatively slender body, slightly depressed head, and conical snout, length of which about 25 percent more than mouth gape. One pair maxillary barbels, enlarged nasal capsules, exceedingly abundant tastebuds and lateral line neuromasts (Reno 1969) and relatively large eyes. Pharyngeal teeth uniformly 0,4-4,0. Origin of dorsal fin directly above to slightly behind insertion of pelvic fins. Anal rays usually eight, sometimes seven. Breast and much of belly naked. No bright colors; body dusky or silvery with randomly distributed melanophores on dorsolateral surface. Nuptial tubercles on pectoral fins. The combination of bulbous snout, small eye, relatively long barbels, and scattered dark spots on back and side distinguishes this chub from other cyprinids in Tennessee River drainage. Its nearest relative in the Tennessee River drainage, the silver chub, is silvery in color and lacks distinctive markings. Recent investigations into systematics of the *Macrhybopsis aestivalis* species complex (speckled chubs), using morphological (Eisenhour 1997) and morphological and genetic characters (C. R. Gilbert and R. L. Mayden, in lit.), reveal a number of evolutionarily distinct taxonomic entities within species as currently recognized. Earlier researchers also recognized much of this diversity. Moore (1950) considered all diversity within M. *aestivalis* complex as subspecies of M. *aestivalis*. Eisenhour (1997) examined morphological variation in the *Macrhybopsis aestivalis* complex and recognizes five species: speckled chub, Rio Grande Basin and Rio San Fernando drainage; prairie chub, upper Red River Basin; burrhead chub, San Antonio, Guadalupe, and Colorado River drainages; peppered chub, upper Arkansas River Basin; and shoal chub, much of Mississippi River Basin and western Gulf Slope rivers. Boschung and Mayden (in press) recognize the shoal chub as a subspecies of the speckled chub, pending the completion of the study by Gilbert and Mayden.

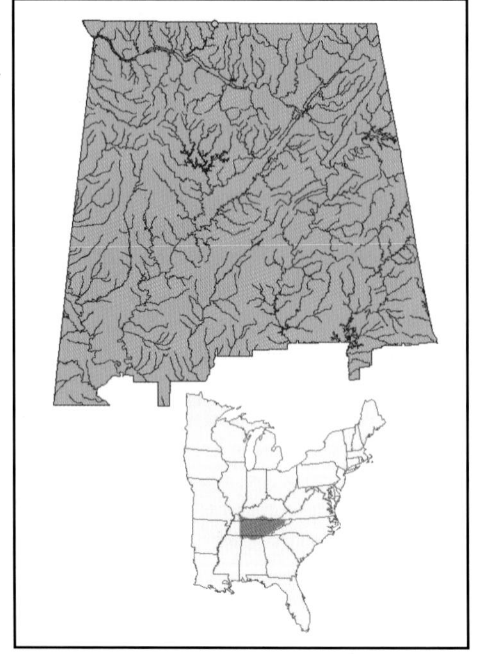

DISTRIBUTION. Tennessee River drainage northward to Ohio River drainage. West of Mississippi River, where taxonomy and distribution not yet resolved.

HABITAT. Gravel riffles and runs of small to large rivers, and sandy substrates in moderate to strong currents. A sedentary species preferring deep water, but can be found in shallow water at night, where they may dive into a sandy substrate when disturbed.

LIFE HISTORY AND ECOLOGY. Numerous tastebuds and lateral-line neuromasts, along with relatively large eyes, account for Shoal Chub's ability to occupy both clear and turbid waters (Moore 1950, Branson 1963). The spawning season is long, from mid-May to late August. Eggs, deposited during midday in deep water, develop as they drift in current. Fertilized eggs transparent, nonadhesive, semibuoyant, and about 2.5 millimeters (0.1 inch) in diameter as they are found suspended in main current of rivers. Yolk-sac fry emerge in about 28 hours and begin actively feeding in another two or three days (Bottrell *et al.* 1964). Growth is rapid, the young reaching about 40 millimeters (1.6 inches) TL by onset of winter. Life expectancy is short, probably only one year, two at most. Shoal Chubs locate food primarily by tastebuds on head, body, and fins. They take sand into the mouth, sort out food, and eject the sand from mouth and gill cavity. Common food items are immature aquatic insects (especially midges), small crustaceans, and plant material (Starrett 1950).

BASIS FOR STATUS CLASSIFICATION. A victim of impoundments, canalization, and channelization; consequently, it has vanished from much of southern bend of the Tennessee River. Very rare and essentially all habitat in Alabama inundated by TVA pools. Known from a single locality in the Elk River near Alabama-Tennessee state line. We believe this population is in danger of extirpation in Alabama.

***Prepared by:* Herbert Boschung and Richard Mayden**

GHOST SHINER

Notropis buchanani Meek

OTHER NAMES. None.

Photo—Richard Mayden

DESCRIPTION. As other members of the *Notropis volucellus* species group, a small (to 75 mm [3 in.] TL) shiner with anterior lateral-line scales much taller than wide, a subterminal mouth, eight anal-fin rays, and pharyngeal teeth 0,4-4,0. Distinguished from all other members of the species group by translucent white overall color with little or no dark pigment, long pointed pelvic fins with tips reaching to, or past, anal fin origin, and lack of, or having an incomplete infraorbital canal. Nuptial males have small tubercles on head, anterior body, and along rays on pectoral fins (Etnier and Starnes 1993; Boschung and Mayden, in press).

DISTRIBUTION. Widespread throughout low-gradient rivers in Mississippi River Basin, Ohio River Basin including Cumberland and Tennessee River drainages, Great Lakes drainage in Ontario, and in Gulf Slope drainages west of Mississippi River to Rio Grande drainage (Page and Burr 1991, Kott and Fitzgerald 2000). Within Alabama, historically found throughout unimpounded main stem of Tennessee River, but today only rarely collected in impounded mouths of large tributaries and in lower flowing reach of Elk River, Lauderdale County.

HABITAT. Typically small to large low-gradient rivers of varying water clarity. Like other members of the species group, typically found in microhabitats of quiet backwater just below, or adjacent to, riffles and runs, or in pools over a wide variety of substrates, including silt, sand, gravel, and bedrock.

LIFE HISTORY AND ECOLOGY. Specimens in spawning condition collected in late April to mid-August. In Missouri, spawning occurs over sluggish riffles of sand or fine gravel. Spawns at two years

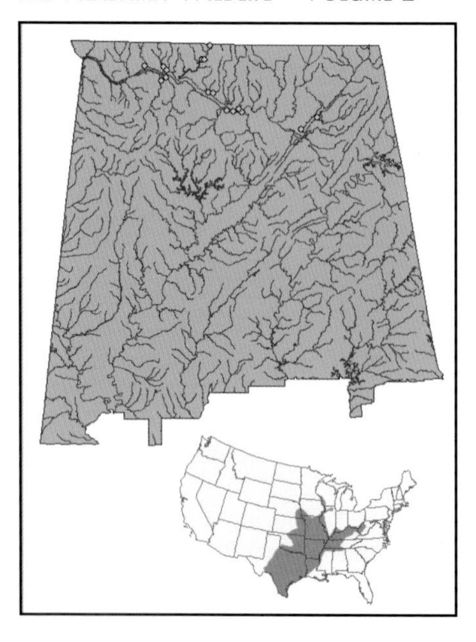

of age and lives to three years. Adults probably feed on aquatic insect larvae and other small invertebrates (Pflieger 1997).

BASIS FOR STATUS CLASSIFICATION. Although widespread, seldom common anywhere, and historically rare in Alabama. Preimpoundment collections by TVA personnel found only 17 specimens in Tennessee River drainage within the state between 1928 and 1938. For next 50 years, routine sampling by TVA and state biologists and ichthyologists failed to record this species. Rediscovered within Alabama in 1991 in Guntersville Reservoir, Marshall County. Since then, specimens have been collected at only a few sites, including impounded mouth of Second Creek and Elk River, and in flowing sections of Elk River (Lauderdale and Limestone Counties). Current threats result from impoundment of most of former habitat in Alabama, altered flow regimens, water quality degradation, and excessive sedimentation.

***Prepared by:* Bernard R. Kuhajda**

DUSKY SHINER
Notropis cummingsae Myers

OTHER NAMES. None.

Photo—Patrick E. O'Neil

DESCRIPTION. Adult size to 72 millimeters (2.8 inches) TL. Possesses wide dark lateral stripe from snout tip and lips to base of caudal fin, where it darkens into a caudal spot that streaks onto caudal fin. Posterior edge of lateral stripe not sharply defined and extends below lateral line. Silver below, dusky yellow-brown above with scales lightly outlined in brown, has a mid-dorsal stripe that is darkest anterior to dorsal fin, and large males have light yellow-orange stripe above lateral stripe. Nuptial males have large tubercles on snout and smaller tubercles elsewhere on head and along fin rays of pectoral, pelvic, and dorsal fins. Possesses slightly subterminal mouth, 10 to 11 anal-fin rays, and pharyngeal teeth 1,4-4,1 (Page and Burr 1991;

Boschung and Mayden, in press).

DISTRIBUTION. Mostly in blackwater Coastal Plain streams along Atlantic Slope from Tar River drainage, North Carolina, to Altamaha River drainage, Georgia. Additional populations in three disjunct areas to the southwest, including St. Johns River, Florida; Aucilla River drainage westward to

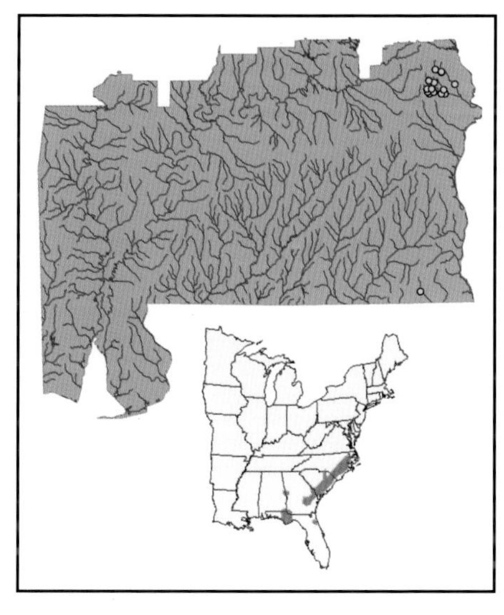

Choctawhatchee River drainage, Florida, Georgia, and extreme southeastern Alabama; and middle Chattahoochee River drainage, Georgia and Alabama. Within Alabama, restricted to single locality in tributary to Chipola River, Houston County (Appalachicola drainage) and found at numerous sites in Uchee Creek System, Russell and Lee Counties (Chattahoochee drainage) (Page and Burr 1991; Mettee *et al.* 1996; Boschung and Mayden, in press).

HABITAT. Clear or tannin-stained streams and small rivers over sand and mud, typically associated with current.

LIFE HISTORY AND ECOLOGY. All life history information from Atlantic Slope populations in South Carolina. Adults associated with current and typically found at head of pools or in eddies, but move into slow or still areas of pools to spawn. Larvae and very young shiners remain in these still pools until they become juveniles (25-30 mm [1-1.2 in.] SL), then move into current. Schools of up to 100 spawn over active nest of redbreast sunfish; nests are saucer-shaped depression in the sand and are guarded and kept silt-free by male sunfish. Spawn from May to August at water temperatures of 27-30°C (81-86°F). During spawning season adults feed heavily on eggs and larvae of redbreast sunfish, whereas they avoid eating their own progeny that are present within nest. Other diet items include aquatic insect larvae, microcrustaceans, and algae (Fletcher 1993).

BASIS FOR STATUS CLASSIFICATION. Always had a limited distribution within Alabama; two disjunct populations at western periphery of species' distribution. Appalachicola River drainage population represented by only a single locality in Chipola River system in Houston County. Chattahoochee River drainage population in Uchee Creek System has many historic localities, but abundance of dusky shiners at these sites declining due to deteriorating water quality and increased sedimentation.

Prepared by: **Bernard R. Kuhajda**

SUCKERMOUTH MINNOW

Phenacobius mirabilis (Girard)

Photo—Patrick E. O'Neil

OTHER NAMES. None

DESCRIPTION. A terete-shaped (up to 90 mm [3.5 in.] TL in Alabama), silvery minnow with a dark yellow to olive-colored lateral stripe ending in a distinct black basicaudal spot. Snout long, conspicuously blunt, and extending beyond an inferior mouth. Eye small, its diameter less than half of snout length (Mettee *et al.* 1996). Dorsal fin has eight rays and its origin is slightly anterior to the pelvic fin origin. A complete, straight lateral line has 41 to 50 scales. Anal fin has seven rays.

DISTRIBUTION. Wide-ranging and often abundant throughout the upper western and central Mississippi River Basin, east to Ohio and West Virginia, and south into Tennessee River drainage in Tennessee and northern Alabama (Etnier and Starnes 1993). A disjunct population occurs in several Gulf Coast drainages from eastern Texas to western Louisiana (Robison and Buchanan 1988). Distribution in Alabama restricted to Bear Creek, a tributary to the Tennessee River, in northwestern Alabama. Has only been collected four times in Alabama. Three collections were taken in 1960s (Wall 1968). A 1993 Geological Survey of Alabama night collection in Pogo Creek produced the first recent record in Alabama (Mettee *et al.* 2002).

HABITAT. Most members of genus *Phenacobius* are silt intolerant, but the suckermouth minnow may be an exception (Etnier and Starnes 1993). Individuals have been collected from small streams to larger rivers having clean gravel, but abundance apparently increases in some silt-laden environments (Robison and Buchanan 1988) in Ohio (Jenkins and Burkhead 1993). Cedar Creek near Pogo, the 1993 collection site, has alternating riffles and pools with a moderate flow and gravel and sand substrates (Mettee *et al.* 1996).

LIFE HISTORY AND ECOLOGY. Unknown in Alabama. Spawning, possibly involving migration, occurs over gravel from April through August in midwestern streams (Robison and Buchanan 1988). Nuptial males occur in Tennessee from May through June and may remain active throughout spawning season (Etnier and Starnes 1993). Females produce more than 1,000 eggs and possibly spawn two or three times in a single season. A bottom feeder; aquatic insect larvae, mainly midges and caddisflies, compose much of diet, although individuals may ingest detritus and other benthic organisms. Becomes sexually mature at age two and may live to age three.

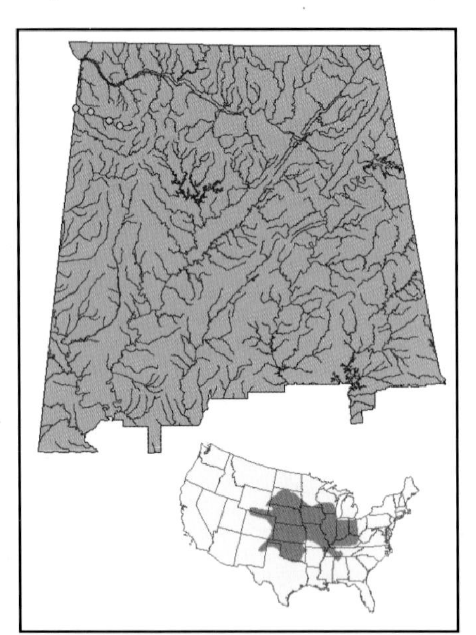

BASIS FOR STATUS CLASSIFICATION. Existing information insufficient to ascertain if this minnow was ever common in Bear Creek. Although widespread and abundant in more northern states, current data suggest it may be on the verge of extirpation from state waters. Population status in Alabama may be adversely affected by increased water temperatures, bed scouring, and associated decreased benthic invertebrate production downstream of several reservoirs in Bear Creek system.

***Prepared by:* Maurice F. Mettee**

STARGAZING MINNOW

Phenacobius uranops Cope

Photo—Patrick E. O'Neil

OTHER NAMES. None

DESCRIPTION. Very elongated, cylindrical body (max. size of adults up to 100 mm [3.9 in.] SL). Long, blunt snout overhangs fairly small, inferior mouth. Elliptical eyes positioned toward top of head. Dorsal area on live individuals dark charcoal to olive, and ventral area white. Complete lateral line, containing 52 to 59 scales, located within a fairly broad lateral band that extends from snout to base of caudal fin. Fins generally rounded (Mettee *et al.* 1996).

DISTRIBUTION. Known from Green, Cumberland, and Tennessee River drainages (Etnier and Starnes 1993). Has been collected at only nine locations in Alabama, most of which are in Shoal Creek (Mettee *et al.* 2002). Historical records include U.S. National Museum records from Cypress Creek and a University of Alabama Ichthyological Collection record from Sugar Creek tributary to Elk River (Boschung 1992).

HABITAT. Most adults in Alabama occur in riffles and runs of moderate-sized streams characterized by a moderate to swift current and gravel, cobble, and silt substrates. Smaller individuals prefer side eddies and pools below riffles with slower current and silt covered substrates. Collected by Geological Survey of Alabama biologists in smaller Shoal Creek tributaries with seines and backpack electrofishing gear and at Goose Shoals in Shoal Creek proper with boat electrofishing gear (Mettee *et al.* 1996).

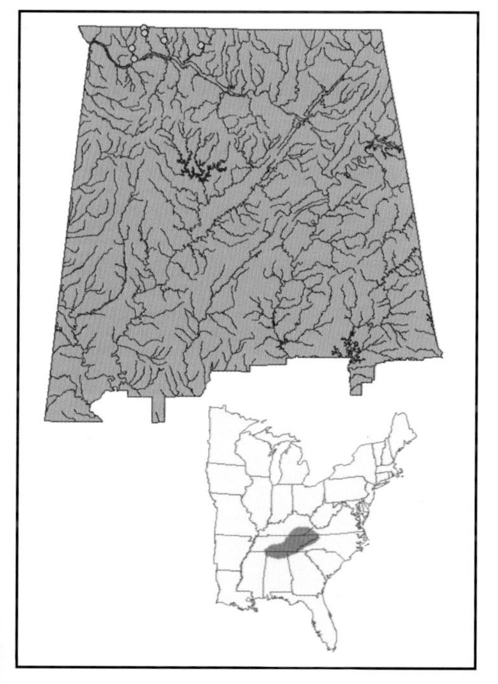

LIFE HISTORY AND ECOLOGY. Unknown in Alabama. Spawning occurs from late April to June in Virginia (Jenkins and Burkhead 1993). Sexual maturity occurs at about one year; life span usually from two to three years. Food items include caddisfly and midge larvae and detritus (Etnier and Starnes 1993).

BASIS FOR STATUS CLASSIFICATION. Populations in Shoal Creek appear stable for present, but current status is unknown in Cypress Creek and Elk River. Stargazing minnows and silver shiners have been collected sympatrically at two or three sites in northern Alabama (Mettee *et al.* 2002). Recent collection of several silver shiners in Sugar Creek suggests stargazing minnows may still occupy this system. Continued survival in Cypress Creek is more remote due to increased nonpoint source pollution.

Prepared by: **Maurice F. Mettee**

BROADSTRIPE SHINER

Pteronotropis euryzonus Suttkus

Photo—Patrick E. O'Neil

OTHER NAMES. None.

DESCRIPTION. Maximum size 54 millimeters (2.1 inches) SL (Suttkus and Mettee 2001). Relatively deep body with broad, dark lateral stripe with narrow orange and green stripes above. Caudal fin orange in life with small clear area in middle in breeding males. Breeding males have darkly colored dorsal fins and more breeding tubercles than females (Suttkus 1955).

DISTRIBUTION. In Alabama, restricted to tributaries of Chattahoochee River from Uchee Creek (Lee County) south to northern Houston County (Suttkus and Mettee 2001). Known from just eight localities in Georgia.

HABITAT. Typically, small headwater streams with clear, sometimes black, water. Often associated with woody debris and vegetation; prefers deeper and swifter water than generally available in habitat (Suttkus 1955, Johnston *et al.* 2001).

LIFE HISTORY AND ECOLOGY. Virtually unknown. Based on spawning behavior of closely related species, most likely broadcasts eggs over vegetation, without parental care. Spawning season in Alabama probably May through July. Probably drift feeders (Mettee *et al.* 1996).

BASIS FOR STATUS CLASSIFICATION. Has suffered a 70 percent distribution reduction in Alabama, largely due to habitat degradation in Uchee Creek (Johnston *et al.* 2001). As a headwater stream specialist, populations in Alabama almost certainly isolated from Georgia populations, and recolonization opportunities may not exist. Therefore, critical to protect species throughout its distribution.

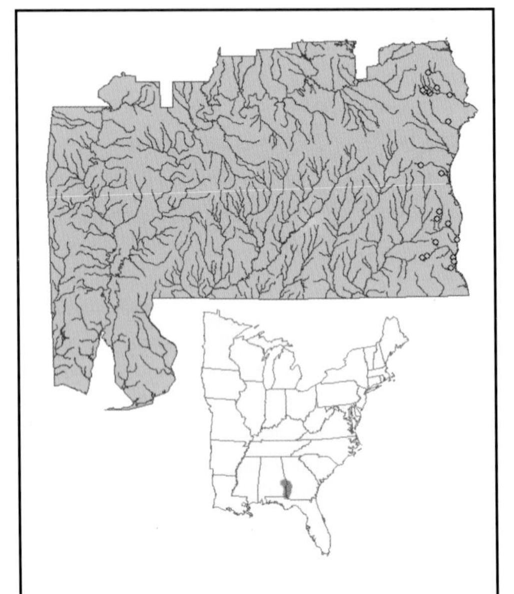

Prepared by: **Carol E. Johnston**

BLUENOSE SHINER

Pteronotropis welaka Evermann and Kendall

OTHER NAMES. None.

Photo—Patrick E. O'Neil

DESCRIPTION. Maximum size 65 millimeters (2.6 inches) SL. Black lateral stripe from chin to caudal fin. Caudal fin with black spot. Breeding males with bright blue pigment on snout, flecks of gold pigment along sides and greatly expanded dorsal and anal fins. Dorsal fin of breeding males black; anal and pelvic fins yellow (Page and Burr 1991).

DISTRIBUTION. St. Johns River, Florida; Gulf Coast drainages, primarily below Fall Line, from Apalachicola system to Pearl River (Page and Burr 1991). In Alabama, known only from 22 sporadically distributed localities in Tombigbee, Cahaba, Alabama, Chattahoochee, and coastal drainages, all below Fall Line (Mettee *et al.* 1996).

HABITAT. Prefer small to medium streams with clear or black water and associated with relatively deep, flowing water with vegetation and sand or muck substrate.

LIFE HISTORY AND ECOLOGY. Breeding season lasts from May to August. Acts as nest associate of sunfishes, broadcasting eggs over nests and benefiting from parental care of host (Johnston and Knight 1999). Males show gradual increase in gonad size and sexually dimorphic characteristics with increasing size. Based on demographic analysis, apparently die after first breeding season (Johnston and Knight 1999). Females typically spawn less than 200 eggs measuring less than one millimeter (0.04 inch) per batch (Johnston and Knight 1999).

BASIS FOR STATUS CLASSIFICATION. Sporadically distributed in Alabama; populations are declining. Short life span and probable limited dispersal ability contributes to vulnerability. Due to increased habitat fragmentation unlikely to recolonize areas once extirpated.

***Prepared by*: Carol E. Johnston**

MOUNTAIN MADTOM

Noturus eleutherus Jordan

Photo—Patrick E. O'Neil

OTHER NAMES. None.

DESCRIPTION. Small (max. size in Alabama 52 mm [2.1 in.] SL), mottled brown, and distinguished by dark saddles or blotches near origin and posterior end of the dorsal fin, and at base of adipose fin. Adipose fin usually has pale white margin. A faint to dark vertical bar occurs at caudal fin base. Caudal fin has two to three crescent-shaped bands and is slightly convex rear margin. Dorsal barbels mottled brown; ventral barbels predominantly white with scattered, usually small, brown spots. Pectoral fin spines have well-developed posterior serrae. Anal fin has 12 to 17 rays (Mettee *et al.* 1996).

DISTRIBUTION. Disjunct populations in White, Ouachita, and Red Rivers west of Mississippi River (Robison and Buchanan 1988). More widespread and sometimes common east of Mississippi River, from Ohio River drainage south through Tennessee River drainage (Etnier and Starnes 1993). Reaches southern limit of distribution in northern Alabama where numbers and distribution are very limited. Geological Survey of Alabama biologists collected the first mountain madtoms taken in Alabama, in Elk River near Alabama-Tennessee state line in 1993 (Mettee *et al.* 1996).

HABITAT. Prefers moderate to swift current over gravelly shoals and riffles of moderate to large rivers; apparently avoids smaller streams. Specimens collected in Elk River came from beneath flattened pieces of bedrock located in a shallow side channel having a swift current, marginal vegetation, coarse gravel, and scattered bedrock.

LIFE HISTORY AND ECOLOGY. Unknown in Alabama. Individuals apparently hide around aquatic vegetation and under bedrock during day and feed on aquatic insect larvae including mayflies, midges, stoneflies, and caddisflies at night. Females ranging from 51 to 59 millimeters (2.0 to 2.3 inches) SL produce 55 to 115 eggs per year in Tennessee (Etnier and Starnes 1993). Spawning occurs in June. A single nest, containing approximately 70 eggs and guarded by a large male, was found under a slab rock in a shaded pool; water temperature was 24°C (75°F). Much of the surrounding substrate consisted of clean fine gravel. Can reach 36-64 millimeters (1.4-2.5 inches) SL in their first year and become sexually mature at 56 millimeters (2.2 inches) SL. Maximum life span estimated to be four to five years (Starnes and Starnes 1985).

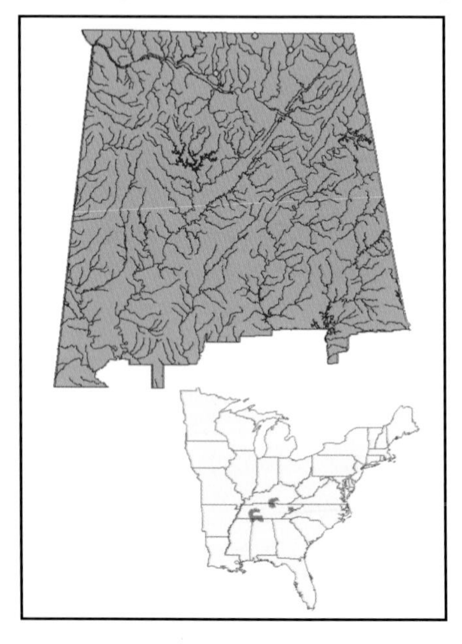

BASIS FOR STATUS CLASSIFICATION. Widespread and locally common in Tennessee, but entire known distribution in Alabama limited to about three miles of Elk River between Alabama Highway 127 and the Alabama-Tennessee state line. This small reach of Elk River is inhabited by more than 55 fish species, including only populations of the shoal chub, bluebreast darter, and federally-listed boulder darter in Alabama. This site is also inhabited by one of Alabama's largest amphibian species, the eastern hellbender. Continued existence of the rich aquatic community in this section of Elk will be directly influenced by controlled discharges from Tims Ford Dam in Tennessee and withdrawal of river water by a local municipality. Any substantial, adverse change in either habitat or water-quality conditions could potentially eliminate several fish species whose present distributions in Alabama are restricted to the Elk River.

***Prepared by:* Maurice F. Mettee**

BRINDLED MADTOM
Noturus miurus Jordan

OTHER NAMES. None.

Photo—Patrick E. O'Neil

DESCRIPTION. A moderately sized (ave. adult about 60 mm [2.4 in.] SL; max. 132 mm [5.2 in.] SL) madtom of subgenus *Rabida*. Mottled with a tan or yellowish ground color, a white to yellowish venter, four dark dorsal saddles, and black blotches on adipose fin and anterior rays of dorsal fin. Mouth subterminal (upper jaw projects slightly beyond lower jaw), premaxillary tooth patch lacks posterior extensions, and a distinct notch separates adipose and caudal fins. Distinguished from congeners in Alabama by black blotches on dorsal and adipose fins that extend to edges of fins, and a subterminal dark band on rounded caudal fin (crescent-shaped bands on caudal fin lacking). Nuptial males develop enlarged muscles on top of head, swollen lips, and enlarged genital papillae.

DISTRIBUTION. A widespread madtom species that occurs in lower half of Mississippi Basin extending east and north into the Ohio, Tennessee, and Cumberland Rivers and southern Great Lakes and west of Mississippi River into the White, Arkansas, and Ouachita Rivers. Also occurs in Pearl River and Lake Pontchartrain drainages of Gulf Coast. In Alabama, known only from Bear Creek system (Tennessee River drainage) (Mettee *et al.* 1996), which is near upstream edge of species' distribution in Tennessee River system (Etnier and Starnes 1993). Records also available for Bear Creek mainstem in Mississippi (Ross 2001) and Elk River system just north of Tennessee-Alabama state line (Etnier and Starnes 1993).

HABITAT. Like its relatives, a bottom-dwelling species that occurs in small to moderate-sized streams, usually in association with slow to moderate current, sandy or gravel substrates, and cover (undercut banks, leaf packs, tree roots, sticks, or logs) (Etnier and Starnes 1993, Mettee *et al.* 1996) During day species retreats to cover, moving into open areas to forage at night.

LIFE HISTORY AND ECOLOGY. Feeds primarily on larvae of bottom-dwelling invertebrates such as midges, black flies, caddisflies, and mayflies (Burr and Mayden 1982). Males reach maturity at about two years (about 50 mm [2 in.] SL), but females may be mature after their first year. Spawns at temperatures of 24-27°C (75.2-80.6°F) in quiet water over soft substrates (Burr and Mayden 1982). Nesting occurs in depressions hollowed out under rocks, woody debris, freshwater mussel shells, or other natural cover, but discarded beverage cans and bottles also are used as nesting sites (Burr and Mayden 1982, Etnier and Starnes 1993, Burr and Stoeckel 1999). Clutch sizes range from 56 to 81. As with other madtoms, male guards and cares for developing eggs, during which time there appears to be little if any feeding by guardian male (Burr and Stoeckel 1999).

BASIS FOR STATUS CLASSIFICATION. Bear Creek system supports only known population within Alabama. Tennessee Valley Authority personnel discovered species in system during rotenone surveys (presumably conducted in 1970s) (Ramsey 1976). At that time, specimens were obtained from Cedar and Little Bear Creeks, both of which are now partially impounded by reservoirs. The 1976 panel on Alabama freshwater fishes considered it a Species of Special Concern (Ramsey 1976), but the freshwater fishes committee of 1986 dropped the species from consideration without comment (Ramsey 1986*a*). Was found only twice in 1993 surveys in Bear Creek (Mettee *et al.* 1996), suggesting recent surface-mining activities, reservoir construction, or other changes in system affected the species. Last recorded in Mississippi portion of Bear Creek in 1990 (S. T. Ross, pers. comm.). Also may persist in Alabama portion of Elk River drainage, but despite repeated collections there, no specimens were recorded (Jandebeur 1972, Etnier and Williams 1989). Likewise, surveys in other tributaries of Tennessee River drainage in Alabama did not yield the species (Jones 2000).

***Prepared by*: Melvin L. Warren, Jr.**

FRECKLEBELLY MADTOM

Noturus munitus Suttkus and Taylor

OTHER NAMES. None.

Photo—Patrick E. O'Neil

DESCRIPTION. A member of subgenus *Rabida*. Small (max. size = 99 mm [3.9 in.] SL) and characterized by short, stout body, large head, and slender caudal peduncle. Eyes relatively large and mouth inferior. Dorsally, body yellow mottled with brown. Head dark brown dorsally with three pronounced dark brown saddles on back. Ventrally, abdomen and head cream, and abdomen sprinkled with dark chromatophores giving species its name. Pelvic, dorsal, and anal fins all have darkly mottled submarginal bands; barbels also mottled. Pectoral spines large and curved with many fine barbs on anterior edge, and less numerous large curved barbs on posterior edge. Anal fin short (12-15 rays), adipose fin narrowly joined to caudal fin base, and caudal fin truncate to slightly rounded (Suttkus and Taylor 1965). There are two dark bands on caudal fin, one near base, and other near fin margin. Other species occurring within Alabama distribution are the speckled, black, tadpole, and freckled madtoms. All are uniformly pigmented dorsally, lack dark saddles, and not easily confused with the frecklebelly madtom.

DISTRIBUTION. Historically known from the Pearl, Bogue Chitto, and Strong Rivers in Mississippi and Louisiana, upper Tombigbee River system, Cahaba River downstream of Fall Line at Centreville, and main channel of Alabama River (Mettee *et al.* 1996). A closely related, undescribed form occurs in Georgia in upper Coosa system. Although once widespread and abundant on expansive gravel shoals that characterized main channel of upper Tombigbee River, species now eliminated there by channelization and impoundment to create Tennessee-Tombigbee Waterway. There are records from four major tributaries of upper Tombigbee River in Alabama and Mississippi: Sipsey River, Luxapallila Creek, Buttahatchie River, and Bull Mountain Creek. Lower section of Sipsey River characterized by high-quality gravel shoals, and population there apparently stable. Although lower reach of Luxapallila Creek has produced several large series of species, portion downstream of Columbus, Mississippi, was channelized for flood control in 1995, eliminating required stable gravel shoal habitat. Channelized stretch upstream of Alabama state line has not produced species. Now probably limited to a short unchannelized stretch of Luxapallila Creek near Steens, Mississippi. Strongest remaining population in upper Tombigbee River system in Buttahatchie River where species has been collected in most stable gravel shoals from first shoal upstream of Columbus Lock and Dam embayment, upstream to Hamilton, Alabama (Shepard *et al.* 1997). In-stream gravel mining in lower reaches of Buttahatchie River has disrupted some shoals. Bull Mountain Creek produced species in 1980-81 at several localities; however, repeated sampling recently at numerous stations throughout creek has failed to produce additional specimens, so may now be extirpated from stream. Population may have been adversely affected when lower section of creek was cut off by construction of Tennessee-Tombigbee Waterway. In Cahaba River, frecklebelly madtoms recently documented from four miles downstream of U.S. Highway 80 bridge in Dallas County upstream to near Fall Line at Centreville, Bibb County (Pierson *et al.* 1989, Shepard *et al.* 1997). Abundance in Cahaba River closely related to condition and stability of gravel shoals. Shoals that are embedded with sand and silt typically not inhabited by species. Frecklebelly madtoms were collected at seven stations in main channel of Alabama River in Wilcox and Monroe Counties by Dr. R. D. Suttkus in 1960s before completion of Jones Bluff, Millers Ferry, and Claiborne Locks and Dams. Recent collections at these localities and others in Alabama River failed to produce species, and it probably no longer occurs there (Shepard *et al.* 1997). Sedimentation during dam construction, dredging, gravel removal, and altered flow regimes may have been responsible for eliminating species from river.

HABITAT. Typically inhabits swift flowing shoals where found in association with stable gravel substrates. Also collected over cobble, boulder, or bedrock substrates covered with river weed. Found in medium to large rivers although individuals have occasionally been taken in smaller streams. Requires clean, stable substrates, and does not occupy shoals that become embedded with sand and silt.

LIFE HISTORY AND ECOLOGY. In Tombigbee River, diet consists principally of aquatic insect larvae including hydropsychids, ephemerellids, simuliids, and chironomids, with a preference for larger prey items (Miller 1984). Spawning occurs in June and July based on gonadal development. Females produce from 50 to 70 mature eggs that apparently are laid in a single clutch (Trauth *et al.* 1981). Spawning behavior unknown.

BASIS FOR STATUS CLASSIFICATION. In Alabama, has disjunct distribution in only three river systems and has been eliminated from main channels of Tombigbee and Alabama Rivers due to impoundment, channelization, and habitat alteration. Stable populations in Alabama now exist only in Cahaba, Sipsey, and Buttahatchie Rivers (Shepard *et al.* 1997). Occurs in main channel shoal habitats. Very intolerant of sedimentation, making it very vulnerable to practices that disturb substrate integrity such as in-stream gravel mining, channelization, and sedimentation due to nonpoint source pollution.

***Prepared by*: Thomas E. Shepard**

HIGHLANDS STONECAT

Noturus sp cf. *flavus*

OTHER NAMES. None

Photo—Patrick E. O'Neil

DESCRIPTION. Largest (adults may reach 200-250 mm [7.9-9.8 in.] SL) madtom in Alabama. Head distinctly flattened. Dorsal area light to dark brown and abdomen white to cream, occasionally with a light scattering of melanophores. Dorsal and adipose fin bases dark becoming pale yellow to white near periphery. Posterior margins of dorsal and anal fins may be darkened. Pelvic fins essentially white with scattered melanophores near base. Premaxillary tooth patch has posterior extensions on either side of mouth, a characteristic not found on any other madtom species in Alabama. Pectoral spines lack posterior serrae. Distinctive color pattern behind head (nape) on Alabama stonecats not found on members of this species elsewhere in its distribution.

DISTRIBUTION. Throughout much of Great Lakes drainage, upper Mississippi Basin, south in Mississippi River into Mississippi, and throughout much of Tennessee River drainage. An undescribed stonecat species occurs in the Cumberland River drainage (Page and Burr 1991). Based on distinct differences in dorsal color pattern, another undescribed population may exist in Shoal Creek and Elk River systems, the only two systems inhabited by this species in Alabama (Mettee *et al.* 1996).

HABITAT. Prefers moderate to large free-flowing streams and rivers having a gentle current, gravel, rubble, and slab rock substrates, and extensive areas of river weed. Some individuals have been taken in very small streams and also in lower Mississippi River proper (Etnier and Starnes 1993).

LIFE HISTORY AND ECOLOGY. Unknown in Alabama. Individuals generally hide under slabrocks during day and emerge to feed on aquatic insect larvae (mayflies, stoneflies, midges, caddisflies) and crayfishes at night. Spawning occurs from April through July in Tennessee (Etnier and Starnes 1993), June in Virginia (Jenkins and Burkhead 1993), and August in Ohio (Walsh and Burr 1985). Water temperatures during spawning period 25°C (77°F) in Tennessee and Virginia, slightly higher in Missouri, and 27.5°C (81.5°F) in Canada. Mature females produce 189 to 570 eggs in Virginia (Jenkins and Burkhead 1993) and 200 to 1,500 eggs in Tennessee (Etnier and Starnes 1993). Eggs generally deposited on the underside of slab rocks and guarded by males. A male may guard from 50 to 500 embryos from different spawns in Virginia (Jenkins and Burkhead 1993). Growth fairly rapid; young fish in Illinois and Missouri streams averaged 49 millimeters (1.9 inches) SL in their first year, 100 millimeters (3.9 inches) SL in their second, and 123 millimeters (4.8 inches) SL in their third year. Largest on record, a 313 millimeters (12.3 inches)TL specimen, collected from Lake Erie. Reaches sexual maturity at age three in Illinois and Missouri. Estimated life span five to six years (Walsh and Burr 1985).

BASIS FOR STATUS CLASSIFICATION. Extirpated from upper Tennessee River drainage in Tennessee although populations still survive in North Fork Holston and Clinch River systems (Etnier and Starnes 1993), Virginia (Jenkins and Burkhead 1993). Elk River population in northern Alabama subject to flow fluctuations from Tims Ford Dam in Tennessee and municipal withdrawals. An accidental discharge from an upstream bicycle factory caused a substantial fish kill in Shoal Creek some years ago, but without long-term adverse effects. Recent sampling by Geological Survey of Alabama biologists revealed the stream is still inhabited by a diverse ichthyofauna consisting of more than 70 species (Mettee *et al.* 2002). However, failure to maintain present water quality and habitat conditions in the Elk River and Shoal Creek could eliminate this species and aquatic communities in these systems.

Prepared by: **Maurice F. Mettee**

SHOAL BASS

Micropterus cataractae Williams and Burgess

OTHER NAMES. None.

Photo—Patrick E. O'Neil

DESCRIPTION. Maximum size 64 centimeters (approx. 25 inches) SL and 3.97 kilograms (8.75 pounds) (Williams and Burgess 1999). Olive colored with five to seven rows of weakly developed spots along side of body below lateral line and several vertical bars along side. Differs from morphologically similar redeye bass in lacking clear margins on tips of caudal fin and a tooth patch on tongue (Williams and Burgess 1999).

DISTRIBUTION. Apalachicola and Chipola Rivers in Georgia, Alabama, and Florida. In Alabama, restricted to Chattahoochee River and major tributaries (Halawakee, Uchee, and Wacoochee Creeks) (Williams and Burgess 1999).

HABITAT. Shoals and riffles of medium rivers and large streams (Mettee *et al.* 1996).

LIFE HISTORY AND ECOLOGY. Like other centrarchids, males build nests for spawning in early spring (April-May). Nests contain 500 to 3,000 eggs, probably spawned by several females. Reach sexual maturity at age three. Diet includes crayfishes, fishes, and insects (references in Williams and Burgess 1999).

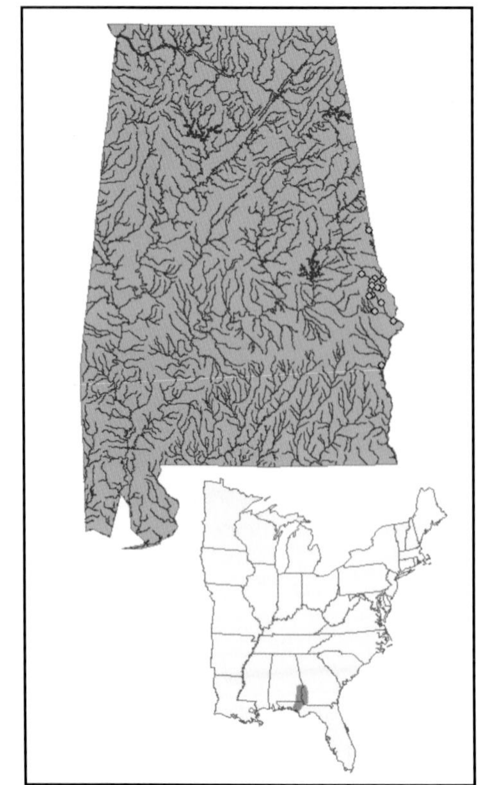

BASIS FOR STATUS CLASSIFICATION. Eliminated from mainstem Chattahoochee in Alabama and restricted to a few major tributaries (Ramsey *et al.* 1972). Reasons for decline include elimination of habitat by impoundment, and habitat degradation due to sedimentation and water withdrawal. Habitat has continued to decrease, and species now rare or absent from Uchee and Halawakee Creeks, once major strongholds in Alabama.

Prepared by: **Carol E. Johnston**

LOCUST FORK DARTER

Etheostoma sp. cf. *bellator*

Photo—Richard Mayden

OTHER NAMES. Warrior Darter.

DESCRIPTION. A member of the snubnose darter group (subgenus *Ulocentra*); all members are smaller (to 7.7 cm [3 in.] TL) darters with extremely blunt snouts, a very narrow or absent premaxillary frenum, broadly joined branchiostegal membranes, and seven to 10 dorsal saddles. Males brightly colored year round, and brilliantly colored in spring. Females brown, straw-colored, and typically lack bright colors, but may have some red in dorsal fins and scattered on their upper sides. Like other members of the *Etheostoma duryi* species group, lacks a premaxillary frenum and has vomerine teeth. Except for blueface darters in Hubbard Creek of Sipsey Fork system, all male snubnose darters in Black Warrior River drainage above the Fall Line, including Locust Fork darters, have a distinctive red spot in first membrane of spinous dorsal fin, a blue-green anal fin, and continuous red-orange ventrolateral coloration (Boschung *et al.* 1992, Suttkus and Bailey 1993). Locust Fork darters distinguished from the other Black Warrior snubnose darters by fixed allelic differences (Clabaugh *et al.* 1996) and by possessing a distinct submarginal red band in spinous dorsal fin and a thin (less than one scale-row wide), cream colored, punctate line along lateral line from opercle to below second dorsal fin origin.

DISTRIBUTION. Restricted to three small eastern tributaries of upper Locust Fork, Blount and Etowah Counties, Alabama (Suttkus and Bailey 1993); all recent collections from Mill Creek, a tributary of Calvert Prong. Recent attempts to capture at other historic collection sites (Sand Valley Creek, tributary to Calvert Prong, 1939; Little Cove Creek, 1969) failed, but no systematic status survey undertaken.

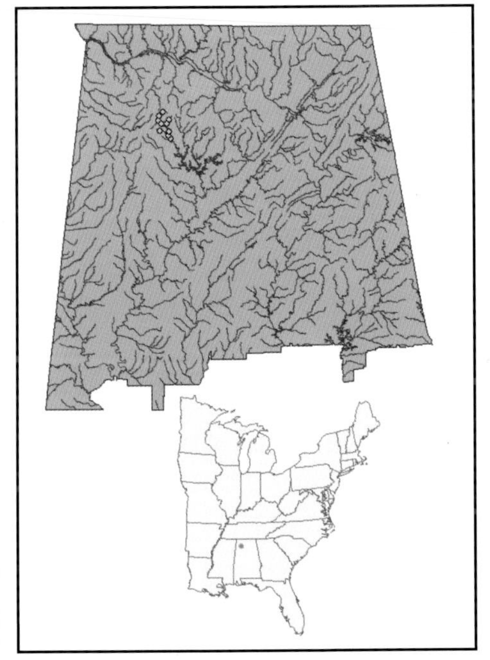

HABITAT. Small streams less than 10 meters (33 feet) wide with overall slow to moderate flow and with substrates of sand, gravel, and cobble. Typically occupy slower, deeper water at head of riffles with larger substrate, but can be present in riffles.

LIFE HISTORY AND ECOLOGY. Relatively unknown, but specimens collected in spawning condition in March. Diet likely includes benthic aquatic invertebrate larvae.

BASIS FOR STATUS CLASSIFICATION. Currently only known from five kilometers (three miles) of Mill Creek, which has sedimentation problems associated with agriculture and urbanization. Collected only once from two historic localities, and recent collections produced no snubnose darters. May warrant higher priority designation in future, but comprehensive survey needed to determine if species persists outside of Mill Creek.

Prepared by: **Bernard R. Kuhajda**

SIPSEY DARTER

Etheostoma sp. cf. *bellator*

OTHER NAMES. Warrior Darter.

Photo—Joseph Tomelleri

DESCRIPTION. A member of the snubnose darter group (subgenus *Ulocentra*); all members are smaller (to 7.7 cm [3 in.] TL) darters with extremely blunt snouts, a very narrow or absent premaxillary frenum, broadly joined branchiostegal membranes, and seven to 10 dorsal saddles. Males brightly colored year round, and brilliantly colored in spring. Females brown, straw-colored, and typically lack bright colors, but may have some red in dorsal fins and scattered on their upper sides. Like other members of the *Etheostoma duryi* species group, lacks a premaxillary frenum and has vomerine teeth. Except for blueface darters in Hubbard Creek of Sipsey Fork system, all male snubnose darters in Black Warrior River drainage above the Fall Line, including Sipsey darters, have a distinctive red spot in first membrane of spinous dorsal fin, a blue-green anal fin, and continuous red-orange ventrolateral coloration (Boschung *et al.* 1992, Suttkus and Bailey 1993). Sipsey darters distinguished from other Black Warrior snubnose darters by fixed allelic differences (Clabaugh *et al.* 1996) and by possessing a distinct submarginal red band in spinous dorsal fin, a distinct cream-colored stripe (one scale-row wide) along lateral line from opercle to below second dorsal fin origin, and red pigment above lateral line greater than one scale row wide and forming a series of rectangles.

DISTRIBUTION. Restricted to Sipsey Fork system, Winston and Lawrence Counties, Alabama. Currently found in main stem of Sipsey Fork from just above Lewis Smith Reservoir embayment to its origin, and in Thompson, Borden (including Flannagin and Braziel), and Caney Creeks, all tributaries of upper Sipsey Fork. Does not occur in Hubbard Creek (Suttkus and Bailey 1993, Kuhajda and Mayden 2002*b*). Although no records in impounded section of Sipsey Fork, likely occurred there before impoundment.

HABITAT. Small to large upland streams five to 25 meters (16 to 82 feet) wide with substrates of sand, gravel, cobble, and boulders. Typically occupy slower, deeper water at head of riffles with larger substrate or at base of riffles/upper end of pools (Kuhajda and Mayden 2002*b*, Powers *et al.* 2003).

LIFE HISTORY AND ECOLOGY. Relatively unknown, but specimens collected in spawning condition in March and April. Diet likely includes benthic aquatic invertebrate larvae.

BASIS FOR STATUS CLASSIFICATION. Currently only known from upper Sipsey Fork and three of its tributaries; populations presumed present in lower Sipsey Fork now extirpated by Lewis Smith Reservoir. Recent collections produced fewer individuals (typically one or two specimens) compared to historic collections. Some sedimentation associated with poor forestry management (clear cuts without appropriately wide streamside management zones) prevalent, especially in tributaries (Kuhajda and Mayden 2002*b*, Powers *et al.* 2003).

Prepared by: **Bernard R. Kuhajda**

BLUEBREAST DARTER

Etheostoma camurum (Cope)

Photo—Patrick E. O'Neil

OTHER NAMES. None.

DESCRIPTION. A member of subgenus *Nothonotus*, distinguished by a somewhat compressed body, deep caudal peduncle, and horizontal banding along posterior two-thirds of body. Adult size in Alabama up to 55 millimeters (2.2 inches) SL. Snout rounded and blunt, with thick lips and a wide frenum connecting upper lip and snout. Seven to 10 obscure saddles on dorsum, and adult males have eight to 12 obscure and vertically elongate lateral bars along sides. Body color olive with greenish cast, speckled with bright crimson spots on sides of males, and brown spots on sides of females. Breast bright green to bluish green on breeding adult males and generally lighter blue on females. Spiny dorsal fin reddish gray with a dark spot anterior and proximal. Both males and females have a dark marginal band and a cream colored submarginal band on median fins. Males and females similarly patterned, but adult males brighter, and nuptial colors generally retained throughout year. Meristic counts include 47 to 70 lateral-line scales; nine to 15 dorsal spines and 10 to 14 rays; two anal spines and six to nine rays; and 13 to 15 pectoral fin rays. Scales absent on cheeks, nape, and breast but present on belly and opercle area (Etnier and Starnes 1993).

DISTRIBUTION. Known in Ohio River Basin from Tennessee and Cumberland Rivers north to upper Allegheny River in Pennsylvania, west to Vermilion River in Illinois and Tippecanoe River in Indiana, and east to Kanawha and Monongahela Rivers in West Virginia (Stauffer 1978). Generally uncommon with a spotty distribution throughout Tennessee; unknown in Alabama until 1993 when discovered in Elk River main channel, Limestone County (Mettee *et al.* 1996). Biologists of Geological Survey of Alabama collected 31 individuals in September 1993 and 18 individuals in October 1993. Known only from about three miles of Elk River main channel from Alabama-Tennessee state line downstream to Alabama Highway 127 bridge.

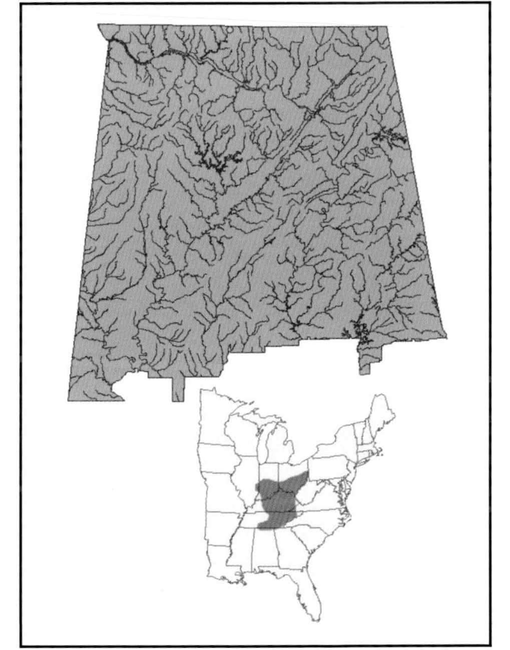

HABITAT. Generally known to occur in main channel of larger rivers and streams in areas with swift current, and substrates of bedrock, cobble, and boulders. Apparently has little affinity for tailwaters and reservoirs and adversely affected by excessive siltation common throughout distribution in Indiana, Illinois, and Ohio. Reportedly increasing in extent and abundance in tailwaters of French Broad and Holston Rivers in Tennessee after measures were taken to improve water quality. In Alabama reach of Elk River, found in shallow bedrock-rubble shoals and cobble-gravel shoals up to 0.5 meters (1.5 feet) deep.

LIFE HISTORY AND ECOLOGY. Spawning occurs throughout distribution from May to early August and likely early May to late June in Alabama. Substantial migrations reported during spring and summer in Ohio streams (Trautman 1981). Fertilization occurs while females partially buried in fine gravel or sand behind large objects in streams such as boulders and areas of bedrock (Mount 1959; Stiles 1972). Diet consists of midge larvae, baetid mayflies, blackflies, and caddisflies that are common inhabitants of riffle and shoal habitats occupied by species (Bryant 1979). In Alabama portion of Elk River, occurs sympatrically with two other *Nothonotus*, the boulder and redline darters.

BASIS FOR STATUS CLASSIFICATION. Has very restricted distribution in Alabama, thought to be limited to about 4.8 kilometers (three miles) of Elk River main channel in Limestone County. Although Alabama population appeared stable in 1993, hydrology of lower Elk River is altered by two significant projects that may have bearing on fish population viability in future. First, water releases upstream from Tims Ford Dam cause substantial variability in stream flows and second, Limestone County has a water intake pipe at upstream end of Mason Island, which is in middle of species' distribution in Alabama.

***Prepared by*: Patrick E. O'Neil**

LIPSTICK DARTER

Etheostoma chuckwachatte Wood and Mayden

OTHER NAMES. None.

DESCRIPTION. A small percid (reaching about 50 mm [2 in.] TL) and member of the greenbreast darter species group. Distinguished from other members by presence of red lips, bright red spots along flanks, a broad red band in anal fin of adult males, and scales on opercles (Wood and Mayden 1993). Named from anglicized version of Creek Indian words for mouth (chuckwe) and red (chattee); refers to bright red lips on mouths of breeding males (Wood and Mayden 1993).

Photo—Malcolm Pierson

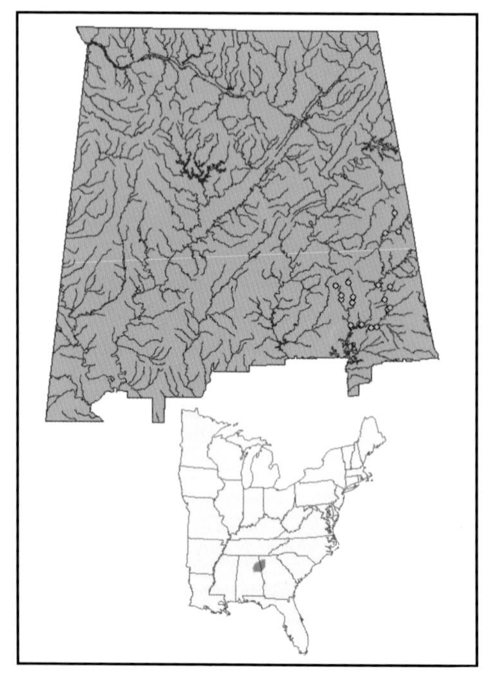

DISTRIBUTION. Known from throughout Tallapoosa River system above Fall Line in Alabama and Georgia (Wood and Mayden 1993). Endemic to Tallapoosa River system of Mobile Basin. Collection records limited almost exclusively to northern Piedmont Upland (Mettee *et al.* 1996).

HABITAT. Adults usually found in riffles and shallow runs of medium to large streams with moderate to swift current. Preferred substrate usually gravel, cobble, and rubble or some combination of these bottom types.

LIFE HISTORY AND ECOLOGY. Populations in Hillabee Creek spawn from late April to late

June with peak activity occurring in May (Orr and Ramsey 1990). Insect larvae (dipterans, ephemeropterans, and plecopterans) accounted for most of diet from Hillabee Creek, Tallapoosa County, Alabama (Orr 1989).

BASIS FOR STATUS CLASSIFICATION. An upper Tallapoosa River system endemic with a very restricted distribution. Two large impoundments (Harris and Martin Hydroelectric Projects) eliminated some riverine habitat and fragmented and isolated populations. Main channel Tallapoosa River habitat downstream of R.L. Harris Hydroelectric Project subjected to regulated river flow from peak hydroelectric discharges. While species continues to be found in good numbers here, unclear what long-term effects these altered flows have on this sensitive riverine species (Pierson 1999).

***Prepared by:* J. Malcolm Pierson**

COLDWATER DARTER

Etheostoma ditrema Ramsey and Suttkus

OTHER NAMES. None.

Photo—Richard Mayden

DESCRIPTION. A member of subgenus *Oligocephalus* attaining a total length of six centimeters (2.4 inches); members variable but all have incomplete lateral line, narrow to moderately joined branchiostegal membranes, and nuptial males with bright red, orange, blue, and sometimes green on body and fins. Females brown and straw colored. A composite of several genetically distinct populations, each with slightly different morphologies, but typically a robust darter with dark brown mottling on a cream to yellow back and sides, three black caudal spots, a suborbital bar, blue and red bands in spinous dorsal fin, and an orange venter that may expand onto lower sides and caudal peduncle as bars. Populations in central Coosa River drainage have fewer unpored scales (12), orange extending onto lower sides, and distinct thin lines on upper sides; those from upper Coosa have more unpored scales (16), less orange on sides, and lack lines on upper sides (Ramsey and Suttkus 1965, Utter 1984, Knott *et al.* 1996).

DISTRIBUTION. Restricted to Coosa River drainage in Alabama, Georgia, and Tennessee above the Fall Line. Two ecological forms readily apparent. Stream form occurs in six tributaries of central Coosa River in four Alabama counties (Chilton, Shelby, Coosa, and Talladega) from Yellow Leaf Creek System, Chilton County, north to Kahatchee Creek System, Talladega County, with most sites within Waxahatchee Creek System in Chilton and Shelby Counties. Second form occurs in springs along upper Coosa River from Tallaseehatchee Creek system, Talladega County, Alabama, to Conasauga River system, Polk County, Tennessee. In Alabama, spring form inhabits 11 springs within seven creek systems and historically inhabited three other sites, all within Talladega, Calhoun, and Etowah Counties (Kuhajda and Mayden 2002*c*).

HABITAT. Stream form inhabits small, vegetated upland streams two to 12 meters (seven to 39 feet) wide with slow to fast current over bedrock, gravel, and cobble substrate. Typically associated with aquatic vegetation, specifically moss, in slow to moderate current in streams six meters (20 feet) wide, but also found associated with detritus. Spring form inhabits vegetated spring pools and runs with mud and silt bottoms. Specimens are collected in aquatic vegetation, especially moss, within spring pools with no or slow flow, but occasionally enter spring runs (Kuhajda and Mayden 2002*c*).

LIFE HISTORY AND ECOLOGY. In Glencoe Spring, Etowah County, spawns from March to September, with peak from April to June, when eggs are attached to vegetation. Life span less than two years. Diet consists of amphipods, chironomids, isopods, and copepods (Seesock 1979). Biology of stream form unstudied, but spawning season probably shorter due to higher water temperatures in summer.

BASIS FOR STATUS CLASSIFICATION. Species complex has ecologically, genetically, and morphologically distinct populations, each with a restricted distribution and threatened habitat in Alabama. Stream form restricted to four counties and six creek systems in a rapidly developing area of Alabama. Sedimentation problems exist in many of these streams, especially within Shelby County. Specimens absent at six historic sites, and stream form has low abundance relative to spring form, which may indicate water quality degradation. Most springs where spring form present have some degree of degradation, even where species is common. Removal of aquatic vegetation and water, excessive sedimentation, and livestock entering springs are common disturbances. This form currently restricted to 11 springs in Alabama and missing from three historic sites (Kuhajda and Mayden 2002*c*).

Prepared by: **Bernard R. Kuhajda**

TUSCUMBIA DARTER

Etheostoma tuscumbia Gilbert and Swain

OTHER NAMES. None.

Photo—Richard Mayden

DESCRIPTION. In monotypic subgenus *Psychromaster*, diagnosed by having scales on top of head and on brachiostegal membranes (that are unconnected), an incomplete lateral line, and one anal spine. Light green to olive brown with golden flecks on dorsum and sides, four to seven irregular brown saddles, dark brown mottling on sides, a suborbital bar, and two basicaudal spots. No sexual dimorphism. Maximum size 65 millimeters (2.6 inches) TL (Etnier and Starnes 1993).

DISTRIBUTION. Restricted to limestone springs and spring runs in southern bend of the Tennessee River at 14 localities in Lauderdale, Colbert, Lawrence, Limestone, and Madison Counties, Alabama. Historic record from one spring in south-central Tennessee, but it, along with six historic sites in

Alabama, inundated with impoundment of the Tennessee River (Etnier and Starnes 1993, Jones *et al.* 1995).

HABITAT. Inhabits limestone springs with aquatic vegetation and fine gravel or sand substrates at temperatures of 15-20°C (59-68°F). Most abundant in springs with diverse aquatic vascular plants, a vegetated buffer zone around spring, and a low diversity of other fishes, but can be present in highly disturbed springs with filamentous algae and centrarchids (Jones *et al.* 1995).

LIFE HISTORY AND ECOLOGY. Ecology of different populations variable. Spawn year round, but peak occurs from either January to March or March to May. Females observed laying eggs in clean gravel or sand substrate in Buffler (King) Spring, Lauderdale County, but attach them to moss in Meridianville Spring, Madison County. Sexual maturity attained at one year of age and three centimeters (1.2 inches) SL, and life span two to three years. Larger females have more eggs, and fecundity ranges from 40 to 100 mature ova. Fertilized eggs hatch in five days at 17°C (63°F). Adult darters active during day in some springs, other populations are active at night, and some are active both day and night. Diet consists of chironomid larvae, amphipods, and snails (Koch 1978, Boyce 1997).

BASIS FOR STATUS CLASSIFICATION. Restricted to 14 springs and spring runs in five counties in Alabama, but most springs have some degradation, including removal of aquatic vegetation and water, excessive sedimentation, livestock entering spring, and small impoundments. Have disappeared from almost half of historic springs; seven sites have been inundated by impoundment of Tennessee River, and have disappeared from an additional five springs due to extensive habitat modification (Jones *et al.* 1995).

Prepared by: **Bernard R. Kuhajda**

BANDFIN DARTER

Etheostoma zonistium Bailey and Etnier

OTHER NAMES. None.

Photo—Patrick E. O'Neil

DESCRIPTION. A member of the snubnose darter group (subgenus *Ulocentra*); all members are smaller (to 77 mm [3 in.] TL) darters with extremely blunt snouts, a very narrow or absent premaxillary frenum, broadly joined branchiostegal membranes, and seven to 10 dorsal saddles. Males brightly colored year round, and brilliantly colored in spring. Females brown, straw colored, and typically lack bright colors, but may have some red in dorsal fins and scattered on upper sides. Like other members of the *Etheostoma duryi* species group, lacks a premaxillary frenum and has vomerine teeth. Male bandfin darters distinguished from all but one member of the species group by possessing three colored and three clear bands in first dorsal fin (may have additional cream color at fin base) and having a continuous well-defined lateral stripe lacking lateral blotches or with blotches restricted to just below stripe (Bailey and Etnier 1988, Etnier and Starnes 1993). Bandfin darters differentiated from blueface darters by possessing a distinctive red spot in first membrane of first dorsal fin, fewer lateral-line scales, and having mostly exposed breast and head scales (Kuhajda and Mayden 1995).

DISTRIBUTION. Coastal Plain streams in Tennessee River drainage from Clarks River, Kentucky, upstream to lower Bear Creek System in Alabama, with an isolated population in Spring Creek and Pleasant Run in upper Hatchie River System, Tennessee (Bailey and Etnier 1988). Within Alabama, found within the Bear Creek System in Pennywinkle and Cripple Deer Creeks and Buck Branch in extreme western Colbert County, with a historic record from Bear Creek proper, which is now impounded by Pickwick Reservoir. Also in Panther Creek, a direct tributary to the Tennessee River in extreme western Lauderdale County (Mettee *et al.* 1996, Kuhajda and Mayden 2002*a*). Populations formerly referred to as bandfin darters in streams on Cumberland Plateau in upper Bear Creek and upper Sipsey Fork (Black Warrior River drainage) systems are actually an undescribed species, the blueface darter (Kuhajda and Mayden 1995).

HABITAT. Small to medium Coastal Plain creeks two to 12 meters (seven to 39 feet) wide with substrates of gravel, sand, silt, mud, and clay with slow to moderate flows. Occupy pools, stream margins, and head and base of riffles over a variety of substrates in slow to moderate current (Carney and Burr 1989, Kuhajda and Mayden 2002*a*).

LIFE HISTORY AND ECOLOGY. Most populations have relatively high abundance; 11 collections in Tennessee and Kentucky averaged 1.14 darters per minute sampling. This contrasts with populations in Pennywinkle and Cripple Deer Creeks in Alabama and Mississippi, which averaged only 0.17 specimens per minute sampling for seven collections, and recent collections have few specimens (less than seven) (Kuhajda and Mayden 2002*a*). Collections from Buck Branch and Panther Creek also are represented by few individuals. In Kentucky, spawns from March to May at 11-17°C (52-63°F), spawning begins at one year of age, and eggs are laid on a variety of habitats in aquaria. Fertilized eggs averaged 1.7 millimeters (0.07 inches) in diameter and hatched in seven days at 20°C (68°F). Life span up to 35 months and principal diet consists of aquatic insect larvae and microcrustaceans (Carney and Burr 1989).

BASIS FOR STATUS CLASSIFICATION. Within Alabama, restricted to less than five kilometers (3.1 miles) of lower Pennywinkle and Cripple Deer Creeks and known from only one site in Buck Branch and Panther Creek in extreme western Colbert and western Lauderdale Counties. Compared with abundance of individuals in populations in Tennessee and Kentucky, are very uncommon, and are becoming rarer. Populations in Bear Creek proper extirpated due to impoundment of Tennessee River. Sedimentation problems exist in all creeks, and clear-cut activity evident along Pennywinkle Creek.

***Prepared by*: Bernard R. Kuhajda**

BLUEFACE DARTER

Etheostoma sp. cf. *zonistium*

Photo—Richard Mayden

OTHER NAMES. Bandfin Darter.

DESCRIPTION. A member of the snubnose darter group (subgenus *Ulocentra*); all members are smaller (to 77 mm [3 in.] TL) darters with extremely blunt snouts, a very narrow or absent premaxillary frenum, broadly joined branchiostegal membranes, and seven to 10 dorsal saddles. Males brightly colored year round, and brilliantly so in spring. Females brown, straw-colored, and typically lack bright colors, but may have some red in dorsal fins and scattered on upper sides. Like other members of the *Etheostoma duryi* species group, lacks a premaxillary frenum and has vomerine teeth. Males distinguished from all but one member of the species group by possessing three colored and three clear bands in first dorsal fin (may have additional cream color at fin base) and having a continuous well-defined lateral stripe lacking lateral blotches or with blotches restricted to just below stripe (Bailey and Etnier 1988, Etnier and Starnes 1993). Blueface darter differentiated from bandfin darter by lack of a distinctive red spot in first membrane of first dorsal fin, possessing more lateral-line scales, and having breast and head scales more deeply embedded (Kuhajda and Mayden 1995).

DISTRIBUTION. Restricted to upland streams on Cumberland Plateau in 10.2 kilometers (6.3 miles) of upper Bear Creek system (Tennessee River drainage) and 5.4 kilometers (3.4 miles) of Hubbard Creek system (Black Warrior River drainage, Sipsey Fork) in Franklin, Lawrence, and Winston Counties, Alabama. Extirpated from half of former distribution in Bear Creek System downstream of, and inundated by, Upper Bear Creek Reservoir in Marion, Franklin, and Winston Counties (Kuhajda and Mayden 2002*a*, *b*).

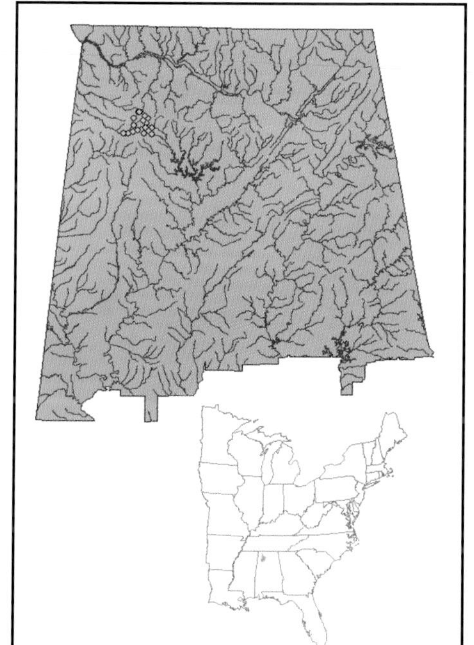

HABITAT. Small upland streams two to 10 meters (seven to 33 feet) wide with predominate cobble and bedrock substrate, but can be found in streams with mostly sand substrate. Prefer slow to moderate current at head and bottom of riffles and in pools over bedrock and cobble, with finer substrates rare; aquatic vegetation, especially moss, may be present (Kuhajda and Mayden 2002*a*, *b*).

LIFE HISTORY AND ECOLOGY. Relatively unknown, but specimens collected in spawning condition from March to May. Species rare to common in Bear Creek system, but typically common in Hubbard Creek system (Kuhajda and Mayden 2002 *a*, *b*). Likely abundance differences attributable to greater habitat degradation in Bear Creek system, but differences in life histories (*i.e.*, fecundity) may be involved. Diet likely includes benthic aquatic invertebrate larvae and microcrustaceans.

BASIS FOR STATUS CLASSIFICATION. Restricted to two creek systems and 16 kilometers (10 miles) within Franklin, Lawrence, and Winston

Counties, Alabama. In Bear Creek system, rare at many sites and found in only half of former distribution due to impoundment of formerly flowing streams, deposition of fine sediments into streambeds, and water quality degradation from reservoir construction, sand and gravel strip mining, and poultry production. Although Hubbard Creek populations stable, and a large portion of watershed lies within Bankhead National Forest, the species is only distributed over 5.4 kilometers (3.4 miles), and Basin and Whitman Creeks have large amounts of sedimentation from poor forestry practices, including no natural riparian buffer zones and highly eroded logging roads (Kuhajda and Mayden 2002 *a,b*).

Prepared by: **Bernard R. Kuhajda**

COAL DARTER

Percina brevicauda Suttkus and Bart

OTHER NAMES. None.

Photo—Malcolm Pierson

DESCRIPTION. Smallest (max. size 50 millimeters [2 in.] SL) member of subgenus *Cottogaster* as well as genus *Percina*. Elongated and slightly compressed, with a blunt snout and an inferior mouth, usually with no premaxillary frenum. Lateral line usually complete with 55 to 60 scales (Suttkus *et al.* 1994). Males typically have seven to 10 modified belly scales; females occasionally develop from one to five modified ventral scales. Has eight to 10 faint dorsal saddles, and eight to 12 oblong or rectangular blotches along lateral line interspersed with another three or four smaller blotches. Spinous dorsal fin darkly pigmented with a basal band and a fainter submarginal band. A suborbital bar present, and snout nearly encircled by preobital bars. Does not develop bright breeding colors; however, nuptial males become darkly pigmented and exhibit blueish iridescence on lower portion of head, opercle, and body to base of pectoral fins (Suttkus *et al.* 1994). Sympatric darter species that might be confused with coal darter include blackbanded and speckled darters. Blackbanded darter has deeper body, lateral blotches expanded dorso-ventrally as bars, and a wide premaxillary frenum. Speckled darter has reddish submarginal band in spinous dorsal fin and V- or W-shaped lateral markings rather than oblong or rectangular lateral blotches.

DISTRIBUTION. An Alabama endemic, known from Cahaba, Coosa, and Black Warrior River systems (Suttkus *et al.* 1994). In Cahaba River, most specimens have come from main channel upstream of Fall Line at Centreville, Bibb County, to near mouth of Little Black Creek, St. Clair County, and lower portion of Little Cahaba River and Six Mile Creek. A few records exist from as far downstream as Heiberger, Perry County (Pierson *et al.* 1989). In Coosa River system, now occurs only in Hatchet Creek where known from a single locality at U.S. Highway 231. In Locust Fork system, occurs in main channel from first shoal upstream of embayment of Bankhead Lake (about 1.6 km [1 mi.] upstream of mouth of Fivemile Creek) upstream to Nectar Bridge, Blount County. Also occurs in lower 6.8 kilometers (4.25 miles) of Blackburn Fork (Shepard *et al.* 2002). There are records from Black Warrior River at Tuscaloosa (1889), before that section was impounded (1895), and from main channel of Coosa River, Talladega County, (1949 and 1950) before impoundment by Logan Martin Dam (1964) (Suttkus *et al.* 1994). These records illustrate that species once had a much more contiguous distribution in Mobile River Basin.

HABITAT. Main channels of small to medium-sized rivers. Most frequently collected in moderate to swift flows over shallow shoals having substrates of gravel, cobble, and sand.

LIFE HISTORY AND ECOLOGY. Little known. Peak spawning probably in May in Cahaba River based on reproductive condition. A spawning aggregation of several hundred individuals in bedrock chutes in Cahaba River observed in mid-May (Suttkus *et al.* 1994). Based on size at maturity and maximum size, most individuals probably reach reproductive maturity at one year of age and few if any survive to spawn a second year (H. L. Bart, Jr., pers. comm.).

BASIS FOR STATUS CLASSIFICATION. Has a disjunct distribution in only three river systems. As a main channel species, vulnerable to chronic water quality and habitat degradation as well as acute impacts such as toxic chemical spills. A single major spill could potentially eliminate this species from Cahaba River, Locust Fork, or Hatchet Creek. As an annual species, poor recruitment in a given year would lead to depressed population levels in following years. Cahaba River system and Locust Fork system are both experiencing increasingly degraded habitat and water quality conditions. Species has only been collected sporadically in Hatchet Creek in recent years and status of population there unclear. In Cahaba River, greatest threats from eutrophication and sedimentation related to urban development in watershed. In Locust Fork system, agricultural activities and surface mining for coal impact habitat and water quality. A water supply reservoir proposed for construction near Jefferson-Blount County line by Birmingham Water Works Board would probably eliminate species from Locust Fork system through inundation of upstream habitat and impacts to downstream habitat related to construction and altered flow regimes.

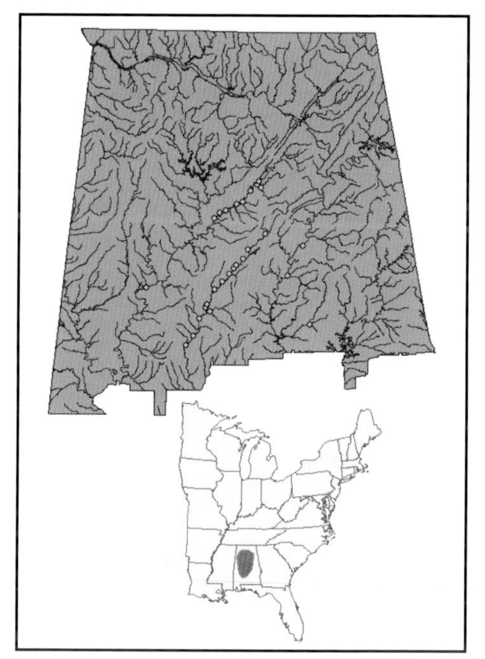

Prepared by: **Thomas E. Shepard**

GILT DARTER

Percina evides (Jordan and Copeland)

OTHER NAMES. None.

Photo—Patrick E. O'Neil

DESCRIPTION. A member of subgenus *Ericosma*, distinguished by russet-orange color of breeding males. Adults range up to 75 millimeters (three inches) SL. Adult males among most colorful Alabama *Percina*; generally dark olive dorsally, fading to lighter yellow-orange shades ventrally. From seven to nine dark, sometimes wide, dorsal saddles extend ventrally as tall blotches along sides. During breeding season these greenish-blue blotches contrast strongly with gilded lateral and ventral regions; blotches often connected forming a continuous lateral band. Base of caudal fin has two oval, orange-yellow spots. Females less colorful and lateral blotches more distinctly formed. Spiny dorsal fin of males has dark orange to amber base and extensive russet-orange color throughout, whereas soft dorsal, anal, and pelvic fins are dusky. Other median fins less colorful. Meristic counts include 51 to 77 lateral-line scales; 10 to 15 dorsal fin spines and 10 to 14 rays; two anal fin spines and five to 10 rays; and 11 to 16 pectoral fin rays. Opercles and nape scaled; breast, belly, and preopercle area naked (Etnier and Starnes 1993).

DISTRIBUTION. Wide distribution, occurring in tributaries of Mississippi River from southwest New York west to Minnesota and south to Alabama, in western tributaries of Mississippi River in Missouri and Arkansas, and in Lake Erie Basin and Maumee River system in Ohio and Indiana (Denoncourt 1969, Page and Burr 1991). In Alabama, occurs in Tennessee River drainage in flowing, main channel reaches of Shoal Creek and Elk River, and in main channel and larger tributaries of Bear Creek system. Prior to 1990s, species known from only a handful of sites in state. Extensive sampling by biologists of Geological Survey of Alabama and Tennessee Valley Authority has since expanded known locations to nine with several additional locations in Shoal Creek system in Tennessee. Shoal Creek at Iron City (Tennessee Highway 227), just north of Tennessee-Alabama state line, yielded 49 individuals during a collection in May 2000.

HABITAT. Clean, flowing streams over gravel and small cobble shoals, most often occurring in deep chutes and runs at head of riffles loads (Hatch 1983, 1986). Believed to avoid riffles with substantial silt and sediment loads.

LIFE HISTORY AND ECOLOGY. Little known in Alabama. Breeding season extends from April through early July in Tennessee, from June to mid-July in Minnesota, and from April to May in Arkansas. An egg burier; spawns over gravel mixed with cobble and boulders in upper reaches of riffles. Feeds on midge larvae, blackflies, caddisflies, mayflies, and occasional snails.

BASIS FOR STATUS CLASSIFICATION. Limited distribution in Alabama and Bear Creek, largest distribution center for this species in state, experiencing degradation of its main channel habitats due to sedimentation, stream bank erosion, and reduced water quality from tailwater releases. Additional sampling in Bear Creek system, both Bear and Cedar Creeks, may yield additional locations. Shoal Creek population appears stable at this time, while Elk River population in Alabama vulnerable due to modified streamflows, water withdrawals, and upstream sedimentation.

***Prepared by:* Patrick E. O'Neil**

COMMON AND SCIENTIFIC NAMES OF PLANTS AND VERTEBRATE ANIMALS MENTIONED IN FISH ACCOUNTS

PLANTS

Algae *Spirogyra*
Bladderwort *Utricularia*
Bur-Reed *Sparganium americanum*
Coontail *Ceratophyllum*
Dock *Rumex*
Hornwort *Ceratophyllum*
Mosses *Fontinalis and/or Fissidens*
Pondweed *Potomogeton foliosus*
River Weed *Podostemum ceratophyllum*
Rushes *Juncus*
Stonewort *Chara*
Watercress *Nasturtium officinale*
Water Milfoil *Myriophyllum*
Water Weed *Elodea*
Water Willow *Justicia virginiana*

FISHES

Southern Cavefish *Typhlichthys subterraneus*

Bigeye Chub *Hybopsis amblops*
Blotched Chub *Erimystax insignis*
Burrhead Chub *Macrhybopsis marconis*
Peppered Chub *Macrhybopsis tetranema*
Prairie Chub *Macrhybopsis australis*
Redeye Chub *Notropis harperi*
Speckled Chub *Macrhybopsis aestivalis*
Silver Chub *Macrhybopsis storeriana*
Bluntnose Minnow *Pimephales notatus*
Pugnose Minnow *Opsopoeodus emiliae*

Banded Darter *Etheostoma zonale*
Blackbanded Darter *Percina nigrofasciata*
Goldstripe Darter *Etheostoma parvipinne*
Greenbreast Darter *Etheostoma jordani*
Muscadine Darter *Percina* sp. cf. *macrocephala*
Redline Darter *Etheostoma rufilineatum*
Redspot Darter *Etheostoma artesiae*
Speckled Darter *Etheostoma stigmaeum*
Warrior Darter *Etheostoma bellator*
Logperch *Percina caprodes*

Spotted Gar *Lepisosteus oculatus*

Black Madtom *Noturus funebris*
Freckled Madtom *Noturus nocturnus*
Speckled Madtom *Noturus leptacanthus*
Tadpole Madtom *Noturus gyrinus*

Mooneye *Hiodon tergisus*

Blackmouth Shiner *Notropis melanostomus*
Coastal Shiner *Notropis petersoni*
Emerald Shiner *Notropis atherinoides*
Mimic Shiner *Notropis volucellus*
Sawfin Shiner *Notropis* sp. cf. *spectrunculus*
Silver Shiner *Notropis photogenis*
Striped Shiner *Notropis chrysocephalus*
Swallowtail Shiner *Notropis procne*
Telescope Shiner *Notropis telescopus*
Weed Shiner *Notropis texanus*

Ohio Shad *Alosa ohioensis*
Skipjack Herring *Alosa chrysochloris*

Redbreast Sunfish *Lepomis auritus*
Redeye Bass *Micropterus coosae*

Note: Common and scientific names of taxa designated as Extinct, Extirpated, Highest, or High Conservation Concern are presented in each account and are not duplicated here.

ALABAMA FISHES WATCH LIST

MODERATE CONSERVATION CONCERN

Taxa with conservation problems because of insufficient data, OR because of two of four of the following: small population; limited, disjunct, or peripheral distribution; decreasing population trend/population viability problem; specialized habitat need/habitat vulnerability due to natural/human-caused factors. Research and/or conservation action recommended.

COMMON NAME	SCIENTIFIC NAME	PROBLEM(s)
Mountain Brook Lamprey	*Ichthyomyzon greeleyi*	Insufficient data
American Brook Lamprey	*Lampetra appendix*	Insufficient data
Alligator Gar	*Atractosteus spatula*	Decreasing population/Habitat vulnerability
Skipjack Herring	*Alosa chrysochloris*	Insufficient data
Bluefin Stoneroller	*Campostoma pauciradii*	Decreasing population/Habitat vulnerability
Bluestripe Shiner	*Cyprinella callitaenia*	Decreasing population/Habitat vulnerability
Blotched Chub	*Erimystax insignis*	Insufficient data
Ribbon Shiner	*Lythrurus fumeus*	Insufficient data
Highscale Shiner	*Notropis hypsilepis*	Decreasing population/Habitat vulnerability
Highland Shiner	*Notropis micropteryx*	Insufficient data
Silver Shiner	*Notropis photogenis*	Insufficient data
Skygazer Shiner	*Notropis uranoscopus*	Decreasing population/Habitat vulnerability
Apalachee Shiner	*Pteronotropis grandipinnis*	Decreasing population/Habitat vulnerability
Blue Sucker	*Cycleptus elongatus*	Insufficient data
Spotted Bullhead	*Ameiurus serracanthus*	Decreasing population/Habitat vulnerability
Southern Cavefish	*Typhlichthys subterraneus*	Small population/Specialized habitat needs
Stippled Studfish	*Fundulus bifax*	Limited distribution/Decreasing population
Marsh Killifish	*Fundulus confluentus*	Decreasing population/Habitat vulnerability
Starhead Topminnow	*Fundulus dispar*	Insufficient data
Saltmarsh Topminnow	*Fundulus jenkinsi*	Decreasing population/Habitat vulnerability
Bayou Killifish	*Fundulus pulverous*	Decreasing population/Habitat vulnerability
Bluefin Killifish	*Lucania goodei*	Insufficient data
Tallapoosa Sculpin	*Cottus* sp. cf. *carolinae*	Insufficient data
Striped Bass	*Morone saxatilis*	Decreasing population/Habitat vulnerability
Banded Sunfish	*Enneacanthus obesus*	Insufficient data
Crystal Darter	*Crystallaria asprella*	Disjunct distribution/Habitat vulnerability
Warrior Darter	*Etheostoma bellator*	Decreasing population/Habitat vulnerability
Fringed Darter	*Etheostoma crossopterum*	Insufficient data
Tuskaloosa Darter	*Etheostoma douglasi*	Limited distribution/Specialized habitat needs
Southern Logperch	*Percina austroperca*	Insufficient data
Freckled Darter	*Percina lenticula*	Disjunct distribution/Habitat vulnerability
Muscadine Darter	*Percina* sp. cf. *macrocephala*	Small population/Limited distribution
Walleye	*Sander vitreus*	Insufficient data

REFERENCES CITED

Albanese, B., and W. T. Slack. 1998. Status of the ironcolor shiner, *Notropis chalybaeus*, in Mississippi. Proc. S.E. Fishes Council 37:1-5.

Aley, T. 1990. Delineation and hydrogeologic study of the Key Cave aquifer, Lauderdale County, Alabama. Ozark Underground Lab. Protein, Missouri. 114 pp.

Anonymous. 1995. Bubbling on the Elk River. Compressed Air Magazine 100:40-43.

Bailey, R. M., and D. A. Etnier. 1988. Comments on the subgenera of darters (Percidae) with descriptions of two new species of *Etheostoma* (*Ulocentra*) from southeastern United States. Misc. Publ. Mus. Zool. Univ. Michigan No. 175:1-48.

_____, and W. J. Richards. 1963. Status of *Poecilichthys hopkinsi* Fowler and *Etheostoma trisella*, new species, percid fishes from Alabama, Georgia and South Carolina. Occas. Papers Mus. Zoology Univ. Michigan No. 630:1-21.

Bart, H. L., Jr., and M. S. Taylor. 1999. A new darter of the subgenus *Fuscatelum* of *Etheostoma* from the upper Black Warrior River system, Alabama. Tulane Studies in Zoology and Botany 31:23-52.

Becker, G. C. 1983. Fishes of Wisconsin. The Univ. of Wisconsin Press, Madison, WI. 1,052 pp.

Bell, D., and T. J. Timmons. 1991. Life history of the brighteye darter *Etheostoma lynceum* (Pisces:Percidae), in Terrapin Creek, Kentucky. Proc. S.E. Fishes Council 23:1-6.

Berry, F. H. 1964. Review and emendation of Family Clupeidae by Samuel F. Hildebrand. Copeia 1964:720-730.

Biggins, R. G. 1984. Endangered and threatened wildlife and plants; final rule reclassifying the snail darter (*Percina tanasi*) from an endangered species to a threatened species. Federal Register 49:27510-27514.

_____. 1988. Endangered and threatened wildlife and plants; determination of threatened status for the boulder darter. Federal Register 53:33996-33998.

_____. 1989. Recovery plan for boulder darter (*Etheostoma* sp.). U.S. Fish and Wildl. Serv., Southeast Region, Atlanta, GA. 15 pp.

_____, and R. B. Eager. 1983. Snail darter recovery plan. U.S. Fish and Wildl. Serv., Southeast Region, Atlanta, GA. 46 pp.

Boschung, H. T. 1992. Catalog of freshwater and marine fishes of Alabama. Bull. Ala. Mus. Nat. Hist. 14. 266 pp.

_____, and R. L. Mayden. In press. Fishes of Alabama. Smithsonian Institution Press, Washington D.C.

_____, _____, and J. R. Tomelleri. 1992. *Etheostoma chermocki*, a new species of darter (Teleosti: Percidae) from the Black Warrior River drainage of Alabama. Bull. Ala. Mus. Nat. Hist. 13:11-20.

_____, and D. Nieland. 1986. Biology and conservation of the slackwater darter, *Etheostoma boschungi* (Pisces: Percidae). Proc. S.E. Fishes Council 4:1-4.

Bottrell, C. E., R. H. Ingersol, and R. W. Jones. 1964. Notes on the embryology, early development and behavior of *Hybopsis aestivalis tetranemus* (Gilbert). Trans. Amer. Micro. Soc. 83:391-399.

Boyce, J. S. 1997. Activity patterns in the Tuscumbia darter, *Etheostoma tuscumbia* (Teleostei: Percidae), with comments on reproductive biology. Unpubl. M.S. Thesis, Univ. of Alabama, Tuscaloosa, AL. 32 pp.

Branson, B. A. 1963. The olfactory apparatus of *Hybopsis gilida* (Girard) and *Hybopsis aestivalis* (Girard) (Pisces: Cyprinidae). J. Morph. 113:215-229.

Bryant, R.T. 1979. The life history and comparative ecology of the darter, *Etheostoma acuticeps*: Tenn. Wildl. Resour. Agency Rept. 79-50. 60 pp.

Burke, J. S., and J. S. Ramsey. 1995. Present and recent historic habitat of the Alabama sturgeon, *Scaphirhynchus suttkusi* Williams and Clemmer, in the Mobile Basin. Bull. Ala. Mus. Nat. Hist. 17:17-24.

Burkhead, N. M., and J. D. Williams. 1992. The boulder darter: a conservation challenge. Endangered Species Tech. Bull. 17: 4-6.

Burr, B. M., and W. W. Dimmick. 1981. Nests, eggs and larvae of the elegant madtom *Noturus elegans* from the Barren River Drainage, Kentucky (Pisces: Ictaluridae). Trans. Ky. Acad. Sci. 42:116-118.

_____, and R. L. Mayden. 1982. Life history of the brindled madtom *Noturus miurus* in Mill Creek, Illinois (Pisces: Ictaluridae). Am. Midl. Natur. 104:198-201.

_____, and J. N. Stoeckel. 1999. The natural history of madtoms (genus *Noturus*), North America's diminutive catfishes. Am. Fish. Soc. Symp. 24:51-101.

Carney, D. A., and B. M. Burr. 1989. Life histories of the bandfin darter, *Etheostoma zonistium*, and the firebelly darter, *Etheostoma pyrrhogaster*, in western Kentucky. Illinois Nat. Hist. Surv. Biol. Notes 134:1-16.

Ceas, P. A., and L. M. Page. 1995. Status surveys of the crown darter (*Etheostoma corona*) and the lollypop darter (*Etheostoma neopterum*) in the Cypress Creek and Shoal Creek systems of Tennessee and Alabama, and the egg-mimic darter (*Etheostoma pseudovulatum*) in the Duck River system of Tennessee. Unpubl. report submitted to Tenn. Wildl. Resour. Agency and U. S. Fish and Wildl. Serv., Office of Endangered Species. Tech. Rept. 1995(14). 43 pp.

Clabaugh, J. P., K. E. Knott, R. M. Wood, and R. L. Mayden. 1996. Systematics and biogeography of snubnose darters, genus *Etheostoma* (Teleostei: Percidae) from the Black Warrior River system, Alabama. Biochem. System. Ecol. 24:119-134.

Coker, R. E. 1930*a*. Keokuk Dam and the fisheries of the upper Mississippi River. U.S. Bureau of Commercial Fisheries Document No. 1063:87-139.

_____. 1930*b*. Studies of common fishes of the Mississippi River at Keokuk. U. S. Bureau of Commercial Fisheries Document No. 1072:164-225.

Cooper, J. E., and R. A. Kuehne. 1974. *Speoplatyrhinus poulsoni*, a new genus and species of subterranean fish from Alabama. Copeia 1974:486-493.

Darr, D. P., and G. R. Hooper. 1991. Spring pygmy sunfish population monitoring. Unpubl. final report prepared for Ala. Dept. Conserv. Nat. Resour., Montgomery, AL. 11 pp.

Denoncourt, R. F. 1969. A systematic study of the gilt darter *Percina evides* (Jordan and Copeland) (Pisces, Percidae). Ph.D. Diss., Cornell Univ., Ithaca, NY.

Eisenhour, D. J. 1997. Systematics, variation, and speciation of the *Macrhybopsis aestivalis* complex (Cypriniformes: Cyprinidae) west of the Mississippi River. Ph.D. Diss. Southern Illinois Univ., Carbondale, IL. 260 pp.

Etnier, D. A. 1975. *Percina (Imostoma) tanasi* a new percid fish from the Little Tennessee River, Tennessee. Proc. Biol. Soc. Wash. 88:469-488.

_____, and W. C. Starnes. 1993. The fishes of Tennessee. The Univ. of Tennessee Press, Knoxville, TN. 681 pp.

_____, and J. D. Williams. 1989. *Etheostoma (Nothonotus) wapiti* (Osteichthyes: Percidae), a new darter from the southern bend of the Tennessee River system in Alabama and Tennessee. Proc. Biol. Soc. Wash. 102:1987-1000.

Evermann, B.W. 1896. Description of a new species of shad (*Alosa alabamae*) from Alabama. Report of the Commissioner of the U.S. Commission of Fish and Fisheries 21:203-205.

_____. 1902. Description of a new species of shad (*Alosa ohiensis*) with notes on other food fishes of the Ohio River. Report of the Commissioner of the U.S. Commission of Fish and Fisheries 27: 273-288.

Federal Register. 1996. Endangered and threatened species: Notice of reclassification of 96 candidate taxa. 60:7457-7463

_____. 2000. Final rule to list the Alabama sturgeon as endangered. 65:26438-26461.

_____. 2001. Endangered and threatened wildlife and plants; rule to list the vermilion darter as endangered. 66:59367-59373.

Fletcher, D. E. 1993. Nest association of dusky shiners (*Notropis cummingsae*) and redbreast sunfish (*Lepomis auritus*), a potentially parasitic relationship. Copeia 1993:159-167.

Fox, D. A., J. E. Hightower, and F. M. Parauka. 2000. Gulf sturgeon spawning migration and habitat in the Choctawhatchee River System, Alabama-Florida. Trans. Am. Fish. Soc. 129:811-826.

Freeman, B. J., M. C. Freeman, and C. Straight. 2002. Status of the undescribed Halloween darter, *Percina* sp. in the Apalachicola River drainage in Alabama and Georgia. Report to USFWS .11 pp.

Freeman, M. C., and B. J. Freeman. 1992. A new *Percina* species in the Apalachicola River basin, Georgia and Alabama. 72nd Annual Meeting Amer. Society of Ichthyologists and Herpetologists, Champaign-Urbana, Urbana, IL.

Gilbert, C. R. 1969. Systematics and distribution of the American cyprinid fishes *Notropis ariommus* and *Notropis telescopus*. Copeia 1969:474-492.

_____. 1980. *Notropis ariommus*, popeye shiner. Pp. 229 *in* D.S. Lee *et al.*, eds. Atlas of North American freshwater fishes. N.C. State Museum of Natural History, Raleigh, NC. 854 pp.

Harris, J. L. 1986. Systematics, distribution, and biology of fishes currently allocated to *Erimystax* (Cyprinidae). Ph.D. Diss., Univ. of Tennessee, Knoxville, TN. 335 pp.

Hatch, J. T. 1983. Life history of the gilt darter, *Percina evides* (Jordan and Copeland), in the Sunrise River, Minnesota. Ph.D. Diss., Univ. of Minnesota, Minneapolis, MN. 162 pp.

_____. 1986. Distribution, habitat, and status of the gilt darter (*Percina evides*) in Minnesota. J. Minnesota Acad. Sci. 51:11-16.

Hickman, G. D., and R. B. Fitz. 1978. A report on the ecology and conservation of the snail darter (*Percina tanasi* Etnier). TVA Tech. Note B28, Norris, TN. 182 pp.

Hildebrand, S. F. 1963. Family Clupeidae. Pp. 257-454 *in* Y. H. Olsen, ed. Fishes of the western North Atlantic. Part 3. Sears Foundation for Marine Research, Yale Univ., New Haven, CN.

Hill, P. L. 1996. Habitat use and life history of the Halloween darter, *Percina* sp., in the upper Flint River system, Georgia. M.S. Thesis, Auburn Univ., Auburn, AL.

Howell, W. M. 1986. Watercress darter. Pp. 5-6 *in* R. H. Mount, ed. Vertebrate animals in Alabama in need of special attention. Ala. Agric. Expt. Sta., Auburn Univ., Auburn, AL.

_____. 1989. Fifth annual population survey of the endangered watercress darter, *Etheostoma nuchale*. Unpubl. U.S. Fish and Wildlife Status Report, Jackson, MS. 42 pp.

_____, and G. Dingerkus. 1978. *Etheostoma neopterum*. A new percid fish from the Tennessee River system in Alabama and Tennessee. Bull. Alabama Mus. Nat. Hist. 3:13-26.

_____, and S. L. Linton. 1996. The fishes of Dugger Mountain, Talladega National Forest, Alabama. 15 pp.

Hubbs, C. 1985. Darter reproductive seasons. Copeia 1985:56-68.

Huff, J. A. 1975. Life history of the Gulf of Mexico sturgeon, *Acipenser oxyrhynchus desotoi*, in the Suwannee River, Florida. Florida Marine Research Publications, Number 16.

Jandebeur, T. S. 1972. A study of fishes of the Elk River drainage in Tennessee and Alabama. Unpubl. M.S. Thesis, Univ. of Alabama, Tuscaloosa, AL. 154 pp.

_____. 1979. Distribution, life history, and ecology of the spring pygmy sunfish, *Elassoma* species. Unpubl. report prepared for Albert McDonald. 22 pp.

_____. 1982. A status report on the spring pygmy sunfish, *Elassoma* sp., in northcentral Alabama. Assoc. S. E. Biologists Bull. 29(2):66 (abstract).

Jenkins, R. E., and N. M. Burkhead. 1984. Description, biology, and distribution of the the spotfin chub, *Hybopsis monacha*, a threatened cyprinid fish of the the Tennessee River drainage. Bull. Ala. Mus. Nat. Hist. 8:1-30.

_____, and _____. 1993. Freshwater fishes of Virginia. American Fisheries Society, Bethesda, MD. 1,079 pp.

Johnston, C. E. 2000. Allopaternal care in the pygmy sculpin (*Cottus pygmaeus*). Copeia 2000:262-264.

_____. 2001. Nest site selection and aspects of the reproductive biology of the pygmy sculpin (*Cottus paulus*) in Coldwater Spring, Calhoun County, Alabama. Ecology of Freshwater Fishes 10:118-121.

_____, M. Castro, and L. Mione. 2001. Status and habitat requirements of the broadstripe shiner (*Pteronotropis euryzonus*) in Alabama. Report to Alabama Dept. Conserv. Nat. Resour. 14 pp.

_____, and W. W. Hartup. 2002. Status survey for *Etheostoma boschungi*, slackwater darter. Report to U.S. Fish and Wildl. Serv., Jackson, MS.

_____, and C. L. Knight. 1999. Life-history traits of the bluenose shiner, *Pteronotropis welaka* (Cypriniformes: Cyprinidae). Copeia 1999:200-205.

_____, and J. R. Shute. 1997. Observational notes on the spawning behavior of the blue shiner (*Cyprinella caerulea*) and the holiday darter (*Etheostoma brevirostrum*), two rare fishes of the Conasauga River, Georgia and Tennessee. Proc. S.E. Fishes Council 35:1-2.

Jones, E. B., III. 2000. Survey of fishes in the southern tributaries of the south bend of the Tennessee River with notes on Wheeler, Wilson, and Pickwick reservoirs. Proc. S.E. Fishes Council 40:1-42.

_____, B. R. Kuhajda, and R. L. Mayden. 1995. Monitoring of Tuscumbia darter, *Etheostoma tuscumbia*, populations in the southern bend of the Tennessee River, Alabama, May 1994-September 1995. Ala. Dept. Conserv. Nat. Resour, Montgomery, AL., and U.S. Fish and Wildl. Serv. Office, Jackson, MS. 63 pp.

Jones, P. W., F. D. Martin, and J. D. Hardy, Jr. 1978. Development of fishes of the Mid-Atlantic bight/An atlas of egg, larval and juvenile stages. Vol. I. Acipenseridae through Ictaluridae. U.S. Fish and Wildl. Serv. Biological Services Program FWS-OBS-78/12. 366 pp.

Knott, K. E., R. L. Mayden, B. R. Kuhajda, and J. Clabaugh. 1996. Systematics and population genetics of the coldwater (*Etheostoma ditrema*) and watercress (*E. nuchale*) darters, with comments on the Gulf darter (*E. swaini*) (Teleostei: Percidae: *Oligocephalus*). Abstract, 76th Annual Meetings of the Amer. Society of Ichthyologists and Herpetologists, New Orleans, LA.

Koch, L. M. 1978. Food habits of *Etheostoma* (*Psychromaster*) *tuscumbia* in Buffler Spring, Lauderdale County, Alabama (Percidae: Etheostomatini). Assoc. S.E. Biologists Bull. 25:56 (abstract).

Kott, E., and D. Fitzgerald. 2000. Comparative morphology and taxonomic status of the ghost shiner, *Notropis buchanani*, in Canada. Environ. Biol. Fishes 59:385-392.

Kuhajda, B. R., and R. L. Mayden. 1995. Discovery of a new species of snubnose darter (Percidae, *Etheostoma*) endemic to the Cumberland Plateau in Alabama. Assoc. S. E. Biologists 42:111-112 (abstract).

_____, and _____. 2001. Status of the federally endangered Alabama cavefish, *Speoplatyrhinus poulsoni* (Amblyopsidae), in Key Cave and surrounding caves, Alabama. Environmental Biology of Fishes 62:215-222.

_____, and _____. 2002*a*. Status survey of the blueface darter, *Etheostoma* sp. cf. *E. zonistium*, in upper Sipsey (Mobile Basin) and Bear Creek (Tennessee River Drainage) of Alabama, final report. U.S. Fish and Wildl. Serv. Office, Jackson. MS. 38 pp.

_____., and _____. 2002*b*. Status survey of the blueface darter, *Etheostoma* sp. cf. *zonistium*, in upper Bear Creek and Hubbard Creek within the Bankhead National Forest in Alabama, final report. U.S. Forest Service Office, Montgomery, AL. 29 pp.

_____, and _____. 2002*c*. Status survey of the coldwater darter, *Etheostoma ditrema*, in Alabama, Georgia, and Tennessee, final report. U.S. Fish and Wildl. Serv. Office, Jackson, MS. 151 pp.

Laurence, G. C., and R.W. Yerger. 1966. Life history of the Alabama shad, *Alosa alabamae*, in the Apalachicola River, Florida. Proc. S. E. Assoc. Game and Fish Comm. 20:260-273.

Leftwich, K. N., and P. L. Angermeier. 1993. Potential use of a closely related surrogate in assessing impacts of stream degradation on endangered fish species. Unpubl. Annual report for Virginia Dept. Game and Inland Fisheries, Richmond, VA. 12 pp.

Lorio, W., ed. 2000. Proceedings of the Gulf of Mexico sturgeon (*Acipenser oxyrinchus desotoi*) status of the subspecies workshop September 13-14, 2000. Mississippi State Univ., Science and Technology Research Center, Stennis Space Center, MS. 20 pp.

Mansueti, A. J., and J. D. Hardy. 1967. Development of fishes of the Chesapeake Bay region: an atlas of egg, larval, and juvenile stages. Part I. Univ. of Maryland, Natural Resources Institute. 202 pp.

Marshall, N. 1947. Studies on the life history and ecology of *Notropis chalybaeus* (Cope). Q. J. Florida Acad. Sci. 9(3-4):163-188.

Mayden, R. L. 1989. Phylogenetic studies of North American minnows, with emphasis on Genus *Cyprinella* (Teleostei: Cypriniformes). Misc. Publ. Univ. Kansas Mus. Nat. Hist. 80:1-189.

Mayden, R. L. 1993. *Elassoma alabamae*, a new species of pygmy sunfish endemic to the Tennessee River drainage of Alabama (Teleostei: Elassomatidae). Bull. Ala. Mus. Nat. Hist. No. 16. 44 pp.

_____, and B. R. Kuhajda. 1989. Systematics of *Notropis cahabae*, a new cyprinid fish endemic to the Cahaba River of the Mobile Basin. Bull. Ala. Mus. Nat. His. 9:1-16.

_____, and _____. 1996. Systematics, taxonomy, and conservation status of the endangered Alabama sturgeon, *Scaphirhynchus suttkusi* Williams and Clemmer (Actinopterygii, Acipenseridae). Copeia 1996:241-273.

McCaleb, J. E. 1973. Some aspects of the ecology and life history of the pygmy sculpin, *Cottus Pygmaeus* Williams, a rare spring species of Calhoun County, Alabama. M.S. Thesis, Auburn Univ., Auburn, AL. 82 pp.

McGregor, S. W., and T. E. Shepard. 1995. Investigations of slackwater darter, *Etheostoma boschungi*, populations, 1992-94. Geological Survey of Alabama Circular No. 184.

McLane, W. M. 1955. The fishes of the St. Johns River system. Ph.D. Diss. Univ. of Florida, Gainesville, FL. 361 pp.

Mettee, M. F., Jr. 1974. A study on the reproductive behavior, embryology, and larval development of pygmy sunfishes of the genus *Elassoma*. Unpubl. Ph.D. Diss. Univ. of Alabama, Tuscaloosa, AL. 130 pp.

_____, P. E. O'Neil, and J. M. Pierson. 1996. Fishes of Alabama and the Mobile Basin. Alabama Geological Surv. Monograph 15. Oxmoor House, Birmingham, AL. 820 pp.

_____, _____, _____, and R. D. Suttkus. 1989. Fishes of the Black Warrior River system in Alabama. Geological Surv. Alabama. Bull. 133. 201 pp.

_____, _____, T. E. Shepard, S. M. McGregor, and W. P. Henderson. 2002. A survey of protected fish species and species of uncommon occurrence in the Tennessee River drainage in north Alabama and northeast Mississippi. Geological Surv. Alabama Bull. 171. 173 pp.

_____, and J. J. Pulliam, III. 1986. Reintroduction of an undescribed species of *Elassoma* into Pryor Branch, Limestone County, Alabama. Proc. S. E. Fishes Council 4:14-15.

_____, and J. S. Ramsey. 1986. Spring pygmy sunfish, *Elassoma* sp. Pages 4-5 in R. H. Mount, ed. Vertebrate animals of Alabama in need of special attention. Ala. Agric. Expt. Station, Auburn Univ., Auburn, AL.

Miller, G. L. 1984. Trophic ecology of the frecklebelly madtom *Noturus munitus* in the Tombigbee River, Mississippi. Am. Midl. Natur. 111:8-15.

Mills, G. M. 1972. Biology of the Alabama shad in northwest Florida. Florida Dept. Nat. Resour. Tech. Series No. 68. 24 pp.

Mount, D. I. 1959. Spawning behavior of the bluebreast darter, *Etheostoma camurum* (Cope). Copeia 1959:240-243.

Moore, G. A. 1950. The cutaneous sense organs of the barbeled minnows adapted for life in the muddy streams of the Great Plains region. Trans. Amer. Microscop. Soc. 69:69-95.

O'Bara, C. J., and D. A. Etnier. 1987. Status survey of the boulder darter. Final report submitted to the U.S. Fish and Wildl. Serv., Asheville, NC.

O'Neil, P. E., S. C. Harris, and M. F. Mettee. 1984. The distribution and abundance of aquatic organisms inhabiting streams draining the Citronelle oil field, Alabama. Geological Surv. Alabama Circ. 99. 26 pp.

_____, M. F. Mettee, S. W. McGregor, T. E. Shepard, and W. P. Henderson. 2000. Life history studies of the Alabama shad (*Alosa alabamae*) in the Choctawhatchee River, Alabama. Geological Surv. Alabama unpubl. report. 21 pp.

Orr, J. W. 1989. Feeding behavior of the greenbreast darter, *Etheostoma jordani* (Teleostei: Percidae). Copeia 1989:202-206.

_____, and J. S. Ramsey. 1990. Reproduction in the greenbreast darter, *Etheostoma jordani* (Teleostei: Percidae). Copeia 1990:100-107.

Page, L. M. 1983. Handbook of darters. TFH Publishers, Inc., NJ.

_____, and B. M. Burr. 1991. A field guide to freshwater fishes, North America north of Mexico. Houghton Mifflin Co., Boston, MA. 432 pp.

_____, P. A. Ceas, D. L. Swofford, and D. G. Buth. 1992. Evolutionary relationships within the *Etheostoma squamiceps* complex (Percidae; subgenus *Catonotus*) with descriptions of five new species. Copeia 1992:615-646.

_____, and P. W. Smith. 1971. The life history of the slenderhead darter, *Percina phoxocephala*, in the Embarras River, Illinois. Illinois Nat. Hist. Surv. Biological Notes 74. 14 pp.

Parauka, F. M., and M. Giorgianni. 2002. Availability of Gulf sturgeon spawning habitat in northwest Florida and southeast Alabama river systems. Final Report, U. S. Fish and Wildl. Serv., Panama City, FL. 77 pp.

Pflieger, W. L. 1975. The fishes of Missouri. First edition. Missouri Dept. Conserv., Jefferson City, MO. 343 pp.

_____. 1997. The fishes of Missouri. Revised edition. Missouri Dept. Conserv., Jefferson City, MO. 372 pp.

Pierson, J. M. 1990. Status of endangered, threatened, and special concern freshwater fishes in Alabama. J. Ala. Acad. Sci. 61:106-116.

_____. 1999. Status of the lipstick darter (*Etheostoma chuckwachatte*) and the undescribed "muscadine darter" (*Percina* (*Alvordius*) sp.) in the Tallapoosa River system of Alabama and Georgia. U.S. Fish and Wildl. Serv. 32 pp.

_____, W. M. Howell, R. A Stiles, M. F. Mettee, P. E. O'Neil, R. D. Suttkus, and J. S. Ramsey. 1989. Fishes of the Cahaba River system in Alabama. Ala. Geological Surv. Bull. 134. 183 pp.

Poly, W. J. 1997. Habitat, diet, and population structure of the federally-endangered palezone shiner, *Notropis albizonatus* Warren and Burr, in Little South Fork (Cumberland River), Kentucky. Unpubl. M.S. Thesis, Southern Illinois Univ. at Carbondale, IL. 178 pp.

Powers, S. L., G. L. Jones, P. Redinger, and R. L. Mayden. 2003. Habitat associations with upland stream fish assemblages in Bankhead National Forest, Alabama. Southeastern Nat. 2:85-92.

_____, and R. L. Mayden. 2002. Threatened fishes of the world: *Etheostoma cinereum* Storer, 1845 (Percidae). Environ. Biol. Fishes 63:264.

Rakes, P. L., and J. R. Shute. 2001. Developing propagation and culture protocols for the Cahaba shiner, *Notropis cahabae*, and the goldline darter, *Percina aurolineata*. Report to the Environmental Protection Agency. 11 pp.

_____, _____, and P. W. Shute. 1999. Reproductive behavior, captive breeding, and restoration ecology of endangered fishes. Environ. Biol. Fishes 55:31-42.

Ramsey, J. S. 1976. Freshwater fishes. Pages 53-65 *in* H. Boschung, ed. Endangered and threatened plants and animals of Alabama. Bull. Ala. Mus. Nat. Hist. No. 2. 92 pp.

_____. 1986*a*. Freshwater fishes. Pages 1-21, *in* R. H. Mount, ed. Vertebrate Animals in Need of Special Attention. Ala. Agric. Expt. Sta., Auburn Univ., Auburn. AL.

_____. 1986*b*. Trispot darter, *Etheostoma trisella*. Pages 20 - 21 in R. H. Mount, ed. Vertebrate animals of Alabama in need of special attention. Ala. Agric. Expt. Sta., Auburn Univ., Auburn, AL.

_____, W. M. Howell, and H. T. Boschung. 1972. Rare and endangered fishes of Alabama. Pages 57-86 *in* J. E. Keeler, ed. Rare and endangered vertebrates of Alabama. Ala. Dept. Conserv. Nat. Resour., Division Game and Fish, Montgomery, AL.

_____, and R. D. Suttkus. 1965. *Etheostoma ditrema*, a new darter of the subgenus *Oligocephalus* (Percidae) from springs of the Alabama River Basin in Alabama and Georgia. Tulane Stud. in Zool. 12:65-77.

Reno, H. W. 1969. Cephalic lateral-line systems of the cyprinid genus *Hybopsis*. Copeia 1969:736-773.

Robison, H. W. 1977. Distribution, habitat notes, and status of the ironcolor shiner, *Notropis chalybaeus* Cope, in Arkansas. Proc. Arkansas Acad. Sci. 31:92-94.

_____, and T. M. Buchanan. 1988. Fishes of Arkansas. Univ. of Arkansas Press, Fayetteville, AR. 536 pp.

Ross, S. T. 2001. The inland fishes of Mississippi. Univ. Press of Mississippi, Oxford, MS. 624 pp.

Ryon, M. G. 1986. Life history and ecology of *Etheostoma trisella* (Pisces: Percidae). Am. Midl. Natur. 115:73-86.

Saylor, C. F. 1980. *Etheostoma cinereum* Storer, ashy darter. Page 635 *in* D.S. Lee *et. al.*, eds. Atlas of North American freshwater fishes. N. C. State Mus. Nat. Hist., Raleigh, NC. 854 pp.

Scott, D. C. 1951. Sampling fish populations in the Coosa River, Alabama. Trans. Am. Fish. Soc. 80:28-40.

Scott, W. B., and E. J. Crossman. 1973. Freshwater fishes of Canada. Fisheries Res. Board Canada Bull. 184. 966 pp.

Seesock, W. E. 1979. Some aspects of the life history of the coldwater darter, *Etheostoma ditrema*, from Glencoe Spring, Etowah County, Alabama. Unpubl. M.S. Thesis, Auburn Univ., Auburn, AL. 70 pp.

Sheldon, A. L., and G. K. Meffe. 1993. Multivariate analysis of feeding relationships of fishes in blackwater streams. Environ. Biol. Fishes 37:161-171.

Shepard, T. E., and B. M. Burr. 1984. Systematics, status, and life history aspects of the ashy darter, *Etheostoma cinereum* (Pisces: Percidae). Proc. Biological Soc. Wash. 97:693-715.

_____, S. W. McGregor, and P. E. O'Neil. 1997. Status survey of the palezone shiner (*Notropis albizonatus*) in the Paint Rock River system 1997. Ala. Geological Surv. contract report to Ala. Dept. Conserv. Nat. Resour. Game and Fish Division, Montgomery, AL. 14 pp.

_____, _____, _____, and M. F. Mettee. 1996. Status survey of the frecklbelly madtom (*Noturus munitus)* in the Mobile River Basin: Summary of results for 1995-96 in the upper Tombigbee, Alabama, and Cahaba River systems. Geological Survey of Alabama, Tuscaloosa, AL. 146 pp.

_____, _____, _____, and _____. 1997. Status survey of the frecklebelly madtom (*Noturus munitus*) in the Mobile River basin, 1995-97. Ala. Geological Surv. open file report. 43 pp.

_____, P. E. O'Neil, S. W. McGregor, and M. F. Mettee. 1995. Status survey of the Cahaba shiner, *Notropis cahabae*, 1992-93. Ala. Geological Surv. Circ. 181. 14 pp.

_____, P. E. O'Neil, M. F. Mettee, Jr., S. W. McGregor, and W. P. Henderson, Jr. 2002. Status surveys of the Cahaba shiner (*Notropis cahabae*), coal darter (*Percina brevicauda*), and Tuskaloosa Darter (*Etheostoma douglasi*) in the Locust Fork and Valley Creek systems, 2001. Ala. Geological Surv. Circ. 201. 19 pp.

Sizemore, D. R., and W. M. Howell. 1990. Fishes of springs and spring-fed creeks of Calhoun County, Alabama. Proc. S. E. Fishes Council 22:1-6.

Smith, C. L. 1985. The inland fishes of New York State. N.Y. Dept. Environ. Conserv., Albany, NY. 522 pp.

Smith, H. M. 1898. Statistics of the fisheries of the interior waters of the United States. Report of the Commissioner of the U.S. Commission of Fish and Fisheries. Part 22. No. 11:489-533.

Smith, P. W. 1979. The fishes of Illinois. Univ. of Illinois Press, Urbana, IL. 314 pp.

Starnes, L. B., and W. C. Starnes. 1985. Ecology and life history of the mountain madtom *Noturus eleutherus* (Pisces: Ictaluridae). Am. Midl. Natur. 106:360-371.

Starnes, W. C. 1977. The ecology and life history of the snail darter, *Percina (Imostoma) tanasi* Etnier. Tenn. Wildl. Resour. Agency Fisheries Res. Report 77-52.

Starrett, W. C. 1950. Food relationships of the minnows and darters of the Des Moines River, Iowa. Ecology 31:216-233.

Stauffer, J. R. 1978. *Etheostoma camurum* (Cope), bluebreast darter. Page. 632 *in* D.S. Lee *et al.*, eds. Atlas of North American freshwater fishes. N. C. State Mus. Nat. Hist., Raleigh, NC. 854 pp.

Stiles, R. A. 1972. The comparative ecology of three species of *Nothonotus* (Percidae: *Etheostoma*) in Tennessee's Little River. Ph.D. Diss., Univ. of Tennessee, Knoxville, TN. 97 pp.

_____. 2000. Preliminary report on the current distribution of the goldline darter, *Percina aurolineata*, in the Cahaba River system of Alabama. Unpubl. report. U.S. Fish and Wildl. Serv., Jackson, MS. 4 pp.

_____. 2002. Final report of a study of the habitat requirements and biology of the pygmy sculpin, *Cottus paulus*, in Coldwater Spring and the feasibility of a transplant into Willett and/or Cabin Club Spring. Unpubl. report. U.S. Fish and Wildl. Serv., Jackson, MS. 4 pp.

_____, and P. D. Blanchard. 2003. Current status of the vermilion darter (*Etheostoma chermocki*) population in Turkey Creek. Report to U.S. Fish and Wildl. Serv. In preparation.

Sulak, K. J. and J. P. Clugston. 1999. Recent advances in the life history of Gulf of Mexico sturgeon, *Acipenser oxyrinchus desotoi*, in the Suwannee River, Florida, USA: A synopsis. J. Appl. Ichthy. 15:116-128.

Suttkus, R. D. 1955. *Notropis euryzonus*, a new cyprinid fish from the Chattahoochee River system of Georgia and Alabama. Tulane Stud. Zool. 3:85-100.

_____, and R. M. Bailey. 1993. *Etheostoma colorosum* and *E. bellator*, two new darters, subgenus *Ulocentra*, from southeastern United States. Tulane Stud. Zool. Bot. 29:1-28.

_____, and D. A. Etnier. 1991. *Etheostoma tallapoosae* and *E. brevirostrum*, two new darters, subgenus *Ulocentra*, from the Alabama River drainage. Tulane Stud. Zool. Bot. 28:1-24.

_____, and M. F. Mettee. 2001. Analysis of four species of *Notropis* included in the subgenus *Pteronotropis* Fowler, with comments on relationships, origin, and dispersion. Geological Surv. Alabama Bull. No. 170. 50 pp.

_____, and W. R. Taylor. 1965. *Noturus munitus*, a new species of madtom, family Ictaluridae, from the southern United States. Proc. Biol. Soc. Wash. 78:169-178.

_____, B. A. Thompson, and H. L. Bart, Jr. 1994. Two new darters, *Percina* (*Cottogaster*), from the southeastern United States, with a review of the subgenus. Occas. Papers Tulane Univ. Mus. Nat. Hist. No. 4. 46 pp.

Swift, C. C. 1970. A review of the eastern North American cyprinid fishes of the *Notropis texanus* species group (subgenus *Alburnops*), with a definition of the subgenus *Hydrophlox*, and materials for a revision of the subgenus *Alburnops*. Ph.D. Diss. Florida State Univ., Tallahassee, FL. 515 pp.

Taylor, W. R. 1969. A revision of the catfish genus *Noturus* Rafinesque, with an analysis of higher groups within the Ictaluridae. U.S. Nat. Mus. Bull. Series 282. 315 pp.

Thomas, D.L. 1970. An ecological study of four darters of the genus *Percina* (Percidae) in the Kaskaskia River, Illinois. Illinois Nat. Hist. Surv. Biol. Notes 70. 18 pp.

Trauth, S. E., G. L. Miller, and J. S. Williams. 1981. Seasonal gonadal changes and population structure of *Noturus munitus* (Pisces: Ictaluridae) in Mississippi. Assoc. S. E. Biologists Bull. 28:66.

Trautman, M. B. 1981. The fishes of Ohio. Second ed. The Ohio State Univ. Press. 782 pp.

U.S. Fish and Wildlife Service. 1975. Endangered and threatened wildlife and plants; amendment listing the snail darter as an endangered species. Federal Register 40:47505-47506.

_____. 1990. Alabama cavefish, *Speoplatyrhinus poulsoni* Cooper and Kuehne 1974 (Second Revision) Recovery Plan. Prepared by J. E. Cooper, North Carolina State Museum of Natural History. Revised by J. H. Stewart, U.S. Fish and Wildl. Serv., Jackson, MS. Atlanta, GA. 17 pp.

_____. 1992. Cahaba shiner (*Notropis cahabae*) recovery plan. Jackson, MS, U.S. Fish and Wildl. Serv. 15 pp.

_____. 1995. Gulf sturgeon recovery plan (prepared with Gulf States Marine Fisheries Commission). Atlanta, GA. 170 pp.

Utter, S. P. 1984. A taxonomic review of the darters referred to as *Etheostoma swaini* and *E. ditrema* (Pisces: Percidae) in the Coosa-Alabama River Drainage. Unpubl. M.S. Thesis, Auburn Univ., Alabama. 115 pp.

Vladykov, V. D., and J. R. Greeley. 1963. Order Acipenseroidei. *In* Fishes of the western North Atlantic. Mem. Sears. Found. Mar. Res. 1(Pt. 3):24-60.

Wall, B. R. 1968. Studies on the fishes of the Bear Creek drainage of the Tennessee River system. M.S. Thesis, Univ. of Alabama, Tuscaloosa, AL. 96 pp.

Walsh, S. J., and B. M. Burr. 1984. Life history of the banded pygmy sculpin, *Elassoma zonatum.* Jordan (Pisces: Centrarchidae), in western Kentucky. Bull. Ala. Mus. Nat. Hist. No. 8. 52 pp.

_____, and _____. 1985. Biology of the stonecat, *Noturus flavus* (Siluriformes, Ictaluridae), in central Illinois and Missouri streams and comparisons with Great Lakes populations and congeners. Ohio J. Sci. 85:85-96.

Warren, M. L., and B. M. Burr. 1998. Threatened fishes of the world: *Notropis albizonatus* Warren, Burr, and Grady, 1994 (Cyprinidae): Environmental Biology of Fishes 51:128.

_____, _____, and J. M. Grady. 1994. *Notropis albizonatus*, a new cyprinid fish endemic to the Tennessee and Cumberland River drainages, with a phylogeny of the *Notropis procne* species group. Copeia 1994:868-886.

_____, _____, S. J. Walsh, H. L. Bart, Jr., R. C. Cashner, D. A. Etnier, B. J. Freeman, B. R. Kuhajda, R. L. Mayden, H. W. Robison, S. T. Ross, and W. C. Starnes. 2000. Diversity, distribution, and conservation status of the native freshwater fishes of the southern United States. Fisheries 25:7- 29.

Wieland, W., and J. S. Ramsey. 1987. Ecology of the muscadine darter, *Percina* sp. cf. *P. macrocephala*, in the Tallapoosa River, Alabama, with comments on related species. Proc. S.E. Fishes Council 17:5-11.

Williams, J. D. 2000. *Cottus paulus*: a replacement name for the pygmy sculpin, *Cottus pygmaeus* Williams 1968. Copeia 2000:302.

_____, and G. H. Burgess. 1999. A new species of bass, *Micropterus cataractae* (Teleostei: Centrarchidae), from the Apalachicola River basin in Alabama, Florida, and Georgia. Bull. Florida Mus. Nat. Hist. 42:81-114.

_____, and G. H. Clemmer. 1991. *Scaphirhynchus suttkusi*, a new sturgeon from the Mobile Basin of Alabama and Mississippi. Bull. Ala. Mus. Nat. Hist. 10:17-31.

Wood, R. M. and R. L. Mayden. 1993. Systematics of the *Etheostoma jordani* species group (Teleostei: Percidae), with descriptions of three new species. Bull. Ala. Mus. Nat. Hist. 16:31-46.

GLOSSARY OF TERMS

Adipose - referring to a fleshy accessory, appendage or membrane; in fish, usually refers to a fin.

Allopatric - a distributional relationship where species with nonoverlapping ranges do not occur together.

Anadromous - fish species that migrate from estuarine or marine areas into fresh water to spawn.

Anterior - at, or near, the front.

Apical gland - a small pimple-like protuberance on the apical lobe of a hydrobiid snail's verge.

Apical lobe - the distal end of a snail's shell.

Aperture - the opening or "mouth" of a snail shell through which the head and foot protrude when the snail is active.

Arcuate - having the form of a bow; curved.

Barbel - a fleshy, tapered sensory organ found around the mouth or on the head of some fishes.

Basal lip - the lip at the base of a snail shell.

Basicaudal - at the base of the caudal fin.

Beak - the raised, inflated, or dome area of the shell along the dorsal margin; used synonymously with umbo.

Benthic - deep or bottom parts of an aquatic environment.

Biangulation - edge along which two surfaces in different planes meet at an angle.

Body whorl - the last complete whorl of a spiral snail shell, measured from the outer lip back to the point immediately above the outer lip; usually the largest whorl of the body.

Branchiostegal membranes - the paired membranes located on the posterioventral (lower-rear) surface of the head, which close the lower opening of the gill cavity.

Bryophyte - member of the division in the plant kingdom of nonflowering plants comprising mosses, liverworts, and hornworts.

Byssus - (*adj*. byssal) strong, silky filament by which mussels attach themselves to rocks and other fixed surfaces.

Callus - a thickened layer of calcareous (composed of calcium) material on a shell secreted by the snail's mantle.

Carina - a sharp spiral edge, ridge, or "keel" on the outer shell margin of a snail shell.

Carinate - having one or more sharp spiral edges, ridges, or keels on the outer shell surface.

Caudal - tail.

Centrarchid - a member of the fish Family Centrarchidae, the sunfishes.

Channelization - see Stream channelization.

Chironomid - a member of the insect Family Chironomidae, the midges or nonbiting midges.

Chromatophore - a cell with pigment that gives color to the surrounding tissue.

Coleopteran - a member of the insect Order Coleoptera, the beetles.

Columella - the internal column around which the whorls of univalve (one piece) shells revolve, as in snails.

Congener - belonging to the same taxonomic genus as another organism.

Conglutinate - a mass of glochidia bound together in a gelatinous/mucous mass, often resembling prey items of potential host fish.

Conspecific - of, or belonging to, the same species.

Copepod - a member of a group of small Crustacea common in marine and fresh waters.

Costa - (*pl.* costae) a transverse rib or rounded ridge of considerable size on the surface of a snail's shell.

Cottid - a member of the fish Family Cottidae, the sculpins.

Crenulate - having a margin with very small, low, rounded teeth.

Cryptic - protective coloration of animals that makes them resemble or blend into their habitats or backgrounds.

Cyprinid - any of numerous often small freshwater fishes of the Family Cyprinidae, which includes the minnows, carps, and shiners.

Demersal - post-spawning, the movement of eggs such that they eventually will sink to, or near, the bottom of a body of water to await larval development.

Demibranch - one freshwater mussel gill; each mussel has two pairs.

Dioecious - having separate sexes; separate male and female individuals.

Dipteran - a member of the insect Order Diptera, including aquatic forms such as mosquitoes, gnats, and midges.

Distal - toward, or near, the outer edge or margin.

Dorsal - on, or pertaining to, the back or dorsum.

Eddies - a water current moving contrary to the direction of the main current, especially in a circular motion.

Endemic - found only in one region to which it is native.

Ephemeropteran - a member of the aquatic insect Order Ephemeroptera, the mayflies.

Excurrent aperture - referring to the opening of a mussel that excretes bodily fluids outside of the shell.

Extant - still living or present.

Fall Line - the boundary between unconsolidated Coastal Plain sediments and the harder rocks of the Appalachian Highlands.

Fork Length - distance from the tip of the snout to the deepest part of the fork of the tail in fish.

Frenum - the fleshy bridge connecting the upper lip to the snout in fish.

Fusiform - tapering at each end; spindle-shaped.

Ganoid scale - a bony scale found in gars that is rhomboid-shaped and hard on the surface.

Globose - globelike or spherical.

Glochidium - parasitic larval stage of freshwater mussels of the Family Unionidae prior to transformation to the juvenile stage; attach to the gills, or other external parts of a host fish.

Gravid - in fish, the reproductive status of females where many of the oocytes are fully or almost fully developed; in mussels, refers to a female that has embryos/glochidia in the marsupium.

Heptageniid - a member of the aquatic insect Family Heptageniidae, mayflies that live in fast-moving streams.

Hermaphrodite - having both male and female reproductive organs.

Heterocercal - caudal fin shape of fish where the upper lobe of tail is larger and longer than the lower lobe.

Humeral spot - in fish, a spot in the shoulder region just behind the operculum (gill cover).

Hydrobiids - a member of the snail Family Hydrobiidae.

Hymenopteran - a member of the most advanced group of insects of the Order Hymenoptera, the ants, bees, wasps, and allies.

Ichthyofauna - fish species of a given area.

Imperforate - refers to a spiral gastropod shell that has no opening or external cavity at the base of the columella.

Incised - grooved or engraved.

Interdentum - a flattened area of the hinge plate of a mussel shell between the pseudocardinal and lateral teeth.

Interradial membrane - membrane between rays in fins of a fish.

Lachrymose - drop-shaped, or tear-drop shaped; usually refers to pustules on a shell.

Lamellate - composed of, or furnished with, thin plates or scales.

Lateral - pertaining to the side.

Lateral line - a canal, usually visible as a dark horizontal line, along the side of a fish that contains pores that open into tubes supplied with sense organs sensitive to low vibrations.

Lentic - pertaining to standing water.

Lira - (*pl.* lirae) a ridge, specifically a spiral ridge on the outer surface of a snail shell.

Long-term brooder - applied to unionid mussels in which glochidia are retained over the winter.

Lotic - pertaining to flowing water.

Macrophyte - member of the macroscopic (visible to the naked eye) plant life especially of a body of water.

Mantle - a membrane that lines the inner surfaces of the shell, envelopes the body, and secretes the substance that forms the shell in mollusks.

Mantle flap display - elongation, or swelling, of the mantle of a mussel to resemble a fish or invertebrate for the purpose of attracting predatory fish close enough to allow parasitic attachment by the mussel's glochidia.

Marsupial swelling - an enlarged or inflated area (often posterior-ventrally) of the female unionid shell that provides space internally for the expansion of gills carrying glochidia.

Melanophore - a color cell with melanin (dark pigmentation).

Meristic - involving modification in number or in geometrical relation of body parts.

Monotypic - a taxon containing only a single form at the lower major taxonomic categories.

Morphometric - measurement of external form.

Nacre - the iridescent, often differently colored, inner layer of a bivalve shell, composed primarily of crystalline calcium carbonate deposited in thin, overlapping layers; mother-of-pearl.

Nematoceran - a member of the insect Suborder Nematocera of the Order Diptera, the true flies.

Neritid-like - having a small spine, large body whorl and large aperture, similar to members of the Family Neritidae.

Neuromast - an exposed receptive cell of the nervous system.

Nucleus - the first formed (earliest) part of beginning of a shell or operculum of a snail.

Nuptial - in fish, a term that refers to being in reproductive condition.

Obtuse - blunt or rounded at the end, not acute or pointed.

Ocellus - a rounded spot contrasting with a lighter or darker color.

Odonate - a member of the insect Order Odonata, the dragonflies and damselflies.

Oligochaete - a member of the Class Oligochaeta, the segmented worms.

Oocyte - a reproductive cell that can develop into a mature ovum or egg.

Opercle - the largest bone of the gill cover.

Operculum - in fish, the proper name for the gill cover; in mollusks, a lid or flap covering an aperture, such as the horny shell cover in snails or other mollusks; a corneous or calcareous plate borne on the dorsal posterior foot of prosobranch snails that closes the aperture when the snail withdraws into its shell.

Oral valve - in fish, a flap of skin along the anterior-dorsal margin of the buccal cavity; helps to regulate flow of water in and out of the mouth.

Ovately conic - egg-shaped cone.

Ovoviviparous - condition in which young snails are formed within an egg, but hatch while still inside the mother snail, from which they emerge as young crawling snails.

Paleomelanin - darkened pigmentation associated with aging.

Pallial line - an indented groove or line approximately parallel with the ventral margin of a bivalve shell that marks the line of muscles attaching the mantle to the shell.

Papilla - in fish, a knoblike protuberance; in mussels, a small finger-like projection posterior along the mantle margin, primarily adjacent to the apertures.

Papillate - bearing papillae.

Parietal - pertains to the inside wall of the shell aperture.

Paucisipiral - in snails, refers to an operculum in which there are few rapidly enlarging spirals, coils, or whorls.

Pectoral - the most anterior or most dorsal paired fins of fish.

Peduncle - a narrow part by which some larger part or the whole body of an organism is attached; in fish, the area between the base of the tail and posterior end of the anal fin.

Pelagic - a term for species that live and roam in open water.

Penultimate - next to last.

Periostracum - the hard chitinous outer covering of the shell of many mollusks that protects them from the erosive action of water.

Pharyngeal - relating to, or found in, the region of the pharynx or throat.

Phylogenetic - a taxonomic arrangement based on natural evolutionary relationships; the arrangement generally begins with the most primitive and concludes with the most advanced forms.

Plica - (*pl.* plicae) a transverse or "vertical" ridge or "rib" on the outer shell surface.

Plication - transverse ridges, typically on the surface of the shell.

Predorsal - in front of the dorsal fin.

Preopercular - referring to the area between the the large membrane bone lying in front of, and parallel to, the opercle of a fish.

Protrusible - having the capability to jut out from the surrounding surface or context.

Pseudocardinal - compact, usually triangular or blade-like structures on the hinge line below the beak of a mussel.

Pupiform - refers to the shape of a snail shell in the form of a moth or butterfly chrysalis.

Pustule(s) - small, usually numerous raised structures on the external part of the shell.

Quadrate - having four sides and four angles; square or rectangular.

Riffle - a stream habitat characterized by shallow depth, broken substrate, and swift flow.

Rimate - refers to a coiled snail shell that has at its base a narrow "umbilical" opening that is partially closed by the expansion of the anterior columellar lip.

Ripe - a state of reproductive readiness when female fish usually extrude eggs and males extrude sperm with light abdominal pressure.

Rotenone - a crystalline insecticide $C_{23}H_{22}O_6$ that is obtained from the roots of several tropical plants and of low toxicity to warm-blooded animals; historically, used extensively to collect fish from waterways.

Rugose - heavily wrinkled or roughened into ridges.

Run - a stream habitat that is transitional between fast, shallow riffles and slow, deeper pools.

Sculpture - the natural surface markings other than those of color, usually found on mussel and snail shells, and that often furnish identifying marks or species recognition.

Scute - an external bony or horny plate, or large scale.

Serrae - in certain fish, refers to the saw toothed-like appearance of the posterior fin spines.

Shagreen - the rough skin of an organism when covered with small close-set tubercles.

Shell disk - in mussels, a general term to describe the broad lateral surface of the shell.

Shoal - a shallow place in a body of water.

Short-term brooder - applied to unionids whose gravid period is short, often only a few months; species spawn, incubate, and release glochidia only during the spring and summer.

Shoulder - refers to the appearance (in outline) of the posterior outer peripheral part of a whorl that is sharply rounded in contrast to the more even curvature of the rest of a snail's shell.

Simuliid - a member of the insect Family Simuliidae, the blackflies.

Spire - the whorls of a snail shell, excepting the last or body whorl; the spire is measured as the distance from the suture where the apertural lip meets the body whorl to the shell apex.

Spiracle - a breathing orifice.

Squamation - the state of being scaly, or the arrangement of scales on an animal.

Standard Length - in fish, the straight distance between the tip of the snout and the base of the caudal fin rays.

Stream channelization - the process of changing wetlands into streams, or straightening the normal meanderings of streams into a straight channel to improve the speed of runoff.

Stria - (*pl.* striae) slight ridges, furrows, or alternating color patterns.

Stygobitic - obligate aquatic cave dweller.

Subadult - an animal, or age class, that resembles an adult but does not breed because of behavioral and/or sexual immaturity.

Subconcentric - having markings, grooves, or ridges that form rings that do not quite have a common center.

Subglobose - not quite having the shape of a globule.

Subnodulous - a small rounded mass of irregular shape.

Suborbital bar - a line or bar situated below the eye.

Subrhomboidal - having generally four distinct sides, two sides being longer than the other.

Substrate - (*pl.* substrata) bottom material in lakes, streams, and rivers.

Subsutural - below a suture.

Sulcus - furrow or groove.

Superconglutinate - packet of glochidia containing the entire annual reproductive output tethered to the excurrent aperture by a hollow mucus tube.

Supralateral - above the side.

Supraorbital - above the eye.

Suture - a spiral seam on a snail shell surface where two adjoining whorls meet.

Sympatric - pertaining to two or more populations that occupy overlapping geographical areas.

Syntopic - a distributional relationship where the ranges of two taxa overlap considerably.

Terete - cylindrical and tapering.

Total Length (TL)- in fish, the distance from the tip of the snout to the posterior end of the longest caudal fin ray.

Trapezoidal - in the shape of a trapezoid.

Trichopteran - a member of the insect Order Trichoptera, the caddisflies.

Truncate - shells or structures appearing squared off, or transversely cut off in a straight line.

Tuberculate - covered with tubercles or rounded knobs.

Tubercule - in fish, a small protuberance, usually pointed and sharp; in snails and mussels, a nodule or small eminence, such as a solid elevation occurring on the shell surface.

Umbilicus - the hollow at the base of the shell of some gastropod (one valve) mollusks.

Umbillicate - refers to a spiral gastropod shell that has an opening or cavity at the base of the columella, and more specifically to one in which the opening is more than a narrow perforation; this cavity is formed in those shells in which the inner sides of the coiled whorls do not join.

Umbo - the oldest part of a bivalve shell, generally in the form of a hump near the hinge of the shell.

Urogenital - the common duct or opening through which urine wastes and reproductive products flow.

Venter - underside of a fish between the pectoral fins and the urogenital opening.

Ventral - on, or pertaining to, the underside or belly.

Ventrolateral - ventral and lateral.

Verge - the male organ of copulation in certain invertebrates.

Viviparid - those who give birth to live young; offspring develop within the mother's body; a member of the snail Family Viviparidae.

Vomerine - referring to a bone of the skull of most vertebrates that is situated below the ethmoid (nasal) region.

LITERATURE USED TO SUPPORT THE GLOSSARY

Many definitions used in the glossary were derived entirely, or partially, from the following sources. Some definitions were formed by combining definitions from various sources. Others were developed by the senior editor.

American® Dictionary of the English Language. © 2000. Fourth Edition. Houghton Mifflin Company.

Burch, J. B. 1989. North American freshwater snails. Malacological Publications. Hamburg, MI. 365 pp.

Howells, R. G., R. W. Neck, and H. D. Murray. 1996. Freshwater mussels of Texas. Texas Parks and Wildlife Press.

Koford, R. R., J. B. Dunning Jr., C. A. Ribic, and D. M. Finch. 1994. A glossary for avian conservation biology. Wilson Bull. 106: 121-137. Jamestown, ND: Northern Prairie Wildlife Research Center Homepage. http://www.npwrc.usgs.govresource/literatr/avian.avian.htm (Version 16JUL97).

Merriam Webster Unabridged Online Dictionary. http://www.dictionary.com Accessed July 2003.

Mettee, M. F., P. E. O'Neil, and J. M. Pierson. 1996. Fishes of Alabama and the Mobile Basin. Oxmoor House, Birmingham. AL. 820 pp.

Parmalee, P. W., and A. E. Bogan. 1998. The freshwater mussels of Tennessee. The Univ. of Tennessee Press. Knoxville, TN. 328 pp.

Romoser, W. S., and J. G. Stoffolano, Jr. 1998. The science of entomology. 4th Ed. McGraw-Hill Dubuque, IA. 605 pp.

About the Editors:

Ralph E. Mirarchi has degrees in biology from Muhlenberg College (1971, B.S.) and in wildlife biology and management from Virginia Polytechnic Institute and State University (1975, M.S.; 1978, Ph.D.). He currently is the William R. and Fay Ireland Distinguished Professor of Wildlife Science in the School of Forestry and Wildlife Sciences at Auburn University, Alabama, where he has taught numerous courses in wildlife science, advised undergraduate wildlife majors, and conducted research on doves and pigeons for the Alabama Agricultural Experiment Station for more than 25 years. He previously served as editor in chief of *The Journal of Wildlife Management* (1992-1993) and as a co-editor, compiler, and author of *Ecology and Management of the Mourning Dove* (1993), an award winning Wildlife Management Institute sponsored text published by Stackpole Books. He also currently serves on the board of Alabama's Forever Wild Land Trust, which purchases wild lands for the people of Alabama to enjoy in perpetuity.

Jeffrey T. Garner has degrees in biology from the University of North Alabama (1987, B.S.) and University of Alabama in Huntsville (1993, M.S.). He currently is the mussel management supervisor for the Alabama Department of Conservation and Natural Resources, Division of Wildlife and Freshwater Fisheries. His research interests over the past 16 years have included distributions of freshwater mussels and snails, as well as various aspects of their life history and ecology. He currently is involved in a variety of mollusk conservation activities across Alabama and serves on The Nature Conservancy of Alabama board of trustees.

Maurice F. Mettee has degrees in biology from Spring Hill College (1965, B.S.) and the University of Alabama-Tuscaloosa (1971, M.S.; 1974, Ph.D.) as well as a degree in secondary education (1968; M.A.) from the same institution. He currently is Chief of the Biological Resources Section of the Geological Survey of Alabama, where he has been employed since 1975. He has conducted research on the distribution, habits, and habitats of the fishes of Alabama for more than 30 years. In addition to numerous technical and peer-reviewed reports, he is senior author of the widely acclaimed Oxmoor House book, *Fishes of Alabama and the Mobile Basin*, a cooperative endeavor of the Geological Survey of Alabama, Division of Wildlife and Freshwater Fisheries, Alabama Department of Conservation and Natural Resources, and the Region 4 Office of the U.S. Fish and Wildlife Service.

Patrick E. O'Neil has degrees in biology (1977; B.S.), vertebrate zoology (1980; M.S.), and civil engineering (Ph.D.) emphasizing water quality and wastewater monitoring from the University of Alabama-Tuscaloosa. He currently is Director of the Water Investigations Program of the Geological Survey of Alabama, where he has been employed since 1979. Since that time he has conducted research on various aspects of Alabama's aquatic fauna and water resources. He is a co-author of the widely acclaimed Oxmoor House book, *Fishes of Alabama and the Mobile Basin*, a cooperative endeavor of the Geological Survey of Alabama, Division of Wildlife and Freshwater Fisheries, Alabama Department of Conservation and Natural Resources, and the Region 4 Office of the U.S. Fish and Wildlife Service.